murach's
R for Data Analysis

Scott McCoy

MIKE MURACH & ASSOCIATES, INC.

3730 W Swift Ave. • Fresno, CA 93722

www.murach.com • murachbooks@murach.com

Editorial team

Author:	Scott McCoy
Editors:	Joel Murach
	Lisa Cooper
Production:	Juliette Baylon

Photo on page 97 by Anne Cooper

Books on SQL and data analysis

Murach's MySQL

Murach's SQL Server for Developers

Murach's Oracle SQL and PL/SQL for Developers

Murach's Python for Data Analysis

Books on programming languages

Murach's Python Programming

Murach's C#

Murach's C++ Programming

Murach's Java Programming

Books on web development

Murach's HTML and CSS

Murach's JavaScript and jQuery

Murach's PHP and MySQL

Murach's ASP.NET Core MVC

For more on Murach books, please visit us at www.murach.com

10 9 8 7 6 5 4 3 2 1
ISBN: 978-1-943873-03-6

Contents

Expanded contents

Section 2 The essential skills for data analysis

Section 4 An introduction to data modeling

Chapter 12   How to work with simple regression models

Chapter 13   How to work with multiple regression models

Appendixes

Introduction

From its start, the R programming language was designed to be used for statistical analysis. Today, it's one of the top languages used by data analysts and statisticians. With this book, you'll learn the R skills you need to become a successful data analyst, even if you're new to programming or have never studied statistics.

What the five sections in this book do

Section 1 gets you off to a fast start. First, you'll learn how to use RStudio, a popular program for coding in R that's available for free. Then, you'll learn the parts of the R language that you need to analyze data. Next, you'll learn how to use R with the tidyverse package to create your first analysis.

Section 2 presents critical skills for descriptive analysis that allow you to understand your data, gain insights, and share your results. That includes how to import and clean data from various sources, prepare that data so it's ready for analysis, and create professional visualizations.

Section 3 contains three complete analyses that show how the skills in the first two sections can be applied to real-world data sets from the fields of political, environmental, and sports analysis. As we see it, you can't master on-the-job skills by working only with pre-prepared sample data sets, so these in-depth analyses make sure that you master the professional skills you need.

Section 4 is an introduction to predictive analysis. This takes descriptive analysis to another level by using statistical models to make predictions. More specifically, this section shows how to use linear regression models to predict continuous numeric values and how to use classification models to predict categorical values.

Section 5 shows how to present an analysis. To do that, you can use R Markdown to convert your analysis into an HTML document, PDF file, or PowerPoint slideshow. This is an important skill because the value of an analysis comes from being able to present the insights gained from it to your target audience, whether that's your boss, your clients, or the general public.

Why you'll learn faster and better with this book

This book is designed to make it as easy as possible for you to learn new skills faster and retain them better. Here are a few of those features:

- All of the information is presented in *paired pages*, with the essential syntax, guidelines, and examples on the right page and clear explanations on the left page. This helps you learn faster by reading less.

- The paired-pages format is ideal for reference when you need to refresh your memory about how to do something.

- The three analyses presented in section 3 use real-world data sets. We believe that studying realistic analyses like these is critical to the learning process.

- The hundreds of short examples present usable code for tasks that you're likely to need for your own analyses.

- The exercises at the end of each chapter provide a way for you to gain valuable hands-on experience without any extra busywork.

The prerequisites

Chapters 1 and 2 present the skills you need to start using R for data analysis. As a result, the only prerequisite for this book is basic computer literacy. However, it's helpful to have some background in statistics.

What software you need

To analyze data with R as shown in this book, you just need to download and install the R language and the free RStudio program. Then, you can install some R packages for data analysis that are freely available. For information about how to do this, see appendix A for Windows or appendix B for macOS.

How our downloadable files can help you learn

You can download all the files that you need for following along with this book from our website (www.murach.com). Again, appendixes A (Windows) and B (macOS) show how to download and install these files. These files include:

- The R scripts for the examples and analyses presented in this book.
- The starting points for the exercises at the end of each chapter.
- The solutions to those exercises.
- The data for the examples, analyses, and exercises presented in this book

After you download these files, you can run the R scripts and experiment with the code to understand it better. In addition, you can copy code from these scripts to use in your own scripts.

If you have any problems with the exercises, you can check the solutions to help you get past any learning blocks, an essential part of the learning process. In some cases, the solutions may show you a more elegant way to handle a problem.

Support materials for instructors and trainers

If you're a college instructor or corporate trainer who would like to use this book for a course, we offer support materials that will help you set up and run your course as effectively as possible. These materials include instructional objectives, test banks, projects, case studies, and PowerPoint slides.

To learn more about our instructor's materials, please go to our instructor's website at www.murachforinstructors.com. Or, if you're a trainer, please go to our retail website at www.murach.com and click on the *Courseware for Trainers* link, or contact Kelly at 1-800-221-5528 or kelly@murach.com.

Please let us know how this book works for you

When we started writing this book, we had three goals. First, we wanted to present the skills that every R data analyst should have. Second, we wanted to do that in a way that works for people without programming experience as well as those who are experienced programmers. Third, we wanted to make this the best on-the-job reference that you've ever used.

Now, we hope that we've succeeded. We thank you for buying this book. And if you have any comments, we would appreciate hearing from you.

Scott McCoy, Author
scott@murach.com

Joel Murach, Editor
joel@murach.com

Section 1

Get started fast

This section gets you off to a fast start with using R for data analysis. First, it shows how to use the RStudio IDE (Integrated Development Environment) to write and run R code. Then, it presents the subset of R coding skills that you need to perform data analysis. Next, it shows how to use the R packages available from the tidyverse to code a simple but complete analysis.

When you complete the chapters in this section, you'll have a subset of the skills that you need for doing analyses of your own. You'll also be able to skip to any chapter in section 2 whenever you need to learn more about a specific phase of data analysis.

Chapter 1

Introduction to RStudio and R

This chapter begins with an introduction to data analysis in general. Then, it introduces the RStudio IDE (Integrated Development Environment) that you can use to write R code. Next, it shows how to get started with programming in R. When you finish this chapter, you'll be ready to learn more about using R for data analysis.

Introduction to data analysis

This chapter begins by providing some perspective on what data analysis is and isn't. Then, it describes five phases that apply to data analysis whether you're using R or any other language.

What data analysis is

Figure 1-1 summarizes the components of *data analysis*. One of the key points is that data analysis includes *data visualization* (or *data viz*) and *data modeling* (or *predictive analysis*). The goal of this book is to teach you a solid set of skills for data analysis, visualization, and modeling.

Data visualization often provides the best, clearest insights into data. A well-constructed graphic can quickly make a point that would otherwise take a paragraph or more to explain. Furthermore, visualization can help identify trends that would be hard to spot just by looking at a table of numbers. For example, the graph in figure 1-1 clearly shows both that child mortality has trended downward for the last century and also that there was a sharp increase during the Spanish Flu pandemic of 1918.

Data modeling involves the use of data for building models. These models can help predict what's going to happen in the future based on the data of the past. For example, looking at this graph you would probably predict that child mortality rates will continue to be low throughout the next decade, but that a pandemic could cause a sudden, unexpected increase. Then, you could use that data to model the expected future population of a country in addition to some worst-case scenarios.

Data analytics is often used as a synonym for data analysis and data visualization. Similarly, *business analytics* refers to data analysis with the focus on business data, and *sports analytics* refers to data analysis with the focus on sports data. Of course, business and sports analytics are just applications of the skills that you'll learn in this book. In fact, your new skills can be applied to any field with data to analyze.

Data science is a field that covers data analysis but also includes advanced skills like *data mining*, *machine learning*, *deep learning*, and *artificial intelligence (AI)*. Although these advanced skills aren't included in this book, they all require a solid set of the essential skills for data analysis and data visualization, which this book covers thoroughly.

Data visualization often provides the best insights into the data

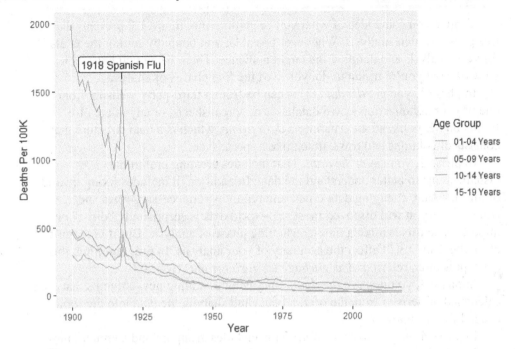

What data analysis includes

- Data analysis
- Data visualization (data viz)
- Data modeling (predictive analysis)

Related terms

- Data analytics
- Business analytics
- Sports analytics
- Data science

- Data mining
- Machine learning
- Deep learning
- Artificial intelligence (AI)

Description

- This book shows how to use the R programming language for *data analysis*, *data visualization* (or *data viz*), and *data modeling* (or *predictive analysis*).
- *Data analytics* is often used as a synonym for data analysis. Similarly, *business analytics* refers to the analysis of business data, and *sports analytics* refers to the analysis of sports data.
- Data analysis is the foundation for advanced skills like data mining, machine learning, deep learning, and artificial intelligence (AI).
- *Data science* is a term that includes both data analysis and advanced skills like data mining, machine learning, deep learning, and AI.

Figure 1-1   What data analysis is

The five phases of data analysis

To give you some idea of what you're getting into, figure 1-2 presents the five phases of data analysis. Whenever possible, you begin by setting the goals for your analysis and defining the target audience. Then, when you have a clear view of what you're trying to do, you start the five phases of analysis.

In phase 1, you *get the data*. This can be from a third-party website, from one of your organization's own databases or spreadsheets, or anywhere else. In this phase, you read the data into a *data frame*, which is a data structure that consists of columns and rows, much like a spreadsheet.

In phase 2, you *clean the data*. That includes creating preliminary visualizations to better understand the data. In addition, it includes fixing invalid or missing data, changing data types, and removing unnecessary rows and columns. As you will discover, most real-world data is surprisingly "dirty," or imperfect, so this is often a time-consuming phase of analysis. But if you don't clean the data, it will affect the accuracy of your analysis. In programming, this concept is often referred to as *garbage in, garbage out* (*GIGO*).

In phase 3, you *prepare the data*. That includes adding new columns that are calculated or derived from the original data and shaping the data into the forms needed for the analysis.

In phase 4, you *analyze the data*. This includes grouping and summarizing the data. It also includes creating data visualizations, because they often provide insights and show relationships that are difficult or impossible to show with tabular data. In addition, it may include data modeling to perform predictive analysis.

In phase 5, you *present the data* in a way that's appropriate for your target audience. That means you enhance your visualizations so they get their points across as clearly and quickly as possible, even to those with no technical or analytical background.

Of course, the divisions between these phases aren't nearly as clear as this figure might make them seem. In fact, there is usually some overlap between the phases. For instance, when you clean or prepare the data, you're already looking ahead to the analyzing phase. And when you analyze the data, you may discover that you need to do more cleaning or preparation.

What you need to do before you start an analysis

Set your goals

- The goals of an analysis can be specific, like trying to confirm or reject a particular hypothesis or trying to answer a particular question.

- The goals of an analysis can also be general, like trying to summarize the main characteristics of the data in order to form new hypotheses that can lead to further analysis. This approach is known as *exploratory data analysis* (*EDA*).

Define your target audience

- The *target audience* includes any other people who you will present your findings to like managers, clients, colleagues, and so on.

The five phases of data analysis

Get the data

- Find the data.

- Load the data into a data frame.

Clean the data

- Create preliminary visualizations to better understand the data.

- Handle invalid or missing values.

- Make sure data types are set correctly.

- Remove unnecessary rows and columns.

Prepare the data

- Add any columns you need.

- Shape the data into the forms you need.

Analyze the data

- Group and summarize the data.

- Create visualizations that provide insights.

- Model the data as part of predictive analysis.

Present the data

- Enhance your visualizations so they're appropriate for your target audience.

Description

- Before you start any analysis, it's generally considered a good practice to set your goals and define the target audience.

- You can divide a typical analysis into five phases like those above. In practice, however, there's almost always overlap between the phases.

Figure 1-2   The five phases of data analysis

How to get started with RStudio

To write R code, you typically use a program called an *Integrated Development Environment (IDE)*. RStudio is the most popular IDE for working with R code, and we recommend using it.

Introduction to RStudio

RStudio is divided into four windows called *panes*. Figure 1-3 shows the default layout of the panes in RStudio.

The top left pane is called the Source pane. You can use it to open, view, and edit source code and data files. When you open RStudio for the first time, it doesn't display this pane. However, you can display it by opening a source file or by creating a new source file.

The bottom left pane is the called the Console pane. You can use this pane to run code and to view the output of the code that you run.

The top right pane is the Environment pane. You can use it to view the data and variables that are currently loaded into RStudio.

The bottom right pane provides several tabs that you can use for different purposes. Typically, you use the Plots tab to view data visualizations like the one shown in this figure. However, you can also use the Help tab to view documentation for R.

When you write R code, it's typically a good practice to save it in a .R file called an *R script*. When you use RStudio to open an R script, it displays the contents in the Source pane. This allows you to edit the script and run its code. When you run the code in a script, RStudio displays the output in the Console pane, and it displays any variables created by the code in the Environment pane.

In addition to .R files, you can open some other types of files in the Source pane. For example, you can open data files such as .csv files.

When you open more than one file, RStudio displays a tab for each open file in the Source pane. Then, you can easily switch between the open files by clicking the appropriate tab. In this figure, for example, you can easily switch between the two scripts that are open in the Source pane by clicking on the tab for the script that you want.

The RStudio IDE

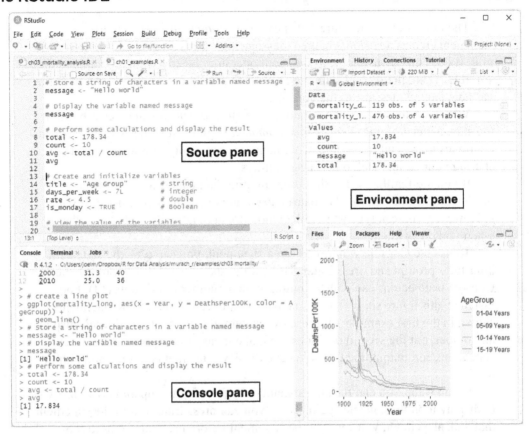

Description

- An *Integrated Development Environment* (*IDE*) is a program that combines several tools for software development.

- RStudio is the most popular IDE for the R programming language.

- RStudio provides four windows called *panes* that you can use to develop R code.

- When you open RStudio for the first time, it doesn't display the Source pane. However, it displays this pane after you open an existing source file or create a new source file.

Figure 1-3   Introduction to RStudio

How to run code in the Console pane

The first example in figure 1-4 shows two *statements* in R. The first statement creates a *variable* named message and stores a value of "Hello world" in it. It does this by using the assignment operator (<-) to assign the value to the variable. Then, the second statement tells R to print the value that's stored in the variable named message to the console.

When coding a script, it's common to include *comments* that describe what the statements do. In R, a *single-line comment* begins with a hash symbol (#) and continues to the end of the line. Comments are especially important if you are sharing your scripts with other programmers.

If you are familiar with other programming languages, note that R doesn't provide multi-line comments. Instead, each line of a comment must start with #. However, you can type #* to tell RStudio to automatically include another #* when you press Enter.

To use the Console window to run a statement, you can type the statement after the > prompt and press Enter. Then, if the statement displays output, the Console pane begins each line of output with a number in brackets ([]). For instance, this figure shows a Console pane that executes the two R statements shown in the first example. Of these two statements, the first one doesn't display any output, but the second one does. To make it easy to differentiate between statements and output, the Console pane displays statements in blue and output in black.

If you want, you can re-run a statement by pressing the up and down arrows to display it on the console again. Then, you can press Enter to run the statement that's displayed. Or, you can edit the statement before pressing Enter to run it. This is often useful when a statement doesn't work the way you want. In that case, you can press the up arrow to display the statement, edit it to fix the problem, and press Enter to run it again.

The code that you type into the Console pane isn't saved for future use. That's why the next figure shows how to use the Source and Console panes to run statements in an R script. This provides a way to save your R statements for future use.

Two R statements

```
message <- "Hello world"
message
```

Two comments added to the statements

```
# store a string of characters in a variable named message
message <- "Hello world"

# display the variable named message
message
```

The Console pane

```
Console   Terminal   Jobs
R 4.1.2 · ~/murach_r/ch01/
> message <- "Hello world"
> message
[1] "Hello world"
> |
```

Description

- A *statement* performs a task such as creating a variable or displaying a variable's contents on the console.
- When writing code, you can add a *comment* that describes a statement or a group of statements.
- A *single-line comment* begins with a hash symbol (#) and continues to the end of the line.
- To run a statement in the Console pane, type it after the > prompt and press Enter.
- If you run a statement that displays output, the Console pane begins each line of output with a number where [1] starts the first line, [2] starts the second line, and so on.
- To display previous statements in the Console pane, press the up and down arrows while in the pane. Then, you can re-run the statement that's currently displayed by pressing Enter. Or, you can edit the statement that's displayed before pressing Enter to run it.

Figure 1-4   How to run code in the Console pane

How to run code in the Source pane

An R script allows you to save your code and run it again later. To open an existing R script, you can select File→Open File from the menus and use the resulting dialog to select the .R file you want to open. In figure 1-5, the tabs at the top of the Source pane show that the user has opened two R scripts. These R scripts contain the code for the examples for two chapters of this book.

To switch between these R scripts, you can click the tab for the file you want. Or, if you want to create your own R script, you can select File→New File→R Script from the menus.

To run the code that's in the Source pane, you can move the cursor to the statement you want to run and press Ctrl+Enter (Windows) or Command+Enter (macOS). When you do that, RStudio moves the statement and any comments associated with it to the prompt in the Console window, runs the statement, and moves the cursor to the next statement. In this figure, the user has executed the first six statements in the script. As a result, the Console pane displays these six statements and their associated comments in blue on the prompt (>). In addition, it displays the output in black prepended by a number in square brackets ([1]).

If you want to run more than one statement, you can select all of the statements that you want to run before pressing Ctrl+Enter (Windows) or Command+Enter (macOS). For example, to run all of the statements in a script, you can press Ctrl+A (Windows) or Command+A (macOS) to select the entire script. Then, you can press the appropriate keystroke combination to run the selected statements. Alternately, you can click the Run button in the upper right of the source pane.

The Source and Console panes

Description

- You can save R code in a .R file known as an *R script*. Then, you can use the Source pane to edit and run the code in the script.
- To open an existing R script, select File→Open File from the menus and use the resulting dialog to select the .R file you want to open.
- To create a new R script, select File→New File→R Script from the menus.
- The Source pane displays the tabs for all open files along the top of the pane.
- To switch between open files, click the tab for the file you want.
- To run a statement in a script, move the cursor to the statement and press Ctrl+Enter (Windows) or Command+Enter (macOS).
- To run multiple lines of code, select the lines and press Ctrl+Enter (Windows) or Command+Enter (macOS), or click the Run button in the upper right of the pane.
- The Console pane displays the output of running any statements in the Source pane.

Figure 1-5   How to run code in the Source pane

How to view variables in the Environment pane

When you create variables in RStudio, they get loaded into *main memory*. Main memory is also referred to as *RAM* (*Random Access Memory*), which is temporary storage for data.

When you close RStudio, you can save these variables in a *workspace image* that's stored in an .RData file. Then, RStudio automatically loads the saved variables from the workspace image the next time you start it. However, if you don't save a workspace image, the variables are lost when you close RStudio.

Once RStudio loads a variable into main memory, you can use it in your scripts and access it from the Console pane. In addition, it becomes visible in the Environment pane. This makes it easy to view all variables that are currently loaded into main memory.

The Environment pane shown in figure 1-6 displays the four variables created by running the beginning of the script shown in the previous figure. This shows the string of characters that's stored in the variable named message as well as the numbers that are stored in the other three variables. Note that RStudio uses quotation marks for the string of characters but doesn't use quotation marks for the numbers.

As you work with RStudio, the Environment pane will eventually become cluttered with variables that you no longer need. This makes it difficult to view the values of the variables you do need. To clear the Environment pane, you can click on the broom icon at the top of the pane. However, this removes all variables from main memory. As a result, you should only do this if you have a way to reload the variables that you need. Fortunately, if you store your code in a script, you can easily reload the variables you need by running some or all of the script.

The Environment pane

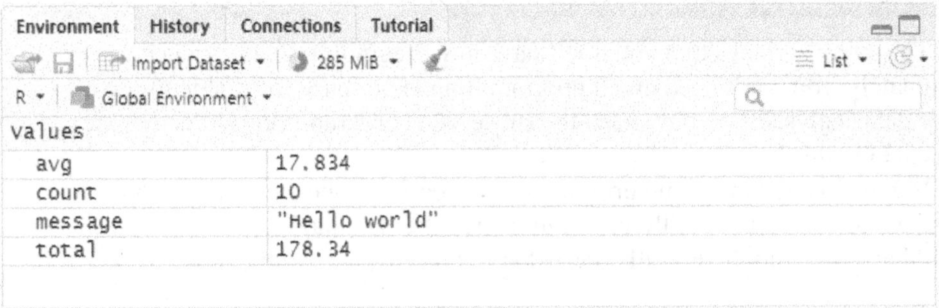

values	
avg	17.834
count	10
message	"Hello world"
total	178.34

Description

- R stores its variables in the computer's *main memory*, which is often referred to as *RAM* (*Random Access Memory*). This type of storage is temporary.
- The Environment pane shows all of the variables that are currently stored in main memory.
- The Environment pane displays the values for strings of characters in double quote marks, but it displays the values for numbers without quote marks.
- To clear the Environment pane, click the broom icon that's on the right side of the toolbar. This removes the variables shown in the Environment pane from main memory.
- When you quit RStudio, you can save the loaded variables in a *workspace image* that's stored in an .RData file. Then, RStudio automatically loads the variables from the workspace image the next time you start it. However, if you don't save a workspace image, the variables are lost when you quit RStudio.

Figure 1-6   How to view variables in the Environment pane

How to get started with R

This book assumes that you don't have any programming experience. Fortunately, you don't need much programming experience to be able to use R for data analysis! You only need the subset of R programming skills that are presented in this book.

If you already have some programming experience, be aware that R has some quirks that make it different from many other languages. As a result, you should at least skim the R skills presented in this chapter and the next.

How to create variables

A *variable* stores data that can change, or *vary*, as code executes. In programming, all variables have an associated *data type*. These data types affect how you work with a variable and what operations you can perform on it. For example, you can perform arithmetic calculations on the double and integer types, but you can't perform them on the character type.

The table at the top of figure 1-7 presents the four most basic data types in R. This table shows the names of the data types as well as some examples of *literal values*, or *literals*, for each type.

To start, the character type can store a string of characters, commonly known as a *string*. A string can include zero or more letters, numbers, or symbols. To code a literal value for a string, you can enclose its characters in double quotes.

After the character type, this figure presents the two *numeric types* available from R, the double and integer types. The double type can store numbers that are positive or negative and with or without a decimal. To code a literal value for the double type, you can just code the number.

The integer type can only store whole (non-decimal) numbers, though the numbers can still be positive or negative. To code a literal value for an integer, you code the number followed by the letter *L*. Here, the letter *L* indicates that the number is an integer, not a double.

The logical type stores a *Boolean value*, which is a value that is either true or false. To create a literal value of the Boolean type, you can code the TRUE or FALSE keywords.

The first example shows how to create a variable for each of these data types using the assignment operator (<-). With modern versions of R, you can also use the equals sign (=) as the assignment operator instead of using the <- operator. However, the equals sign works a little differently, and most R programmers use the <- operator to maintain compatibility with older versions of R. To make it easier to type the assignment operator, RStudio provides a keyboard shortcut for Windows (Alt+-) and macOS (Option+-).

The second example shows how to view the values stored in each variable. To do that, you can enter the name of the variable to display its value on the console. If you forget the type of a variable, you can use the typeof() function to check it as shown in the third example in this figure. To do this, you code a call to the typeof() function by coding the function's name, an opening parenthesis, the variable that you want to check, and a closing parenthesis.

The four basic data types in R

Type	Name	Literal values
character	String	"x", "abc", "123", "this is a character string"
integer	Integer	1L, -3L, 5L
double	Double	12.3, 5.0, -3.562, 7
logical	Boolean	TRUE, FALSE

How to create and initialize variables

```
title <- "Age Group"
days_per_week <- 7L
rate <- 4.5
is_monday <- TRUE
```

How to view the value of a variable

```
title
[1] "Age Group"
days_per_week
[1] 7
rate
[1] 4.5
is_monday
[1] TRUE
```

How to view the type of a variable

```
typeof(title)
[1] "character"
typeof(days_per_week)
[1] "integer"
typeof(rate)
[1] "double"
typeof(is_monday)
[1] "logical"
```

Description

- A *variable* stores data that can change, or *vary*, as code executes.
- To create a variable and assign it an initial value, you can code the name of the variable, followed by the *assignment operator* (<-), followed by a *literal value*. Then, R automatically determines the data type based on the literal value.
- In RStudio, the keyboard shortcut for the assignment operator is Alt+- (Windows) or Option+- (macOS).
- You can use the typeof() function to check the data type of a variable.
- The character type stores a string of characters, commonly known as a *string*.
- Since the double and integer types provide a way to store numbers, they are known as *numeric types*.
- The logical type stores a *Boolean value*, which is a value that is either true or false.

Figure 1-7   How to create variables

How to work with variables

Now that you know how to create a variable and assign a value to it, you're ready to learn how to work with a variable. The first example in figure 1-8 begins by assigning a value of 4.5 to a variable named rate1. Then, it assigns a new value of 5.3 to this variable. This shows that you can change the value of a variable by assigning a literal value to it.

After changing the value of the rate1 variable, this example creates a variable named rate2 and assigns it to rate1. This shows that you can change the value of a variable by assigning another variable to it.

It's also possible to change the value of a variable by assigning an *expression* to it. For example, the next figure shows how to create arithmetic expressions that you can assign to a variable.

In most cases, when you assign a new value to a variable, you assign a value of the same data type. For instance, the first example creates a variable of the double type and then assigns other values of the double type to that variable.

However, it's possible to assign a value of another data type to a variable as shown by the second example. Here, the first statement shows that the variable named rate1 stores a value of the double type. Then, the second statement assigns a value of the integer type to that variable. To do that, it uses a literal variable of 10L. Next, the final two statements show that the rate1 variable has been changed to the integer type and it now stores an integer value of 10.

The third example shows that you must use the correct case when referring to variable names. In this example, R can't find a variable named Rate1 because variable names are *case-sensitive*. As a result, Rate1 is not the same as rate1.

When you have multiple words in a variable name, you can make the variable name easier to read by using underscores to separate words. This is known as *snake case* because it looks a little like a snake. Alternately, you can use capital letters to start each new word. This is known as *camel case* because the capital letters look a little like a camel's hump. A third possibility is to use periods to separate each word. However, this generally isn't considered a best practice, so we don't recommend this approach.

It's generally considered a best practice to pick one naming convention and use it consistently. This book uses snake case for most variable names. However, some programmers prefer camel case. If you're joining a new team, it's a good idea to ask about naming conventions prior to writing new code.

How to change the value of a variable

```
rate1 <- 4.5
rate1
[1] 4.5
rate1 <- 5.3
rate1
[1] 5.3
rate2 <- 6.8
rate1 <- rate2
rate1
[1] 6.8
```

How to change the data type for a variable

```
typeof(rate1)
[1] "double"
rate1 <- 10L
typeof(rate1)
[1] "integer"
rate1
[1] 10
```

Code that causes an error because of incorrect case

```
Rate1
Error: object 'Rate1' not found
```

Common conventions for coding variable names

```
days_per_week <- 7L      # snake case
daysPerWeek <- 7L        # camel case
days.per.week <- 7L      # period separators
```

Description

- To change the value of a variable, you can assign a literal value, another variable, or an expression such as one of the arithmetic expressions described in the next figure.

- You can assign a value of any data type to a variable, even if that variable has previously been assigned to a value of a different data type.

- Because variable names are *case-sensitive*, you must be sure to use the correct case when coding the names of your variables.

- When you have multiple words in a variable name, you can make the variable name easier to read by using underscores to separate words (*snake case*), by using capital letters to start each new word (*camel case*), or by using periods to separate words.

- It's generally considered a best practice to pick one naming convention and use it consistently.

Figure 1-8   How to work with variables

How to code arithmetic expressions

Figure 1-9 shows how to code *arithmetic expressions*. These expressions consist of two or more *operands* that are operated upon by *arithmetic operators*. The operands in an expression can be numeric variables or numeric literals. This might sound complicated, but it's just the same math you're familiar with from daily life.

The first table in this figure shows the arithmetic operators available in R. The first four operators work as they do in normal mathematics. However, the integer division (**%/%**) and modulo (**%%**) operators may be new to you. The integer division operator performs normal division on the operands but it drops the decimal point from the result and returns an integer (whole number). In contrast, the modulo operator returns only the remainder of the integer division operator.

The second table shows some specific examples of how these operators work. For the most part, this is simple arithmetic. Note that the division operator always returns a double value. However, the integer division operator truncates the result and returns an integer, and the modulo operator returns an integer for the remainder of the integer division. As a result, 5 divided by 3 with regular division is 1.666667, 5 divided by 3 with integer division is 1, and 5 divided by 3 with modular division is 2.

When an expression includes two or more operators, the *order of precedence* determines which operators are applied first. For order of precedence, R uses *PEMDAS* (*Parentheses, Exponents, Multiplication, Division, Addition, Subtraction*). As a result, if you need to override the default order of precedence, you can use parentheses as shown by the third table. Then, R performs the operations in the innermost sets of parentheses first, followed by the expressions in the next set of parentheses, and so on. This works the same as it does in basic algebra.

The arithmetic operators

Operator	Name	Description
+	Addition	Adds two operands.
-	Subtraction	Subtracts the right operand from the left operand.
*	Multiplication	Multiplies two operands.
/	Division	Divides the left operand by the right operand. The result is always a double.
%/%	Integer division	Divides the left operand by the right operand and drops the decimal portion of the result.
%%	Modulo/Remainder	Divides the left operand by the right operand and returns the remainder.
^	Exponentiation	Raises the left operand to the power of the right operand.
**	Exponentiation	Raises the left operand to the power of the right operand.

Examples with two operands

Example	Result
5 + 3	8
5 – 3	2
5 * 3	15
5 / 3	1.666667
5 %/% 3	1
5 %% 3	2
5 ^ 3	125
5 ** 3	125

Examples that show how the order of precedence works

Example	Result
3 + 4 * 5	23
(3 + 4) * 5	35

Description

- An *arithmetic expression* consists of one or more *operands* that are operated upon by *arithmetic operators*.
- If you use multiple arithmetic operators in one expression, R uses *PEMDAS* (*Parentheses, Exponents, Multiplication, Division, Addition, Subtraction*) to control the *order of precedence*. As a result, you can use parentheses to override the sequence the other arithmetic operations.

Figure 1-9   How to code arithmetic expressions

How to use arithmetic expressions in statements

Now that you know how to code arithmetic expressions, figure 1-10 shows how to use these expressions with variables and assignment statements. The first example shows how to use the multiplication operator in an R statement, and the second example shows how to use the integer division and modulo operators.

The third example uses the addition and multiplication operators. This example also uses parentheses to clarify the order of operations. Since multiplication is performed before addition, these parentheses aren't necessary for the code to calculate the perimeter correctly. However, they make the code easier to read by clarifying the order of operations.

The fourth example shows how to calculate the decade for a year. Here, the first statement stores the year 1992 in a variable. Then, the second statement uses integer division to divide by 10. This leaves an integer value 199. Next, the second statement multiplies 199 by 10 to get 1990, which is the decade for 1992. To clarify the order of operations, the expression encloses the division operation in parentheses. As it turns out, the parentheses aren't necessary for this expression. However, they make the code clearer.

The fifth and sixth examples show how to use the constant named pi. This constant is useful for working with circles and arcs. Here, the fifth example displays the constant to show that it stores the value of pi up to six decimal digits. Then, the sixth example use pi to calculate the area of a circle. To do that, it multiplies pi by the radius squared.

Unlike most other languages, R doesn't protect its constants from being changed. This means you could assign a value like 3 to pi to use that value until you remove it from memory. It's best to avoid overwriting constants because other programmers expect them to have the usual value. Instead, you can create your own variable to hold the modified value.

The seventh example shows that you can use logical values in arithmetic expressions. This works because the TRUE keyword evaluates to a value of 1, and the FALSE keyword evaluates to a value of 0. As a result, if you have multiple logical values, you can get a count of how many are TRUE by calculating the sum of all logical values.

The pi constant

Constant	Description
pi	The mathematical constant named pi that's approximately equal to 3.14159.

Calculate rate per 100,000 people

```
rate <- 0.0198
rate_per_100K <- rate * 100000          # rate_per_100K is 1980
```

Calculate hours and remaining minutes

```
minutes <- 127
hours <- minutes %/% 60                 # hours is 2
minutes <- minutes %% 60                # minutes is now 7
```

Calculate the perimeter of a rectangle

```
width <- 4.25
length <- 8.5
perimeter <- (2 * width) + (2 * length)  # perimeter is 25.5
```

Calculate the decade for a year

```
year <- 1992
decade <- (year %/% 10) * 10            # decade is 1990
```

Display the value of the constant named pi

```
pi
[1] 3.141593
```

Use pi to calculate the area of a circle

```
radius <- 10
area <- pi * radius ^ 2                 # area is 314.1593
```

Use arithmetic on logical values

```
true_count <- FALSE + TRUE + FALSE + TRUE + FALSE
true_count
[1] 2
```

Description

- It's common to use an assignment statement to assign the result of an arithmetic expression to a variable.
- R contains some pre-defined constants, like pi, which you can use in arithmetic expressions.
- Unlike some other languages, constants in R are not protected. This means you could assign a value like 3 to pi and use that value until you remove it from memory.

Figure 1-10   How to use arithmetic expressions in statements

How to interpret error messages

When you run a statement, R may encounter an error if you made any coding mistakes such as misspelling a variable name. There are many different types of errors. Regardless of the type, R prints an error message to the console. This message is designed to help you find the source of the error so you can correct it. The example in figure 1-11 shows two common errors in R.

The first error indicates that R can't find an object named numdays. Programmers often introduce this type of error by misspelling a variable name, not capitalizing a variable name correctly, or forgetting to create the variable in the first place. The error message doesn't tell you which one of these possibilities caused the error, so you have to think critically to determine the cause of the error. In this case, the error is due to forgetting to code the underscore in the num_days variable.

The second error occurs when you pass a string to an arithmetic operator. You may think that this is a silly error to make, but it's a common error. That's because the data for an analysis doesn't always get imported with the correct data type. As a result, you may think that you are working with numeric data, but the data may actually be stored with the character type.

Regardless of the source of the error, you often have to do a little digging to find the solution. To start, you can read the error message carefully and think about why the error might be occurring. In many cases, this is enough to solve the problem. However, if the error message doesn't make sense to you, you can enter it into a search engine to learn more about it and to get ideas for how to fix it. Also, if you're working on a team, your team members can be a valuable source of information.

Everyone accidentally makes errors when they code. The important thing is learning how to find and fix the errors you make.

Two common error messages in R

The "object not found" error

```
numdays
Error: object 'numdays' not found
```

The "non-numeric argument to binary operator" error

```
length <- "8.5"
width <- 4.24
area <- length * width
Error in length * width : non-numeric argument to binary operator
```

Description

- When R encounters an error, it prints an error message to the console to help you determine what went wrong.

- "Not found" errors typically occur when a programmer misspells the name of a variable or a function or uses incorrect capitalization.

- "Non-numeric" error messages typically occur when the programmer does not notice a number is being stored as a string.

Figure 1-11   How to interpret error messages

Perspective

Now that you've completed this chapter, you should be able to use RStudio to run R code, and you should have some basic skills for working with R like creating variables and using the arithmetic operators to make calculations. In the next chapter, you'll learn some more R skills that you often need for data analysis.

Summary

- *Data analysis* (or *data analytics)* includes *data visualization* (or *data viz*) as well as *data modeling* (or *predictive analysis*).

- The term *data science* includes data analysis as well as advanced topics such as data mining, machine learning, deep learning, and artificial intelligence (AI).

- With *exploratory data analysis* (*EDA*), the goal is to summarize the main characteristics of the data in order to form new hypotheses that can lead to further analysis.

- Data analysis can be divided up into five phases: *get, clean, prepare, analyze,* and *present* the data. However, there's often overlap between these phases.

- If you don't clean the data for an analysis, it will affect the accuracy of the analysis. In programming, this is often referred to as *garbage in, garbage out* (*GIGO*).

- An *Integrated Development Environment* (*IDE*) is a program that combines several tools for software development.

- RStudio is the most popular IDE for writing R code. It's divided into four windows called *panes* that you can use to develop R code.

- A *statement* performs a task such as creating a variable or displaying a variable on the console.

- A *comment* typically describes a statement or a group of statements.

- A *single-line comment* begins with a hash symbol (#) and continues to the end of the line.

- R stores its variables in the computer's *main memory*, which is often referred to as *RAM* (*Random Access Memory*). This type of storage is temporary.

- When you quit RStudio, you can save the loaded variables in a *workspace image* that's stored in an .RData file.

- You can save R code in a .R file known as an *R script*.

- A *variable* stores data that can change, or *vary*, as code executes.

- To create a variable and assign it an initial value, you can code the name of the variable, followed by the *assignment operator* (<-), followed by a *literal value*. Then, R automatically determines the data type based on the literal value.

- The character data type stores a string of characters, commonly known as a *string*.

- Since the double and integer data types provide a way to store numbers, they are known as the *numeric types*.

- The logical data type stores a *Boolean value*, which is a value that is either true or false.

- Because variable names are *case-sensitive*, you must be sure to use the correct case when coding the names of your variables.

- When you have multiple words in a variable name, you can make the variable name easier to read and understand by using underscores to separate words (*snake case*) or by using capital letters to start each new word (*camel case*).

- An *arithmetic expression* consists of one or more *operands* that are operated upon by *arithmetic operators*.

- If you use multiple arithmetic operators in one expression, R uses *PEMDAS* (*Parentheses, Exponents, Multiplication, Division, Addition, Subtraction*) to control the *order of precedence*.

- When R encounters an error, it displays an error message that describes the error. You can use this error message to find and fix the error.

Before you do the exercises for this chapter

Before you do any of the exercises in this book, you need to install R and RStudio. You also need to download the source code for this book. Appendixes A (Windows) and B (macOS) provide step-by-step instructions for how to do this.

Exercise 1-1   Get started with R and RStudio

This exercise is designed to get you started with RStudio and some of the basic coding skills for R. In addition, it's designed to reassure you that you can have fun experimenting with code, and that your experiments won't cause any errors that can't be quickly and easily fixed.

Use RStudio and run some code from the Console pane

1. Start RStudio.
2. Enter the following code in the Console pane:
   ```
   string_var <- "Testing 1 2 3"
   ```
   You should now see your variable in the Environment pane along with its value.
3. Display your variable's value in the console by typing its name in the Console pane:
   ```
   string_var
   ```
4. Experiment with creating variables of different types and assigning them different values as described in figures 1-7 and 1-8. Create at least one Boolean variable, one integer, one double, and one string. When you do, make sure to try using the keyboard shortcut to type the assignment operator.
5. Use the typeof() function to check the types of your variables.
6. Clear RStudio's memory by clicking the broom icon in the Environment pane. Notice that your variables no longer appear in this pane.
7. Enter the name of one of your variables into the console. Note that RStudio now gives you a "not found" error.
8. Congratulate yourself for successfully programming in R!

Use RSudio to open, run, and edit an R script

9. Open the file named exercise_1-1.R using either File→Open File in RStudio or locating the file on your computer and double-clicking it. If you followed the instructions in the appendix, the file should be located in this folder:
   ```
   Documents/murach_r/exercises/ch01
   ```

10. Click the Run button in the Source pane. This should run the first statement in the script. Note that comments are printed to the console but otherwise don't do anything.

11. Press Ctrl+Enter (Windows) or Command+Enter (macOS). This should run the second statement in the script.

12. Keeping running statements until you have run every line of code in the script. As you do, notice how variables are added to the Environment pane and changed.

13. Examine the code in the Source pane and note how it uses some of the arithmetic operators you learned about in figure 1-9 to calculate percentages. Compare the first calculation, which uses two statements to divide and multiply, to the second calculation, which uses one statement with parentheses.

14. Add a calculation for the number of pirates who chose neither R nor C as their favorite letter and store this number in a new variable. You can name this variable whatever you want.

15. Print a message and the value of your new variable to the console.

Experiment on your own

16. If you have the time and interest, further experiment with RStudio. Try clicking on the different tabs in each pane.

17. Note that RStudio has many options and views to make coding and data analysis as easy as possible while significantly minimizing the possibility of errors that you can't quickly undo.

Chapter 2

More skills for working with R

R is a complex language that provides many features for working with statistics and graphics. As a result, it can take a long time to master the R language. Fortunately, you don't need to master all features of the R language to get started with data analysis. Instead, you just need to learn the R skills presented in the first two chapters of this book.

How to use functions

A *function* is a reusable unit of code that performs a specific task. R provides many built-in functions that you'll use throughout this book.

How to call functions

When you *call* a function, you code the function name followed by a set of parentheses. Within the parentheses, you code any arguments that you want to pass to the function. An *argument* is a value that the function uses to perform its task. If there is more than one argument, you separate the arguments with commas.

A function uses *parameters* to define the arguments that it accepts. Parameters can be required or optional. Optional parameters have default values that are used if you don't specify a value for them. In practice, many programmers use the terms parameter and argument interchangeably. However, this book uses argument to describe a value that's passed to a function and a parameter to describe part of the function's definition.

To show how to call a function, figure 2-1 presents one of R's built-in functions, seq(). This function generates a sequence of numeric values that starts at one value (the from argument) and goes to another value (the to argument). By default, the numbers in this sequence are incremented by a value of 1. However, you can change that by passing the optional third argument (the by argument).

The table that summarizes the seq() function uses boldface to indicate the parts that you must code exactly as shown. In this case, that's just the name of the function and its parentheses. You can substitute your own values for the parameters that aren't boldfaced. In this case, those values are the arguments that you specify within parentheses for the function.

The seq() function requires two arguments named from and to. If you pass two values to a function without specifying the names of the parameters, R assumes that you have coded the values in the order that they are defined in the function. That is, it assumes you have coded the from value first and the to value second. This is known as using *positional arguments*.

However, you can explicitly specify the value that should be used for each argument by coding the parameter name, an equals sign (=), and the value. This is known as using *named arguments*. If you use this approach, it doesn't matter which order you code your arguments in.

The last two examples show how to use the optional parameter named by. This parameter instructs the seq() function to increment the numbers in the sequence by the value of the by parameter. Both examples set the by parameter to a value of 2. As a result, the seq() function increments each number by 2 when it generates the sequence.

The syntax for calling any function

```
function_name(arguments)
```

The seq() function

Function	Description
seq(from, to, by)	Returns a sequence of numbers from one number to another number. By default, it increments each number by 1. To change that, you can specify an amount for the optional by parameter.

How to call a function

With positional arguments
```
seq(1, 10)
[1]  1  2  3  4  5  6  7  8  9 10
```

With named arguments
```
seq(from = 1, to = 10)
[1]  1  2  3  4  5  6  7  8  9 10
```

With named arguments in a different order than the parameters
```
seq(to = 10, from = 1)
[1]  1  2  3  4  5  6  7  8  9 10
```

With the optional third parameter
```
seq(1, 10, 2)
[1] 1 3 5 7 9
```

With the optional third parameter as a named argument
```
seq(from = 1, to = 10, by = 2)
[1] 1 3 5 7 9
```

Description

- A *function* is a reusable unit of code that performs a specific task. Many functions are built into the R language.
- To *call a function*, you code the name of the function followed by a pair of parentheses. Within the parentheses, you code any *arguments* that you want to pass to the function, and you separate multiple arguments with commas.
- A function uses *parameters* to define the arguments that it accepts. Parameters can be required or optional. Optional parameters have default values that are used if you don't specify a value for them.
- You can call functions with *positional arguments* or *named arguments*. Positional arguments must be coded in the same order as defined by the function. Named arguments can be coded in any order.
- In practice, the terms *parameter* and *argument* are often used interchangeably.

Figure 2-1   How to call functions

How to use functions to work with strings

Now that you understand how to call a function, you're ready to learn how to use functions to work with data that's stored with the character type. The character type stores strings of characters commonly known as *strings*. This data type can store zero or more letters, numbers, and symbols.

When working with strings, an *index* is the numeric position of a character in a string. In R, indexes start with the number 1. As a result, you can use 1 to access the first character in a string, 2 to access the second character, and so on. If you are familiar with other programming languages, you are probably used to indexes starting with 0 instead of 1. As a result, it might take you a while to get used to this difference.

The table at the at the top of figure 2-2 summarizes several common functions for working with strings. Here, the paste() function uses an ellipsis (...) for its first parameter. This indicates that you can pass one or more strings to the paste() function.

The first example uses the paste() function to combine strings. To start, the example defines three variables that store two strings and a number. Here, the first statement uses double quotes to create a string, and the second statement uses single quotes. There are pros and cons to each technique, and you may see both in other programmers' code, but this book typically uses double quotes to create strings. The third statement assigns a number to create a value of the double type, which is a numeric type.

After defining the three variables, the first example shows how to use the paste() function to combine these variables into a single string. The first paste() function doesn't specify a value for the separator so it uses the default value, a single space. The second paste() function uses the sep parameter to separate the two strings by a comma and a space. And the third paste() function automatically converts a number to a string. That's what typically happens when you try to pass a number to a parameter that expects a string. However, in some cases, this might cause an error.

The second example uses the grep() function to check if a string contains a *substring*. Here, the first argument specifies the substring to search for and the second argument specifies the string. If the function finds the substring, it returns a value of 1, which is equivalent to a Boolean value of TRUE. By default, this function uses a case-sensitive comparison. However, if you want to ignore case in the comparison, you can set the ignore.case parameter to TRUE.

The third example uses the substr() function to extract a substring from a string. This function takes three arguments: the string, the start index, and the stop index. Here, the code extracts two numbers from the string and stores them in variables named min_age and max_age.

The fourth example uses the sub(), gsub(), toupper(), and tolower() functions to modify a string. These functions operate on the col_name variable and store the string that's returned in the same variable. As a result, they modify the original variable by overwriting it. Here, the first sub() function replaces the first space character with an underscore character, the second sub() function replaces

Some functions for working with strings

Function	Description
`paste(..., sep)`	Combines two or more strings into one string. Each combined string is separated by the specified separator string.
`substr(str, start, stop)`	Extracts part of a string from the start index to the stop index.
`grep(pattern, str, ignore.case)`	Returns 1 if the pattern is in the string. By default, the comparison is case-sensitive. To ignore case, you can set the optional ignore.case parameter to TRUE.
`sub(pattern, replacement, str)`	Replaces the first found pattern with the replacement string.
`gsub(pattern, replacement, str)`	Replaces all instances of a pattern with the replacement string.
`toupper(str)`	Converts the string to uppercase.
`tolower(str)`	Converts the string to lowercase.

How to combine strings

```
first_name <- "Bob"                          # double quotes
last_name <- 'Smith'                          # single quotes work too
age <- 40

paste(first_name, last_name)                 # returns "Bob Smith"
paste(last_name, first_name, sep = ", ")     # returns "Smith, Bob"
paste(first_name, "is", age, "years old.")   # returns "Bob is 40 years old."
```

How to check if a string contains a substring

```
state <- "Maine CD-1"
grep("CD-", state)                       # returns 1, which is equivalent to TRUE
grep("cd-", state)                       # does not return 1
grep("cd-", state, ignore.case = TRUE)   # returns 1, TRUE
```

How to get substrings

```
col_name <- "01-04 Years"
min_age <- substr(col_name, 1, 2)    # returns "01"
max_age <- substr(col_name, 4, 5)    # returns "04"
```

How to modify a string

```
col_name <- "05-08 Years"
col_name <- sub(" ", "_", col_name)    # returns "05-08_Years"
col_name <- sub("-", "_", col_name)    # returns "05_08_Years"
col_name <- gsub("0", "", col_name)    # returns "5_8_Years"
col_name <- toupper(col_name)          # returns "5_8_YEARS"
col_name <- tolower(col_name)          # returns "5_8_years"
```

How to nest function calls

```
name <- "years"
toupper(substr(name, 1, 1))            # "Y"
```

Description

- R provides many functions for working with *strings* of characters.
- In R, the first character in a string has an index of 1, the second has an index of 2, and so on. This is different from many other programming languages, which start at 0.

Figure 2-2   How to use functions to work with strings

the dash character with an underscore character, and the gsub() function removes both zeroes by replacing them with an empty string. Then, the toupper() function converts the letters in the string to uppercase, and the tolower() function converts the letters in the string to lowercase.

The fifth example in figure 2-2 shows how to *nest* function calls. When you nest functions, you code one function as the argument for another function. You can do this as many times as you want as long as each inner function returns the correct data type for the outer function. In this example, the substr() function is nested within the toupper() function to get a capital letter for the first letter in the name variable.

How to use functions to work with numbers

The table at the at the top of figure 2-3 summarizes some common functions for working with numbers. To start, the first example shows how to use the round() function to round numbers. The first round() function doesn't pass a second argument that specifies the number of decimal digits. As a result, the round() function rounds the number to the nearest whole number. However, the second and third round() functions specify the number of decimal digits. As a result, the round() function rounds to the nearest number with the specified number of decimal digits.

The first example also shows how to use the floor(), ceiling(), and trunc() functions to round or truncate numbers. You can use ceiling() to round up (towards the ceiling), you can use floor() to round down (towards the floor), and you can use trunc() to truncate the decimal digits without rounding. These three functions don't allow you to specify the number of decimal digits as they always return whole numbers.

The second example shows how to use the abs() function to get the absolute value of a number. This removes the negative sign from negative numbers and leaves positive numbers unchanged.

The third example shows how to use functions to work with square roots and logarithms. As you would expect, the sqrt() function returns the square root of the given argument, and the log() and log10() functions return the natural logarithm and base 10 logarithm for the given argument. You may occasionally need these functions for data modeling as described in section 3 of this book.

Some functions for working with numbers

Function	Description
round(num, digits)	Rounds the number to the nearest number with the specified number of digits after the decimal point.
floor(num)	Rounds the number down to the nearest integer.
ceiling(num)	Rounds the number up to the nearest integer.
trunc(num)	Truncates the decimal part of a number.
abs(num)	Returns the absolute value of the number.
sqrt(num)	Returns the square root of the number.
log10(num)	Returns the base 10 logarithm of the number.
log(num)	Returns the natural logarithm of the number.

How to round and truncate numbers

```
percent = 10.193456
round(percent)          # returns 10
round(percent, 1)       # returns 10.2
round(percent, 2)       # returns 10.19

floor(percent)          # returns 10
ceiling(percent)        # returns 11
trunc(percent)          # returns 10
```

How to get the absolute value of a number

```
abs(-7.77)              # returns 7.77
```

How to work with square roots and logarithms

```
sqrt(16)                # returns 4
log10(100)              # returns 2
log(2.718*2.718)        # returns 1.999793 (approximately 2)
```

Description

- R provides many functions for working with numbers.

Figure 2-3   How to use functions to work with numbers

How to work with data structures

Data structures provide structures for storing data. Every type of data structure has its advantages and disadvantages, and R provides many types of built-in data structures. This chapter describes three data structures that are commonly used in data analysis: vectors, data frames, and lists.

How to work with vectors

Vectors store a collection of items of the same data type. Figure 2-4 begins by showing how to create a vector. To do that, you can code the c() function. Within the parentheses, you can code any number of items separated by commas. Then, the c() function *combines* these items into a vector.

If you want to create a vector for a range of numeric values, you can generate the vector by using the seq() function. Or, if you prefer, you can use the *colon operator* (:) to generate a vector of values starting at the operand on the left side of the colon and ending at the operand on the right side of the colon, incremented by 1.

In a vector, all of the values must have the same data type. If you attempt to store values of different types in a vector, R *coerces* the values into a single data type. For example, if you try to store numbers and strings in a vector, R coerces the numbers to the character type and stores all items with the character type. Since this can lead to unexpected results, it's best to avoid mixing data types in a vector whenever possible.

To access items in a vector, you code the name of the vector followed by a set of brackets ([]). Within the brackets, you can code one or more indexes. To select a single item, you can provide a single index. For example, an index of 1 selects the first item, 2 selects the second item, and so on.

To select all items except a specified item, you can code a negative index. For example, an index of -1 selects all items except the first item. To select multiple values, you can pass a vector of indexes. To do that, it's common to use colon operator (:) to generate a range of indexes. However, if you need more flexibility, you can code the vector that contains the indexes.

A function for creating a vector

Function	Description
c(...)	Combines the specified arguments to form a vector.

How to create a vector

A vector of numbers
```
nums_of_week <- c(1,2,3,4,5,6,7)
```

A vector of strings
```
days_of_week <- c("Mon","Tues","Wed","Thurs","Fri","Sat","Sun")
```

A vector that stores a range of numbers
```
ids <- seq(1, 10)
```

Another vector that stores a range of numbers
```
ids <- 1:10
```

Code that attempts to store items of different types in a vector
```
mixed <- c(1,"Tues",3,"Thurs",5,"Sat",7)
mixed
[1] "1"      "Tues" "3"      "Thurs" "5"      "Sat"    "7"
```

How to access the items in a vector

The item at the specified index
```
days_of_week[1]
[1] "Mon"
```

All items except the item at the specified index
```
days_of_week[-1]
[1] "Tues"  "Wed"    "Thurs" "Fri"    "Sat"    "Sun"
```

All items in the range of indexes
```
days_of_week[1:5]
[1] "Mon"    "Tues"   "Wed"    "Thurs" "Fri"
```

All items in the vector of indexes
```
days_of_week[c(1,3,5)]
[1] "Mon" "Wed" "Fri"
```

Description

- A *vector* can be used to store multiple related items in a single data structure.
- If you try to store items of different data types in a vector, R *coerces* the items into a single type.
- To access the items in a vector, you use a number known as an *index*.
- To access items stored in a vector, you code the name of the vector followed by a set of brackets. Within the brackets, you can code a single index, a negative index, a range of indexes, or a vector of indexes.

Figure 2-4   How to work with vectors (part 1)

Part 2 of figure 2-4 presents some additional skills for working with vectors. To start, the first example shows how to create a vector of temperatures that range from 77.4 degrees to 101.3 degrees. Then, the second example shows how to use the length() function to get the number of items in the vector and how to use the typeof() function to display the data type of the items stored in the vector.

A *named vector* works much like a regular vector. However, it provides a name for each of its items. This allows you to store some additional data (the names) in the vector.

To create a named vector, you can call the c() function. Within the parentheses for this function, you code the items for the vector, using commas to separate each item. For each item, you code the item's name, the parameter assignment operator (=), and the item's value. For instance, the third example creates a named vector that provides the names for each item. So, the item named "Mon" has a value of 99.8, the item named "Tues" has a value of 97.2, and so on.

When you display a named vector on the console, R displays its names as well as the values for each item. If you only want to display the names for a vector, you can pass the vector to the names() function as shown in the fourth example.

You can also use the names() function to add or modify the names in an existing vector as shown in the fifth example. To do that, you can use the names() function to assign a vector of names to the names area. The vector of names must be the same length as the vector you want to add the names to.

Once you create a named vector, you can use indexes or names to access its items as shown by the sixth example. In this example, both statements access the second item, but the first statement uses an index to get the second item while the second statement uses a name to get the second item.

The syntax for adding names to items in a vector

```
c(name1 = item1, name2 = item2, ...)
```

Functions for working with vectors

Function	Description
length(vec)	Returns the number of items in the vector.
names(vec)	Returns a vector of names for a named vector or NULL for a vector that doesn't use names.

A vector of temperatures

```
temps <- c(99.8,97.2,77.4,101.3)
temps
[1]  99.8  97.2  77.4 101.3
```

How to get the length and type of a vector

```
length(temps)
[1]  4
typeof(temps)
[1] "double"
```

How to create a named vector

```
temps <- c("Mon" = 99.8, "Tue" = 97.2,
           "Wed" = 77.4, "Thu" = 101.3)

temps
  Mon   Tue   Wed   Thu
 99.8  97.2  77.4 101.3
```

How to get the names for a named vector

```
names(temps)
[1] "Mon" "Tue" "Wed" "Thu"
```

How to add or modify the names of an existing vector

```
names(temps) <- c("M","Tu","W","Th")
```

Two ways to get values from a named vector

```
temps[2]      # by index
  Tu
97.2
temps["Tu"]   # by name
  Tu
97.2
```

Description

- A *named vector* provides names for its items. This allows you to access each item by its name or its index.
- Names can be added to a vector when you create it or after you create it.

Figure 2-4   How to work with vectors (part 2)

How to work with data frames

Vectors work well for collecting related values of a single data type. However, when analyzing data, you often want to store multiple vectors together. To do this, you can use a *data frame*. Data frames store *tabular data*, which is data that is organized into a table. Data frames are made up of *rows* (also called *observations*) and *columns* (also called *variables*).

Figure 2-5 begins by summarizing the data.frame() function that you can use to create a data frame. To do that, you code the data.frame() function. Then, within its parentheses, you can code as many columns as you want, using commas to separate each column. For each column, you typically code the column name, the equals operator (=), and a vector that provides the values for the column.

The first example shows how to create a data frame that stores information about each month. To start, this example creates three vectors. The first vector stores the month's name, the second stores the number of days in the month, and the third stores the season for the month. Then, the data.frame() function uses these three vectors to provide the data for the three columns in this data frame and names the columns Month, NumDays, and Season.

When you display a data frame, the console includes an extra column to the left. This column contains the index for each row, starting from 1. This figure shows that the indexes range from 1 to 5 for the first five rows.

The second example shows how to select the vector for a column. To do that, you typically use the $ operator to specify the name of the column. However, that technique doesn't work if the name of the column contains spaces. In that case, you need to use the double brackets ([[ ]]) to specify the name or index for the column.

The third example shows how to select columns, rows, and values from a data frame. To do that, you can begin by coding the name of the data frame followed by a set of brackets ([ ]). Within the brackets, you can code indexes or names that select the columns and rows that you want.

To select columns only, you can code the index or name for the column. When you use this technique, it selects the entire column including its name, not just the vector that stores the values for the column. To select rows only, you can code the index or name for the row followed by a comma. You can think of this as leaving the column argument blank to mean that you want all columns. To select a subset of columns and rows, you can code the indexes or names for the rows, a comma, and the indexes or names for the columns.

Once you select part of the data frame, you can assign a new value to it as shown in the fourth example. The first statement assigns a value of 29 to the second row and second column. This replaces the value of 28 at this location in the original data frame. In other words, it replaces the number of days for February. Then, the second statement assigns a value of "Autumn" to rows 9 through 11 for the column named Season.

A function for creating a data frame

Function	Description
data.frame(...)	Creates a data frame by combining the specified vectors to form the columns. Typically, you specify a column name before each vector.

How to create a data frame

```
# create three vectors for the data frame
months <- c("Jan","Feb","Mar","Apr","May","June","July","Aug","Sept",
            "Oct","Nov", "Dec")
num_days <- c(31,28,31,30,31,30,31,31,30,31,30,31)
seasons <- c("Winter","Winter","Spring","Spring","Spring","Summer",
             "Summer","Summer","Fall","Fall","Fall","Winter")

# combine the vectors to make the data frame
calendar_df <- data.frame(
  Month = months, NumDays = num_days, Season = seasons)

calendar_df
   Month NumDays Season
1    Jan      31 Winter
2    Feb      28 Winter
3    Mar      31 Spring
4    Apr      30 Spring
5    May      31 Spring
...
```

How to select the vector for a column

```
calendar_df$Month          # by name using $ operator
calendar_df[["Month"]]     # by name using double brackets
calendar_df[[1]]           # by index using double brackets
```

How to select columns, rows, and values

```
calendar_df[1]             # select 1st col by index
calendar_df["Month"]       # select 1st col by name
calendar_df[1,]            # select 1st row, all columns
calendar_df[2:4,1:2]       # select the 2nd to 4th rows, 1st to 2nd columns
calendar_df[4,2]           # select 4th row, 2nd column
```

How to assign new values to selected cells

```
calendar_df[2,2] <- 29
calendar_df[9:11,"Season"] <- "Autumn"
```

Description

- A *data frame* provides a way to store data in a table.
- In a data frame, the *rows* are known as *observations*, the *columns* are known as *variables*, and the individual data points are known as *values*.
- In a data frame, you can uniquely identify each column by its column name or index, and you can uniquely identify each row by its index.

Figure 2-5   How to work with data frames

Unfortunately, using the bracket selector often results in code that's hard to read and understand. That's why this book largely avoids using it. Instead, this book typically uses the select() and filter() functions described in section 2. However, many programmers use the bracket selector. As a result, when you review code from other programmers, you're bound to come across it.

How to work with lists

A *list* is a special type of vector that can store values of different data types, including other lists. This makes lists optimal for storing data that doesn't fit well in a data frame.

Figure 2-6 begins by summarizing two functions that are commonly used with lists. To start, you can use the list() function to create a list. To do that, you call the list() function. Within its parentheses, you can code as many items as you want, separated by commas. For each column, you can optionally provide a name for the item by coding its name and the equals operator (=) before the item. This works similarly to supplying the names for columns in a data frame.

The first example shows how to create a list that stores three items that contain information about each month. The first item doesn't have a name and stores a string, "Misc Calendar Data". The second item has a name of headers and stores a vector. And the third item has a name of rows and stores another list. This list contains two more lists that provide data for two rows. Here, each row uses more than one data type.

To display the structure of this list on the console in a compact way, this example passes the list to the str() function. The resulting output shows the structure of the list. In particular, it shows that the list has three items where the first item is a string, the second item is a character vector that contains three strings, and the third item is a list that contains two more lists that contain three items each.

To retrieve an item from a list, you can use the bracket selector as shown in the second example. However, this retrieves the item, not the item's value. That's why the second example displays the name of the second item, which is $headers, before displaying the value of the item, which is the vector of strings.

To retrieve an item's value, you need to code two sets of brackets as shown in the third example. This example retrieves the vector that's stored as the value of the second item, and it doesn't retrieve the name of this item.

Both the second and third examples use an index to select the second item in the list. However, if an item has a name, you can use its name to select it as shown by the fourth example. Here, the first statement uses a name to get the vector that's stored as the value of the second item. The second statement gets the item for the first row in the rows item. And the third statement gets the value for the second column of the first row.

Two functions for working with lists

Function	Description
list(...)	Creates a list by combining the specified items. Optionally, you can supply a name for an item.
str(object)	Displays the structure of the object in a compact way.

How to create a list and display its structure

```
calendar_list <- list(
  "Misc Calendar Data",
  headers = c("Month","NumDays","Season"),
  rows = list(
    list("Jan",31,"Winter"),
    list("Feb",28,"Winter")
  ))
str(calendar_list)
List of 3
 $ : chr "Misc Calendar Data"
 $ headers: chr [1:3] "Month" "NumDays" "Season"
 $ rows    :List of 2
  ..$ :List of 3
  .. ..$ : chr "Jan"
  .. ..$ : num 31
  .. ..$ : chr "Winter"
  ..$ :List of 3
  .. ..$ : chr "Feb"
  .. ..$ : num 28
  .. ..$ : chr "Winter"
```

How to use single brackets to select an item

```
calendar_list[2]                        # get the 2nd item (headers)
$headers                                # The item's name
[1] "Month"    "NumDays" "Season"       # The item's value
```

How to use double brackets to select an item's value

```
calendar_list[[2]]                      # get the value of the 2nd item (headers)
[1] "Month"    "NumDays" "Season"       # The item's value
```

How to use names or indexes to select nested items

```
calendar_list[["headers"]]         # get value for headers item
calendar_list[["rows"]][1]         # get the first item in rows
calendar_list[["rows"]][[1]][[2]]  # get value for row 1, column 2
```

Description

- A *list* is a type of vector that can store multiple items with different data types, including nested lists.

- With a list, use single brackets ([]) to select an item and double brackets ([[]]) to select an item's value. If an item has a name, you can select it by its name or its index.

Figure 2-6   How to work with lists

How to add values to data structures

After creating a vector or list, you might want to add more values to it. Similarly, after creating a data frame, you might want to add more rows or columns to it. Fortunately, R provides functions that make this easy to do. The table at the top of figure 2-7 summarizes some of these functions.

To add values to a vector or list, you can use the append() function. It has two required arguments, the vector or list and the values you want to add. If you want to add more than one value, you can collect the values in another vector or list as shown in the first and second examples. By default, the append() function appends the values to the end of the data structure. However, you can use the optional third parameter to insert the values at the specified the index.

To append a row to a data frame, you can use the rbind() function. This function takes two or more data structures and combines them by row. If a row uses different data types to store its data, it typically makes sense to use a list to store the data as shown in the third example. That way, the row can use different data types for each column. However, if a row only contains columns of the same data type, a vector works equally well.

The cbind() function works similarly to rbind() but it appends columns, not rows. Also, cbind() typically takes vectors as arguments, not lists. When you add columns, you can optionally name them. To do that, you can use the same syntax as for naming a column when you use data.frame() to create a data frame. For instance, the fourth example provides a name of AvgTemp for the column that's appended to the data frame.

Functions for adding values to data structures

Function	Description
`append(x, values, after)`	Appends the given values to a vector or list. To append more than one value, use a vector or list. By default, values are added to the end of the vector or list, but you use the after argument to insert the value at the given index.
`rbind(df, ...)`	Short for row bind. Appends one or more vectors or lists to a data frame as rows.
`cbind(df, ...)`	Short for column bind. Appends one or more vectors to a data frame as columns.

How to add values to a vector

```
numbers <- c(1,2,3,4)
numbers <- append(numbers, 5)              # numbers is 1,2,3,4,5
numbers <- append(numbers, c(6,8))         # numbers is 1,2,3,4,5,6,8
numbers <- append(numbers, 7, after = 6)   # numbers is 1,2,3,4,5,6,7,8
```

How to add values to a list

```
letters <- list("a","b","c")                        # letters is a,b,c
letters <- append(letters, "d")                     # letters is a,b,c,d
letters <- append(letters, c("e","h"))              # letters is a,b,c,d,e,h
letters <- append(letters, list("f","g"), after = 5) # letters is a,b,c,d,e,f,g,h
```

How to add rows to a data frame

```
calendar_df <- rbind(calendar_df,
                 list("NewJan", 31, "Bonus"),
                 list("NewFeb", 29, "Bonus"))
...
11     Nov     30   Fall
12     Dec     31 Winter
13 NewJan      31   Bonus
14 NewFeb      28   Bonus
```

How to add columns to a data frame

```
avg_temp <- c(52.3,56.1,62,68.9,75.9,80.7,81.9, 82.7,78.5,70.7,61.1,54.3,0,0)
calendar_df <- cbind(calendar_df, AvgTemp = avg_temp)
  Month NumDays Season AvgTemp
1   Jan      31 Winter    52.3
2   Feb      28 Winter    56.1
3   Mar      31 Spring    62.0
...
```

Description

- To add more than one value at a time to a vector or list, you can pass a vector or a list to the append() function.

Figure 2-7   How to add values to data structures

How to code Boolean expressions

Boolean expressions (or *conditional expressions*) evaluate to either TRUE or FALSE. They are typically based on a comparison between two or more values. You use these expressions in if statements, and you can pass them as arguments to some functions.

How to use the relational operators

Figure 2-8 shows how to code Boolean expressions that use *relational operators*. To start, the table summarizes the six relational operators. Then, the examples show how these operators work.

For instance, the first expression evaluates to TRUE if the variable named num_days is equal to 30. Otherwise, this expression evaluates to FALSE. Similarly, the second expression evaluates to TRUE if the season variable is equal to "Winter". Otherwise, it evaluates to FALSE.

The rest of the examples work similarly, although they use the other operators. They also show that you can compare a variable to a literal value such as 30 or to another variable such as max_days.

The last example shows that you can use an arithmetic expression as an operand in a Boolean expression. This works because the arithmetic expression evaluates to a value that can be used as an operand.

When you're coding Boolean expressions, you need to make sure that your expressions compare numbers with numbers and strings with strings. If you compare a number with a string, it doesn't cause an error to occur, but it might not yield the result you expect. On the other hand, if you compare an integer with a decimal number, the values are compared numerically, so the expression should yield the result you expect.

When you use the equal to operator (==), you need to remember to code two equals signs, not one. That's because R uses one equals sign to assign a value, not to compare two values. This is a common mistake when you're learning to program, and it causes a syntax error.

Last, due to the way computers store floating-point numbers, double values aren't always exact. As a result, you shouldn't use any of the operators that check for equality (==, !=, >=, or <=) when comparing them. Instead, you should use the greater than (>) and less than (<) operators.

Relational operators

Operator	Name	Description
==	Equal to	Returns TRUE if both operands are equal.
!=	Not equal to	Returns TRUE if the left and right operands are not equal.
>	Greater than	Returns TRUE if the left operand is greater than the right operand.
<	Less than	Returns TRUE if the left operand is less than the right operand.
>=	Greater than or equal to	Returns TRUE if the left operand is greater than or equal to the right operand.
<=	Less than or equal to	Returns TRUE if the left operand is less than or equal to the right operand.

Boolean expressions

```
num_days == 30          # variable equal to numeric literal
season == "Winter"      # variable equal to string literal

num_days != 28          # variable not equal to a numeric literal

distance > 5.6          # variable greater than a numeric literal
num_days < max_days     # variable less than a variable

num_days >= max_days    # variable greater than or equal to a variable
num_days <= max_days    # variable less than or equal to a variable

rate / 100 >= 0.1       # expression greater than or equal to a literal
```

Description

- You can use the *relational operators* to create a *Boolean expression* (or *conditional expression*) that compares two operands and returns a Boolean value of TRUE or FALSE.

- Because the double data type doesn't store floating-point numbers as exact values, you shouldn't use any of the operators that check for equality to compare them.

- When coding Boolean expressions, be sure to compare numbers with numbers and strings with strings. Mixed expressions might not work the way you expect them to.

Figure 2-8   How to use the relational operators

How to use the logical operators

A *compound conditional expression* uses the *logical operators* shown in figure 2-9 to combine two or more Boolean expressions. If you use the AND operator, the compound expression only returns TRUE if both expressions are TRUE. If you use the OR operator, the compound expression returns TRUE if either expression is TRUE. If you use the NOT operator, the value returned by the expression is reversed.

Below the table of operators, this figure presents their *order of precedence*. This defines the order in which the operators are evaluated for compound expressions that use more than one logical operator. This means that NOT operators are evaluated before AND operators, which are evaluated before OR operators. Although this is normally what you want, you can override this order by using parentheses.

The expressions in this figure show how these operators work. To start, the first expression uses the AND operator to connect two Boolean expressions. As a result, it evaluates to TRUE if both the expression on its left *and* the expression on its right are TRUE. Similarly, the second expression uses the OR operator to connect two Boolean expressions. As a result, it evaluates to TRUE if either the expression on its left *or* the expression on its right is TRUE.

The third and fourth expressions show how to use the NOT operator to reverse the value of an expression. As a result, these expressions evaluate to TRUE if the age variable is *not* greater than or equal to 65. In this case, using the NOT operator is okay, but often the NOT operator results in code that's difficult to read. That's why it's a good practice to rewrite your code so it doesn't use the NOT operator. To do that, you could code this expression:

```
age < 65
```

This code is shorter, simpler, and easier to read.

The next four expressions show that compound conditions aren't limited to two expressions. For instance, the fifth expression uses two AND operators to connect three Boolean expressions, and the sixth expression uses two OR operators to connect three expressions.

You can also mix AND and OR operators as shown in the seventh and eighth expressions. In the seventh expression, the parentheses *clarify* the sequence of operations since R would perform the AND operation first anyway. In the eighth expression, though, the parentheses *change* the sequence of operations. As a result, R evaluates the two expressions connected by the OR operator before evaluating the rest of the expression.

Logical operators

Operator	Name	Description
&	AND	Returns a TRUE value if both expressions are TRUE.
|	OR	Returns a TRUE value if either expression is TRUE.
!	NOT	Reverses the value of a Boolean expression (TRUE becomes FALSE and FALSE becomes TRUE).

Order of precedence

1. NOT operator
2. AND operator
3. OR operator

Boolean expressions that use logical operators

```
# The AND operator
age >= 65 & city == "Chicago"

# The OR operator
age >= 65 | city == "Greenville"

# The NOT operator
!age >= 65

# The NOT operator with parentheses to clarify sequence of operations
!(age >= 65)

# Two AND operators
age >= 65 & city == "Greenville" & state == "SC"

# Two OR operators
age >= 65 | age <= 18 | status == "retired"

# AND and OR operators with parens to clarify the sequence of operations
(age >= 65 & status == "retired") | age < 18

# AND and OR operators with parens to change the sequence of operations
age >= 65 & (status == "retired" | state == "SC")
```

Description

- You can use the *logical operators* to join two or more Boolean expressions. This can be referred to as a *compound conditional expression*.

- The *order of precedence* determines the sequence in which the logical operators are executed. However, you can use parentheses to clarify or change that sequence.

Figure 2-9   How to use the logical operators

How to work with control structures

Like all programming languages, R provides *control structures* that let you control how the code is executed. These control structures include if statements, for loops, and functions.

How to code if statements

An *if statement* lets you control the execution of statements based on the results of a Boolean expression. Figure 2-10 shows how to code an if statement. Each if statement must start with an *if clause*. The if clause can be followed by one or more *else if clauses*, but they are optional. An *else clause* can be included at the end of the if statement, but it is also optional.

The examples in this figure show how this works. Here, the first example shows an if statement with only an if clause. To code the if clause, you code the if keyword followed by a set of parentheses and a set of braces. Within the parentheses, you code a condition. Within the braces, you code one or more statements that are executed if and only if the condition evaluates to TRUE. In this example, if the days variable is less than 30, the if clause sets the month variable to "February". Otherwise, this if statement doesn't do anything.

The second example shows an if statement with an else clause. Coding an else clause works similarly to coding an if clause, except that you just code the else keyword followed by braces. There is no need for parentheses and a condition because it executes only if all prior clauses are false. In this example, if the size variable is greater than 100, the if clause sets the desc variable to "Large". Otherwise, the else clause sets the desc variable to "Small".

The third example shows the same code for the second example after the braces have been removed and both clauses have been coded on the same line. This works because the braces for a clause are optional if the clause only contains a single statement. In general, it's considered a good practice to code the braces for a clause. That way, if you need to add more statements to a clause, you're ready to do it. It also makes your code easier to read.

The fourth example shows an if statement with two else if clauses and an else clause. Coding an else if clause works like coding an if clause, except that you start with the else if keywords. As a result, for this example, if the size variable is greater than 100, the if clause sets the desc variable to "Large". Otherwise, R evaluates the first else if clause. Then, if the size variable is greater than 70, the first else if clause sets the desc variable to "Medium Large". Otherwise, R evaluates the second else if clause. Then, if the size variable is greater than 30, the second else if clause sets the desc variable to "Medium Small". Otherwise, the else clause sets the desc variable to "Small".

Note that the sequence of the conditions matters when you use else if clauses. For instance, the fourth example wouldn't work correctly if you reversed the three conditions in the if and else if clauses. In that case, the condition in the if clause would check if the size is greater than 30. As a result, R would never evaluate the else if clauses.

An if statement

```
if (days < 30) {
  month <- "February"
}
```

An if statement with an else clause

```
if (size > 100) {
  desc <- "Large"
} else {
  desc <- "Small"
}
```

An if statement with no braces that performs the same task

```
if (size > 100) desc <- "Large" else desc <- "Small"
```

An if statement with two else if clauses and an else clause

```
if (size > 100) {
  desc <- "Large"
} else if (size > 70) {
  desc <- "Medium Large"
} else if (size > 30) {
  desc <- "Medium Small"
} else {
  desc <- "Small"
}
```

Description

- An *if statement* always contains an *if clause*. In addition, it may optionally contain one or more *else if clauses* and one *else clause*.

- When an if statement is executed, the condition in the if clause is evaluated first. If it is true, the statements in the if clause are executed and the if statement ends. Otherwise, if there are additional clauses, the condition in the next clause is evaluated.

- If the condition in the next clause is true, the statements in that clause are executed and the if statement ends. This continues until the condition in a clause is true or an else clause is reached.

- The statements in the else clause are executed if and only if the conditions in all of the preceding clauses are false.

- An if statement can only execute the statements in one clause. After that, R executes the code after the if statement.

Figure 2-10   How to code if statements

How to code nested if statements

To get the logic for an if statement right, you may sometimes need to code one if statement within a clause of another if statement. The result is known as *nested if statements*.

The first example in figure 2-11 shows how to use a nested if statement to determine the number of points for a shot in a basketball game. Here, the outer if statement checks if the shot_result variable is equal to "made". If so, the nested if statement checks if the shot is a 3-point shot. If so, it assigns a value of 3 to the points variable. Otherwise, it checks if the shot is a 2-point shot. If so, it assigns a value of 2 to the points variable. On the other hand, if the shot_result variable isn't equal to "made", the shot must have been missed. In that case, the else clause in the outer if statement assigns a value of 0 to the points variable.

The second example gets the same results as the first example, but it doesn't use nested if statements. Instead, it uses compound conditions in the if and else if clauses. Some programmers find this if statement easier to read and understand since all of the conditions are coded at the same level.

However, the second example duplicates the condition that checks the shot_result variable. This code duplication makes your code more difficult to maintain. If, for example, you need to change the name of the shot_result variable, you'd have to make that change in two places in the second if statement. By contrast, you'd only have to change the code in one place in the first example. Furthermore, it's debatable whether the second if statement is any easier to read than the first statement.

A nested if statement

```
if (shot_result == "made") {
  if (shot_type == "3pt shot") {
    points <- 3
  } else if (shot_type == "2pt shot") {
    points <- 2
  }
} else {
  points <- 0
}
```

Another way to get the same result

```
if (shot_result == "made" & shot_type == "3pt shot") {
    points <- 3
} else if (shot_result == "made" & shot_type == "2pt shot") {
    points <- 2
} else {
  points <- 0
}
```

Description

- It's possible to code one if statement within a clause of another if statement. The result is known as *nested if statements*.

- In some cases, you can use the logical operators to get the same results that you get with nested if statements.

Figure 2-11   How to code nested if statements

How to code for loops

Figure 2-12 shows how to use *for statements* to define *for loops* that run once for each item in a collection of items. The first example shows how this works by using the print() function to print the integers from 1 to 3 to the console. To do that, the code begins with the for keyword followed by a pair of parentheses and a pair of braces. Within the parentheses, the code defines a variable named i, followed by the in keyword, and a vector of numbers from 1 to 3. Within the braces, the code contains a print() function that prints the variable named i to the console.

When R executes this for statement, it stores the first number in the vector in the variable named i. Then, for each iteration of the loop, this variable receives the next number in the vector of numbers. When the statement finishes looping through all of the numbers in the vector, it ends.

Incidentally, it's common to use a variable name of i for numbers that can be integers or indexes. However, you can use any name you want. For instance, this example would work equally well if this variable name was changed from i to num.

The second example uses a for loop to get a sum of the numbers in the vector named grades. Before the loop, this example creates a variable named total and assigns it a value of 0. Then, this code defines a for loop that loops through each grade in the grades vector. Each time through the loop, the statement within the loop's braces adds the next grade to the total variable. To do that, it codes the total variable on both sides of the assignment operator. That way, the loop adds each grade to the total so far instead of overwriting the value of the total variable with the grade. When the loop ends, this code displays the total variable on the console.

The third and fourth examples show how to use a loop to modify some or all of values in a vector. To do that, you must use an index to access the value in the vector. Then, you can assign a new value to that index.

The third example starts by defining a for loop that loops through every index in the grades vector. To do that, it begins at the first index and uses the length() function to get the value for the last index. Then, within the braces for the loop, the code uses the index to get the value at the specified index and add a value of 10 to it. This increases each grade by 10 points.

The fourth example works much like the third example. However, it uses a nested if statement to check if each value is greater than 100. If so, it sets the value at that index to 100. As a result, this modifies all grades higher than 100 but doesn't modify the other grades.

A function for printing data to the console

Function	Description
print(val)	Prints the specified value to the console.

A for loop that prints the numbers 1 through 3 to the console

```
for (i in 1:3) {
  print(i)
}
[1] 1
[1] 2
[1] 3
```

A for loop that sums the numbers in a vector

```
grades <- c(100,90,110,70)
total <- 0
for (grade in grades) {
  total <- total + grade
}
total
[1] 370
```

A for loop that modifies all values in a vector

```
grades <- c(100,90,110,70)
for (i in 1:length(grades)) {
    grades[i] <- grades[i] + 10
}
grades
[1] 110 100 120  80
```

A for loop that modifies some values in a vector

```
grades <- c(100,90,110,70)
for (i in 1:length(grades)) {
  if (grades[i] > 100) {
    grades[i] <- 100
  }
}
grades
[1] 100  90 100  70
```

Description

- A *for statement*, also known as a *for loop*, provides a way to loop through every value in a vector.
- If you need to modify the values in the vector, you can use an index to access the value in the vector. Then, you can assign a new value to the value at that index.

Figure 2-12   How to code for loops

How to define functions

Earlier in this chapter you learned how to use some of the functions that are available as part of the R language. However, there will be times when you want to perform a task, but you can't find an existing function that works for it. In that case, you can define your own function. The examples in figure 2-13 show how to do that.

The first example shows how to define a function that prints a greeting to the console. This function doesn't define any parameters or return any data. Instead, it contains a single statement that prints "Greetings!" to the console. To do that, this example defines the function by coding the function keyword followed by an empty set of parentheses and a set of braces. Within the braces, this example codes a single statement that prints "Greetings!" to the console.

After defining the function, the code assigns it to the print_greeting variable that provides the name of the function. Then, this example calls the print_greeting() function to show that it prints "Greetings!" to the console.

The second example defines a function that has a single parameter named size and returns a string. To define the parameter, this example codes the name of the parameter within the parentheses for the function. Then, within the braces, the code uses an if statement to set a variable named desc based on the value of the size parameter. Next, it uses the return() function to return the desc variable to the calling code.

After defining the get_desc() function, the second example calls that function and passes an argument of 101. This returns a string of "Large".

The third example defines a function that has two parameters and returns a number. To define the parameters, the example separates the names of the parameters with a comma. Then, within the braces, a nested if statement sets the number of points based on the two parameters. Next, it returns the number of points.

After defining the get_points() function, the third example calls that function and passes arguments for a made 3-point shot. This returns a numeric value of 3.

Since a function ends when it calls the return() function, you can't use the return() function to return multiple values. However, you can store multiple values in a data structure and then return that data structure. For example, if you wanted to return multiple numeric values, you could store them in a vector and then return the vector.

A function for returning data from a function

Function	Description
return(value)	Returns the specified value to the calling code.

A custom function

```
print_greeting <- function() {
  print("Greetings!")
}
print_greeting()
[1] "Greetings!"
```

A function that defines a parameter and returns a value

```
get_desc <- function(size) {
  if (size > 100) {
    desc <- "Large"
  } else {
    desc <- "Small"
  }
  return(desc)
}
get_desc(101)
[1] "Large"
```

A function that defines two parameters and returns a value

```
get_points <- function(shot_type, shot_result) {
  if (shot_result == "made") {
    if (shot_type == "3pt shot") {
      points <- 3
    } else {
      points <- 2
    }
  } else {
    points <- 0
  }
  return(points)
}
get_points("3pt shot", "made")
[1] 3
```

Description

- To define a custom function, code the function keyword, a set of parentheses, and a set of braces. Optionally, code parameters within the parentheses. Within the braces, code one or more statements.

- When you define a function, you can assign it to a variable that provides the name of the function.

- Calling the return() function exits the function. To return more than one value, you can return a data structure such as a vector.

Figure 2-13   How to define functions

Perspective

Now that you've completed this chapter, you should understand some core R skills like calling functions as well as defining your own functions. You should understand how to work with data structures such as vectors, data frames, and lists. And you should understand how to work with control statements such as if statements and for loops.

As you progress through this book, you'll see how you can use these skills in the context of data analysis. In addition, you'll learn about other R skills that are commonly used in data analysis.

Summary

- A *function* is a reusable unit of code that performs a specific task. Many functions are part of the R language.

- To *call a function*, you code the name of the function followed by a pair of parentheses. Within the parentheses, you code any *arguments* that you want to pass to the function, and you separate multiple arguments with commas.

- A function uses *parameters* to define the arguments that it accepts. Parameters can be required or optional. Optional parameters have default values that are used if you don't specify a value for them.

- You can call functions with *positional arguments* or *named arguments*. Positional arguments must be coded in the same order as defined by the function. Named arguments can be coded in any order.

- R provides many functions for working with *strings* of characters.

- You can use a *data structure* to organize your data.

- A *vector* can be used to store multiple items of the same data type in a single data structure.

- If you try to store items of different data types in a vector, R *coerces* the items into a single type.

- To access the items in a vector, you can use a number known as an *index*.

- To create a vector for a range of numeric values, you can use the *colon operator* (:) to generate the values for the vector.

- A *named vector* provides names for its items. This allows you to access each item by its index or its name.

- A *data frame* provides a way to store data in a table.

- In a data frame, the *rows* are known as *observations*, the *columns* are known as *variables*, and the individual data points are known as *values*.

- A *list* is a type of vector that can store multiple items with different data types including nested lists.

- You can use the *relational operators* to create a *Boolean expression* (or *conditional expression*) that compares two operands and returns a Boolean value of TRUE or FALSE.

- You can use the *logical operators* to join two or more Boolean expressions. This can be referred to as a *compound conditional expression*.

- The *order of precedence* determines the sequence in which the logical operators are executed. However, you can use parentheses to clarify or change that sequence.

- *Control structures* let you control how the code is executed.

- An *if statement* always contains an *if clause*. In addition, it may contain one or more *else if clauses* and one *else clause*.

- It's possible to code an if statement within a clause of another if statement. The result is known as *nested if statements*.

- A *for statement*, also known as a *for loop*, provides a way to loop through every value in a vector.

Exercise 2-1   Creating and editing data structures

This exercise is designed to make you feel comfortable working with functions, data structures, and control structures. So, we encourage you to experiment with each step and make sure you understand the process.

Create a data frame for a multiplication table

1. Create a new script by selecting File→New File→R Script and save it with a name of exercise_2-1.R.

2. In the new script, create a vector named ones that contains the numbers from 1 to 5.

3. Create a vector named twos that contains the numbers from 2 to 10 counting by 2 like this:

    ```
    twos <- c(seq(from = 2, to = 10, by = 2))
    ```

4. Create three more vectors, named threes, fours, and fives, each counting by 3, 4, and 5 respectively until they reach five numbers.

5. To confirm that you've correctly created each of your vectors, check their values in the Environment pane.

6. Create a data frame named times_table that combines your five vectors. Include a name for each column like One, Two, and so on.

7. Display your data frame in the Console pane by typing its name. Compare the console display with how the data frame's data is displayed in the Environment pane.

8. Create another vector for sixes. This one should go up to 6 times 6 (36) and contain six values total.

9. Use rbind() to add the sixes row to your multiplication table. Then, view it. Notice the last value, 36, isn't included because the data frame only has five columns.

10. Use cbind() to add the sixes vector as a column named Six. The multiplication table should now be complete up to 6 times 6.

11. Use brackets to select the value for 6 times 5 that's at the sixth row and the fifth column. When you do, create two statements where the first selects the column by index and the second selects it by name.

Define and call a function

12. Define a function named divide_and_round() that accepts two numbers as arguments.

13. Inside the function's braces, divide the first number by the second, round the result to two decimal places, and return the result to the calling code.

14. Test your function with some different numbers. Try using named arguments to pass the second number first and make sure the function still works correctly.

Create another data frame

15. Create a vector of pirate captain names. You can use any names you like, but make sure you have at least three names in your vector.

16. Create a second vector of crew sizes. This vector should contain the number of crew members for each captain you named in the previous step. Each crew size should be between 1 and 7.

17. Create a data frame named ships where the first column stores the captain's name and the second column stores the crew sizes. Give the first column a name of Captain and the second column a name of CrewSize.

18. Display the data frame on the console. It should have two columns and at least three rows.

19. Use the single bracket selector to select the column named Captain. Note that it selects the item for column, not the value for the column, which is a vector.

20. Use the $ selector to select the vector for the column named Captain.

21. Use the rbind() function to add a row for another captain and crew size. When you do, use a list to store the data for the row.

Use control structures to work with a data frame

22. Code a for loop that loops through all values in the vector for the CrewSize column.

23. Within the for loop, code a statement that prints each crew size value to the console. Then, run the loop to make sure it works correctly.

24. Within the loop, modify the code so it uses an if statement that checks if the crew size for the captain is greater than 5. If so, it should print a line to the console with the crew size and its description (large) like this.

    ```
    [1] "7 is a large crew size."
    ```

25. Run the loop and make sure it works correctly.

26. After the if clause, add an else if clause to the if statement that checks if the crew size is greater than 3. If so, print a line to the console with the crew size and its description (medium).

27. Run the loop and make sure it works correctly.

28. After the else if clause, add an else clause that prints a line to the console with the crew size and its description (small).

29. Run the loop and make sure it works correctly. It should display one line for each row with a description that matches the crew size something like this:

    ```
    [1] "7 is a large crew size."
    [1] "4 is a medium crew size."
    [1] "2 is a small crew size."
    ```

Chapter 3

How to code your first analysis

The goal of this chapter is to give you a taste of how data analysis works. In addition, it's designed to introduce you to some of the most important R packages for working with data analysis. To do that, this chapter presents the R code for a simple but complete analysis of child mortality data. The code for this analysis uses a collection of packages known as the tidyverse.

An introduction to the analysis

Before presenting the code for the analysis of child mortality data, this chapter shows how to get ready for an analysis. In particular, it shows how to familiarize yourself with the data, set the working directory, and load packages of prewritten code.

The child mortality data

The data for the child mortality analysis is stored in an Excel spreadsheet named child_mortality_rates.xlsx. Figure 3-1 begins by showing the path to this file. This assumes that you've downloaded the files for this book and put them in your Documents directory as described in appendix A (Windows) or B (macOS).

The second example shows the mortality data after it has been opened in Excel. This shows that the data is stored in five columns. The first column stores the year. The next four columns store the death rates per 100,000 people for each age group.

As you examine this data, you can begin to think of questions that you might want to explore. If you're operating in a scientific capacity, you might want to create an inference (educated guess) about the data. Then, you can attempt to validate or disprove the inference. Even if you're just exploring the data and looking for interesting patterns, it's still worth making a list of questions to consider as you go through your analysis.

An Excel file that contains the child mortality rates

`/Documents/murach_r/data/child_mortality_rates.xlsx`

The spreadsheet when opened in Excel

	A	B	C	D	E	F	G	H
1	Year	01-04 Years	05-09 Years	10-14 Years	15-19 Years			
2	1900	0.019838	0.004661	0.002983	0.004848			
3	1901	0.01695	0.004276	0.002736	0.004544			
4	1902	0.016557	0.004033	0.002525	0.004215			
5	1903	0.015421	0.004147	0.002682	0.004341			
6	1904	0.015915	0.00425	0.003052	0.004714			
7	1905	0.014989	0.003963	0.002798	0.004393			
8	1906	0.0158	0.003774	0.002722	0.004452			
9	1907	0.014683	0.003656	0.002658	0.004377			
10	1908	0.013968	0.003542	0.002479	0.003977			
11	1909	0.013489	0.003302	0.002305	0.003636			
12	1910	0.013973	0.003484	0.002359	0.003719			
13	1911	0.01176	0.0031	0.002222	0.00366			
14	1912	0.010941	0.002875	0.002022	0.003472			
15	1913	0.011934	0.003177	0.002148	0.003603			

Sheet1   ⊕

Inference

- Child mortality has steadily declined over the years.

The goals of this analysis

- Examine the child mortality data to identify trends.
- Identify potential reasons for changes in child mortality.
- Visualize the child mortality data.

Description

- The data for the Child Mortality analysis is stored in an Excel spreadsheet.
- The rates stored in this spreadsheet are per 100,000 people.

Figure 3-1   The child mortality data

How to set the working directory

Before you can get the data, you need to set the *working directory*. This is the directory that R uses when it reads or writes files.

The first example in figure 3-2 shows how to get the working directory with the getwd() function. Using getwd() can help you determine if you need to set the working directory or if the working directory is already set to the directory you want. Here, the working directory is set to the Documents directory. This is the default directory for both Windows and macOS. Of course, this path varies depending on your name and operating system.

The second example shows how to use the setwd() function to set the working directory to a relative path, which is a path based on your current directory. Here, the example sets the directory to the murach_r/ch03 child directory of the Documents directory. Then, the third example uses the parent directory symbol (..) to move back to the parent directory.

The fourth example shows how to specify an absolute file path, which is a path that begins at the root directory and specifies the entire path. Here, the path begins with a front slash (/). As a result, it starts from the root directory for the current hard drive.

The fifth example shows how to use the tilde symbol (~) to set the working directory to the murach_r/data subdirectory of the Documents directory on a Windows system. If you followed the instructions in the appendixes to store the murach_r directory in the Documents directory, this is a good way to set your working directory.

The sixth example shows how to use the tilde symbol to set the working directory to the murach_r/data directory on macOS. This works much like the fifth example. However, on macOS, the tilde symbol refers to the directory for the user, not the Documents directory. As a result, this example needs to include the Documents directory in the path.

To execute code that gets or sets the working directory, you typically enter it directly into the Console pane of RStudio, not the Source pane. That way, if you share the script for your analysis with a colleague, it doesn't set their working directory.

When working on an analysis, you may want to set the working directory to the same directory as the R script that's currently open in the Source pane. To do that quickly, you can open the script in RStudio and select Session→Set Working Directory→To Source File Location. This should execute the appropriate setwd() function in the console.

The functions for getting and setting the working directory

Function	Description
getwd()	Gets the path to the working directory.
setwd(dir)	Sets the working directory to the specified directory.

How to get the working directory

```
getwd()
[1] "C:/Users/joelm/Documents"
```

How to specify a relative path

```
setwd("murach_r/examples/ch03")
getwd()
[1] "C:/Users/joelm/Documents/murach_r/examples/ch03"
```

How to use the parent directory symbol (..)

```
setwd("..")
getwd()
[1] "C:/Users/joelm/Documents/murach_r/examples"
```

How to specify an absolute path

```
setwd("/Users")
getwd()
[1] "C:/Users"
```

How to specify the murach_r/data directory on Windows

```
setwd("~/murach_r/data")
getwd()
[1] "C:/Users/joelm/Documents/murach_r/data"
```

How to specify the murach_r/data directory on macOS

```
setwd("~/Documents/murach_r/data")
getwd()
[1] "/Users/joelm/Documents/murach_r/data"
```

Description

- The *working directory* is the directory that R uses when it reads or writes files.
- A *relative path* specifies a path that's relative to the working directory.
- An *absolute path* specifies the entire path to the file.
- When separating directories, you must use the front slash (/), not the backslash (\).
- You can use the tilde symbol (~) to access the Documents directory (Windows) or the user directory (macOS).
- To set the working directory to the same directory that stores the R script, open the script in RStudio and select Session→Set Working Directory→To Source File Location. This should execute the appropriate setwd() function in the console.

Figure 3-2   How to set the working directory

How to work with packages

In R, a *package* contains prewritten code that you can use to save time and work. This prewritten code can include functions, sample data, and documentation. For example, the tidyverse package, or just *tidyverse*, is a collection of R packages designed for data science.

Before you can use the tidyverse, you need to make sure it's installed on your computer. If you followed the instructions in appendix A (Windows) or B (macOS), the packages that you need for this book should already be installed on your system. The procedures in these appendixes use the install.packages() function shown in the first example to install the tidyverse package. However, if you need to update a package, you can use the update.packages() function as shown in the second example.

When you install or update a package, you only need to do it once on each computer. However, to use a package in an analysis, you need to load the package at the beginning of each session. In other words, you need to load the package each time you restart RStudio.

To load a package, you can run the library() function shown in the third example. Since you use the library() function to load packages, packages are sometimes referred to as *libraries*. In practice, you can use these terms interchangeably. However, *package* is more technically correct in most cases.

When you load the tidyverse package, it displays a message that includes the names and version numbers of the eight packages that are part of the *core tidyverse*. These packages include the tibble, dplyr, tidyr, and ggplot2 packages that are used by the Child Mortality analysis presented in this chapter.

After displaying the list of packages that are loaded, the tidyverse message includes some information about conflicts. For example, the first message shows that the filter() function in the dplyr package "masks" the filter() function in the stats package, which is one of R's base packages. This means that you can call the filter() function from the dplyr package normally like this:

```
filter()
```

However, to call the filter() function from the stats package, you need to qualify it with the name of the package like this:

```
stats::filter()
```

After displaying the list of conflicts, the tidyverse message may include some warning messages. Since these messages are warnings, not errors, you can often ignore them. Or, if they bother you or cause problems, you can search the internet to learn more about them. In many cases, you can get rid of warning messages by updating R, RStudio, or the packages that you're using.

The tidyverse package includes many packages that aren't part of the core tidyverse. These packages are installed as part of the tidyverse, but they aren't loaded automatically when you load the tidyverse package. As a result, if you want to use them, you need to use the library() function to load them separately. For instance, in the third example, the second statement loads the readxl package. This is necessary because that package isn't part of the core tidyverse. However, it's used by the Child Mortality analysis presented in this chapter.

Four of the eight packages in the core tidyverse

Package	Description
tibble	Contains the code for working with a tibble, which is a special type of data frame that's optimized for working with data analysis.
dplyr	Contains functions that allow you to manipulate the data in a data frame in a consistent way that yields code that's easy to read and understand.
tidyr	Contains functions that allow you to tidy your data by performing tasks like changing its shape.
ggplot2	Contains functions that use the Grammar of Graphics (GG) to plot data to create data visualizations.

One of the many tidyverse packages that's not part of the core tidyverse

Package	Description
readxl	Contains functions for reading data from Excel spreadsheets into a tibble.

How to install a package

```
install.packages("tidyverse")
```

How to update a package

```
update.packages("tidyverse")
```

How to load packages

```
library("tidyverse")
-- Attaching packages ------------------------------------- tidyverse 1.3.1 --
✓ ggplot2 3.3.5      ✓ purrr    0.3.4
✓ tibble  3.1.6      ✓ dplyr    1.0.8
✓ tidyr   1.2.0      ✓ stringr 1.4.0
✓ readr   2.1.2      ✓ forcats 0.5.1
-- Conflicts ------------------------------------- tidyverse_conflicts() --
x dplyr::filter() masks stats::filter()
x dplyr::lag()    masks stats::lag()
library("readxl")
```

Description

- In R, a *package* contains prewritten code that can include functions, sample data, and documentation.

- The *tidyverse* is a collection of R packages designed for data science. When you load the tidyverse package, it only loads the *core tidyverse*. If necessary, you can load additional tidyverse packages that aren't part of the core tidyverse.

- You only need to install a package once on your computer, but you need to load the package each time you restart RStudio.

Figure 3-3   How to work with packages

How to get and examine the data

The first step to starting a new analysis is to get the data. For this analysis, you can get the data by reading it from an Excel spreadsheet into a *tibble*, which is a special type of data frame. Like a data frame, a tibble consists of *columns* and *rows*. However, tibbles are newer than data frames and have been optimized for data analysis.

Once the data has been loaded into a tibble, you can clean the data. For dirty data, cleaning can be a tedious and time-consuming task. Fortunately, for this analysis, the data is already clean. As a result, you don't need to clean it. Instead, you can focus on examining the data to better understand it.

How to read the data into a tibble

To read the data from the child mortality spreadsheet into a tibble, you need to load the readxl package as described in the previous figure. After you load this package, you can use the read_excel() function to read the data from the spreadsheet into a tibble as shown in figure 3-4. Here, the first example shows how to read the first sheet of the spreadsheet into a tibble named mortality_data. This works because the working directory was set correctly earlier in this chapter. In addition, the second argument uses a value of 1 to specify that the function should read the first sheet in the Excel file.

After you read the data into the tibble, you can view the tibble as shown by the second example. Here, the console output shows that the tibble has 119 rows and 5 columns. In addition, it shows the name and data type for each column. In this case, every column uses the double type. It might be more accurate for the Year column to use the integer type, but the double type works fine for the purposes of the analysis presented in this chapter.

Note that the column names for the age groups are each surrounded by tick marks (`` ` ``). This shows that the column names contain spaces. In general, renaming columns to remove spaces makes them easier to work with when writing the code for an analysis. However, as it turns out, this isn't necessary for the analysis presented in this chapter.

By default, printing a tibble to the console displays the first ten rows of a tibble as shown in this figure. That's usually enough rows to give you a good idea of the data that's stored in the tibble. However, if you want to select a smaller number of rows from the top or bottom of a tibble, you can use the functions presented in the next figure to do that.

A function for reading the data from an Excel file

Function	Description
read_excel(path, sheet)	Reads the data in specified sheet of the Excel spreadsheet at the specified path and returns that data as a tibble.

How to read the data from the Excel file

```
mortality_data <- read_excel("../../data/child_mortality_rates.xlsx", sheet = 1)
```

How to view the tibble

```
mortality_data
# A tibble: 119 x 5
    Year `01-04 Years` `05-09 Years` `10-14 Years` `15-19 Years`
   <dbl>         <dbl>         <dbl>         <dbl>         <dbl>
 1  1900        0.0198       0.00466       0.00298       0.00485
 2  1901        0.0170       0.00428       0.00274       0.00454
 3  1902        0.0166       0.00403       0.00252       0.00422
 4  1903        0.0154       0.00415       0.00268       0.00434
 5  1904        0.0159       0.00425       0.00305       0.00471
 6  1905        0.0150       0.00396       0.00280       0.00439
 7  1906        0.0158       0.00377       0.00272       0.00445
 8  1907        0.0147       0.00366       0.00266       0.00438
 9  1908        0.0140       0.00354       0.00248       0.00398
10  1909        0.0135       0.00330       0.00231       0.00364
# ... with 109 more rows
```

Description

- A *tibble* is special type of data frame that's optimized for data analysis. Like a data frame, a tibble consists of *columns* and *rows*.

- The read_excel() function is in the readxl package. Since this package isn't part of the core tidyverse package, you need to load it even if you've already loaded the tidyverse package.

- If you get an error like "`path` does not exist: 'child_mortality_rates.xlsx'", make sure you have set the working directory correctly.

Figure 3-4   How to read the data into a tibble

How to select the top and bottom rows

In the previous figure, you learned how to displays the first ten rows of the tibble. However, if you just want to select a specified number of rows from the top or bottom of a tibble, the head() and tail() functions allow you to do so.

The examples in figure 3-5 show how to use the head() and tail() functions. Here, the head() function selects the first six rows of the tibble. That's because these functions select six rows by default. However, you can supply a second argument to specify the number of rows to select. For example, the tail() function uses a second argument of 3 to select only the last three rows of the tibble.

As you review this data, note that the head() and tail() functions select the specified rows and return a tibble that has fewer rows than the original tibble. For example, the tibble returned by the head() function has six rows and five columns. Similarly, the tibble returned by the tail() function has three rows and five columns.

If you look at the data selected by the head() and tail() functions, you can already begin to do some analysis. For instance, you can note that the mortality rates for the later years (2016 through 2018) are much lower than the rates for the earlier years (1900 through 1905). That's probably what you expected, but it's nice to verify that the data supports this.

Functions for selecting the top and bottom rows

Function	Description
head(data, n)	Returns the specified number of rows from the top of the data set. By default, the n parameter is set to 6.
tail(data, n)	Returns the specified number of rows from the end of the data set. By default, the n parameter is set to 6.

How to select the first six rows

```
head(mortality_data)
# A tibble: 6 x 5
   Year `01-04 Years` `05-09 Years` `10-14 Years` `15-19 Years`
  <dbl>         <dbl>         <dbl>         <dbl>         <dbl>
1  1900        0.0198       0.00466       0.00298       0.00485
2  1901        0.0170       0.00428       0.00274       0.00454
3  1902        0.0166       0.00403       0.00252       0.00422
4  1903        0.0154       0.00415       0.00268       0.00434
5  1904        0.0159       0.00425       0.00305       0.00471
6  1905        0.0150       0.00396       0.00280       0.00439
```

How to select the last three rows

```
tail(mortality_data, 3)
# A tibble: 3 x 5
   Year `01-04 Years` `05-09 Years` `10-14 Years` `15-19 Years`
  <dbl>         <dbl>         <dbl>         <dbl>         <dbl>
1  2016      0.000253      0.000122      0.000146      0.000512
2  2017      0.000243      0.000116      0.000155      0.000515
3  2018       0.00024      0.000115      0.000149      0.000492
```

Description

- The head() and tail() functions are part of the base R language.

Figure 3-5   How to select the top and bottom rows

How to view summary statistics

The summary() function provides a quick way to view summary statistics for the columns in your data set. How this function works depends on the type of data in each column. In figure 3-6, each column uses the double type. Since the double type is a numeric type, the summary() function displays the minimum, first quartile, median, mean, third quartile, and maximum for each passed column.

For a numeric column, the summary numbers include the *five-number summary* (minimum, first quartile, median, third quartile, and maximum). This provides a concise summary of the distribution of the values in the column. In addition, the summary numbers include a sixth value for the average (mean) of the column.

The first example shows how the summary() function works on all columns in the tibble. However, you may want to only summarize a certain column or set of columns. If so, you can nest the select() function inside the summary() function.

The second example shows how to use the select() statement. To start, you pass the name of the tibble. Then, you list the names of the columns you want, separated by commas. If a column name contains a space, you must enclose the name in quotation marks. To select every column except the specified columns, you can use the subtraction operator (-) to remove the specified columns as shown by the third statement.

The third example shows how to use summary() with select() to summarize just the data for the "01-04 Years" age group. This displays a subset of the information displayed by the first example, making it easier to focus on the statistics for the specified column.

If you look at the mean for the four age groups, you might be surprised to see that the average rate for the fourth age group (15-19) is higher than the average rate for the second (5-9) and third (10-14) groups. During your analysis, this might lead you to do more digging into why that is.

A function for viewing summary data in a tibble

Function	Description
summary(data)	Returns summary statistics for all columns in the data set. For a numeric column, these statistics include the *five-number summary* (mimimum, first quartile, median, third quartile, and maximum) as well as the averge (mean).

How to view summary statistics for all columns

```
summary(mortality_data)
      Year          01-04 Years          05-09 Years          10-14 Years
 Min.   :1900    Min.   :0.0002400    Min.   :0.000114    Min.   :0.0001390
 1st Qu.:1930    1st Qu.:0.0005065    1st Qu.:0.000240    1st Qu.:0.0002720
 Median :1959    Median :0.0010910    Median :0.000484    Median :0.0004460
 Mean   :1959    Mean   :0.0038323    Mean   :0.001173    Mean   :0.0009377
 3rd Qu.:1988    3rd Qu.:0.0057730    3rd Qu.:0.001989    3rd Qu.:0.0015890
 Max.   :2018    Max.   :0.0198380    Max.   :0.004661    Max.   :0.0037510
  15-19 Years
 Min.   :0.0004480
 1st Qu.:0.0008545
 Median :0.0010690
 Mean   :0.0017737
 3rd Qu.:0.0028365
 Max.   :0.0077740
```

A function for selecting columns from a tibble

Function	Description
select(data, col1, ...)	Returns a tibble for the specified columns. If a column name contains spaces, you must enclose its name in quotes.

How to select columns

```
select(mortality_data, Year)                 # selects the Year column
select(mortality_data, Year, "01-04 Years")  # selects two columns
select(mortality_data, -Year)                 # selects all columns except Year
```

How to view summary statistics for a selected column

```
summary(select(mortality_data, "01-04 Years"))
  01-04 Years
 Min.   :0.0002400
 1st Qu.:0.0005065
 Median :0.0010910
 Mean   :0.0038323
 3rd Qu.:0.0057730
 Max.   :0.0198380
```

Description

- The the summary() function is part of the R lanaguage, and the select() function is in the dplyr package.

Figure 3-6   How to view summary statistics

How to prepare and analyze the data

Once you import and clean your data, you can move on to preparing and analyzing the data. When preparing data, it's common to combine several columns into two new columns, where one of the new columns holds the original column names and the other column holds the values from the original columns. This is known as *melting* the data.

How to melt the data

Melting the data converts *wide data* to *long data*. Long data is typically preferred by R because it works better with its plotting functions. So far in this chapter, you've been working with a tibble named mortality_data that stores wide data that's 5 columns wide and 109 rows long. However, the tibble named mortality_long that's shown at the bottom of figure 3-7 is only 3 columns wide but 476 rows long.

To melt data, you can use the pivot_longer() function as shown in the second example. This function takes four arguments. The data argument specifies the tibble that contains the columns to melt. The cols argument specifies a vector of the column names to be melted. In this example, it specifies the names of the four columns that store the death rates for each age group.

The names_to and values_to parameters specify the names for the new columns. In this example, the names_to argument specifies a name of AgeGroup, and the values_to argument specifies a name of DeathRate. This results in data that's in the long form where the AgeGroup column contains a string that identifies the age group, and the DeathRate column contains a number that specifies the death rate.

The third example shows the long data that results from melting the mortality data. The second column now stores a string that identifies the age group category while the third column stores a double value for the death rate. This format makes it easier to add calculated columns and plot data.

In this figure, the long data is *tidy* because it stores the values for the death rate in a single column, and it stores the strings for the age group in a single column. In other words, it stores each *variable* (AgeGroup and DeathRate) in a column, it stores each *observation* in a row, and each cell only stores one value. By contrast, the wide data stores the death rate variable in multiple columns, which is *messy*, and it stores the age group data as a column name, which is also messy. This results in four death rate observations in each row.

The wide data (messy)

```
mortality_data
# A tibble: 119 x 5
    Year `01-04 Years` `05-09 Years` `10-14 Years` `15-19 Years`
   <dbl>         <dbl>         <dbl>         <dbl>         <dbl>
1  1900        0.0198       0.00466       0.00298       0.00485
2  1901        0.0170       0.00428       0.00274       0.00454
3  1902        0.0166       0.00403       0.00252       0.00422
4  1903        0.0154       0.00415       0.00268       0.00434
5  1904        0.0159       0.00425       0.00305       0.00471
...
```

A function for melting data

Function	Description
pivot_longer(data, cols, names_to, values_to)	Combines (melts) the specified columns into two columns where the first column stores the column names and the second column stores the column values.

How to melt the columns to create long data

```
mortality_long <- pivot_longer(mortality_data,
    cols = c("01-04 Years", "05-09 Years", "10-14 Years", "15-19 Years"),
    names_to = "AgeGroup", values_to = "DeathRate")
```

The long data (tidy)

```
mortality_long
A tibble: 476 x 3
    Year AgeGroup    DeathRate
   <dbl> <chr>           <dbl>
1  1900 01-04 Years    0.0198
2  1900 05-09 Years    0.00466
3  1900 10-14 Years    0.00298
4  1900 15-19 Years    0.00485
5  1901 01-04 Years    0.0170
...
```

Description

- When you *melt* the data in a tibble, you combine two or more columns into just two columns: one that contains the names of the combined columns and one that contains the related values. In other words, you convert *wide data* to *long data*.

- *Tidy data* stores each *variable* in a column, each *observation* in a row, and each cell only stores one value. By contrast, *messy data* sometimes uses multiple columns to store the same variable or uses column names to store variables. This often results in multiple observations per row.

- The pivot_longer() function is in the tidyr package.

Figure 3-7   How to melt the data

How to add, modify, and rename columns

Sometimes, you may want to add a new column to your data frame by performing a calculation on existing data. Performing a calculation on an existing column to form a new column is sometimes known as *transforming* the column.

To transform a numeric column, you can use the mutate() function and any of the arithmetic operators to modify the column's values as shown in figure 3-8. The mutate() function takes at least two arguments. The first is the tibble or data frame you're working with. The second is the name of the column you want to mutate and what the new values should be. You can mutate multiple columns at once by separating them with commas as shown in the third example.

The first example multiplies the DeathRate column by 100,000. This calculates the number of deaths per 100,000 people and assigns the result of the expression to the same column, DeathRate. In this example, the mutate() function returns the mortality_long tibble that's passed as the data argument but with the modified DeathRate column. Then, the code assigns this tibble to the same variable, overwriting the original version of the tibble.

The second example shows how to create a column for the decade that each row belongs to. To do this, the code uses integer division to divide by 10 to remove the last digit from the year column. Then, it multiplies that result by 10 to add 0 as the last digit of this number. Next, this code assigns the result to a column named Decade. Since this column doesn't already exist in the tibble, this adds the column to the tibble.

The third example combines the first and second examples. Both techniques yield the same result, but most programmers prefer to combine the statements so that the code is shorter and the mutate() function only needs to run once.

The fourth example renames the DeathRate column to DeathsPer100K. It does this by using the rename() function. Note that the new name for the column is coded on the left of the equals sign, and the old name is coded on the right.

The last example shows the data after the new columns have been added. This shows that the DeathsPer100K column stores the number of deaths per 100,000 people, and the Decade column stores the decade for the row. These examples should give you an idea of how you can add calculated columns to a tibble.

Functions for modifying columns in a tibble

Function	Description
`mutate(data, col = val, ...)`	Modifies the specified column by assigning a new value to it. If the column name already exists, it modifies the existing column. Otherwise, it adds a new column.
`rename(data, new = old, ...)`	Renames the specified colum by providing a new name for it.

How to modify the values in the DeathRate column

```
mortality_long <- mutate(mortality_long,
                    DeathRate = DeathRate * 100000)
```

How to add a caculated column for the decade

```
mortality_long <- mutate(mortality_long,
                    Decade = (Year %/% 10) * 10)
```

How to modify two columns in a single statement

```
mortality_long <- mutate(mortality_long,
                    DeathRate = DeathRate * 100000,
                    Decade = (Year %/% 10) * 10)
```

How to rename the DeathRate column

```
mortality_long <- rename(mortality_long, DeathsPer100K = DeathRate)
```

The tibble with the calculated columns

```
mortality_long
# A tibble: 476 x 4
    Year AgeGroup    DeathsPer100K Decade
   <dbl> <chr>             <dbl>  <dbl>
 1  1900 01-04 Years       1984.   1900
 2  1900 05-09 Years        466.   1900
 3  1900 10-14 Years        298.   1900
 4  1900 15-19 Years        485.   1900
 5  1901 01-04 Years       1695    1900
 6  1901 05-09 Years        428.   1900
 7  1901 10-14 Years        274.   1900
 8  1901 15-19 Years        454.   1900
 9  1902 01-04 Years       1656.   1900
10  1902 05-09 Years        403.   1900
# ... with 466 more rows
```

Description

- The mutate() and rename() functions are in the dplyr package. These functions return a modified version of the tibble that's passed as the data argument.

Figure 3-8   How to add, modify, and rename columns

How to save a tibble as an RDS file

If you're working on a large analysis, you may want to save your work. R has a special file format named RDS that you can use to save a single R data object such as a tibble. When you save an R data object as an RDS file, it keeps all of its formatting. Then, when you read the RDS file later, it restores the object.

Figure 3-9 shows two functions for saving and reading RDS files. The saveRDS() function takes two arguments, the object to be saved and the name and path of the file to save it in. Typically, the file name should have the .rds extension. If necessary, you can indicate parent directories with two periods (..).

The readRDS() function only needs one argument, the filename with or without a path, and you can assign the object that it returns to a variable as shown in the second example. Here, the file is read into a variable named mortality_long.

Functions for saving a tibble to a file

Function	Description
saveRDS(object, file)	Saves the specified object to the specified file.
readRDS(file)	Returns the object stored in the specified file.

How to save a tibble to a file...

In the current working directory
```
saveRDS(mortality_long, "mortality_long.rds")
```

In a relative directory
```
saveRDS(mortality_long, "../../data/mortality_long.rds")
```

How to read a tibble from a file...

In the current working directory
```
mortality_long <- readRDS("mortality_long.rds")
```

In a relative directory
```
mortality_long <- readRDS("../../data/mortality_long.rds")
```

Description

- R has its own data format named RDS for storing a single R data object in a file. Files that store an R data object in the RDS format typically use an extension of .rds.
- By default, the saveRDS() and readRDS() functions use the current working directory. If you want to use another directory, you can use a relative path or an absolute path.
- The saveRDS() and readRDS() functions are part of the R language.

Figure 3-9   How to save a tibble as an RDS file

How to calculate summary columns

Sometimes, you may want to make a calculation based on groups of data. For example, you might want to calculate the mean number of deaths per 100,000 people for each decade. One way to do this is to combine the group_by() and summarize() functions.

The first example in figure 3-10 uses a different syntax for calling a function than you've seen before. This syntax uses the *pipe operator* (%>%) to *pipe* the output of one function into the input of another. In this example, coding the mortality_long variable would normally display the data for this tibble on the console. However, since the pipe operator is coded after this variable name, the tibble is used as the first argument of the group_by() function.

The first two lines in the example work the same as coding this statement:

```
group_by(mortality_long, Decade)
```

So why would you want to use the pipe operator? In short, it makes it easier to pass data through a series of functions to get to some result. This makes your code more concise by making it unnecessary to define variables to store intermediate forms of the data. It also makes it unnecessary to nest several function calls, which can become unwieldy.

When you first start using the pipe operator, you may find that typing the three characters for the pipe operator is unwieldy. That's why RStudio provides the keyboard shortcut shown in this figure for the pipe operator. With a little practice, using this keystroke shortcut becomes second nature and makes it easy to enter the pipe operator.

After the pipe operator, this example codes the group_by() function. This function creates groups in the data, which allows you to apply summary functions based on those groups. In this case, the code creates groups based on the Decade column. As a result, it groups all rows with the decade 1900, all rows with decade 1910, and so on. This doesn't change the existing data. It just adds information to the tibble's structure about how to group the data.

After grouping the data, this code pipes it into the summarize() function. This function allows you to create summary columns by applying a function to the rows in each group.

To create a new summary column, you code the new column's name, the assignment operator (=), and the statistical function that you want to use. In this example, the code uses summarize() to calculate two columns. The first column contains the mean of the DeathsPer100K column for each decade, and the second counts the number of rows for each decade.

The resulting tibble displays both the grouping column (Decade) and the columns calculated by the summarize() function (MeanDeaths and Count). This shows that the average number of deaths decreases dramatically from early decades to later ones. In addition, the decade of 2010 only has 36 rows, not 40 like the other decades. That's because the data set only contains information to the year 2018, so the decade of 2010 doesn't include the four age groups for the year 2019.

Functions for calculating summary columns in a tibble

Function	Description
group_by(data, col1, ...)	Groups the data by the specified column or columns.
summarize(data, col = fun, ...)	Summarizes the data by adding a column for each specified statistical function.
n()	A statistical function that returns the number of rows for each group.
mean(col)	A statistical function that returns the mean (average) of the specified column for each group.

How to calculate summary columns

```
mortality_long %>%
  group_by(Decade) %>%
  summarize(MeanDeaths = mean(DeathsPer100K),
            Count = n())
# A tibble: 12 x 3
   Decade MeanDeaths Count
    <dbl>      <dbl> <int>
 1   1900       669.    40
 2   1910       525.    40
 3   1920       366.    40
 4   1930       243.    40
 5   1940       134.    40
 6   1950       81.9    40
 7   1960       70.2    40
 8   1970       62.9    40
 9   1980       48.8    40
10   1990       41.1    40
11   2000       31.3    40
12   2010       25.0    36
```

Description

- When summarizing grouped data, it's common to use the *pipe operator* (%>%) to pipe the data into the group_by() and summarize() functions.
- In RStudio, the keyboard shortcut for the pipe operator is Ctrl+Shift+M (Windows) or Command+Shift+M (macOS).
- The group_by(), summarize(), and n() functions are in the dplyr package.
- The the mean() function is part of the R lanaguge.

Figure 3-10   How to calculate summary columns

How to create a line plot

Visualizing data is a crucial part of analyzing data. That's because there's often too much data to analyze without visualizing it. For example, you can learn some things about the child mortality data by viewing it and calculating some summary statistics. However, visualizing this data is typically the best way to analyze it. To do that, you can create a plot like the *line plot* shown in figure 3-11.

To create plots with R, you typically use the ggplot2 package. This package is part of the core tidyverse, so it's loaded automatically when you load the tidyverse package. Alternatively, you can load the ggplot2 package separately.

The first example shows the code that creates the line plot. Like all plots created with ggplot2, this code begins with a call to the ggplot() function. Then, it passes the data for the plot as the first argument. Next, it passes a call to the aes() function to set the *aesthetic mappings* of the plot. In this case, the aes() function specifies that the Year column should be plotted on the x axis, the DeathsPer100K column should be plotted on the y axis, and a different color should be used for each value in the AgeGroup column.

The ggplot() function creates an empty plot known as the base plot that contains the x and y axes and their labels. However, it doesn't draw the line plot. To do that, this code uses the addition operator (+) to add the geom_line() function to the base plot. This function draws the lines on the plot.

When you use RStudio to run this code, it displays the line plot in the Plots tab. In this figure, the plot that's displayed makes it easy to see the spike representing the Spanish Flu of 1918 as well as the general downward trend in deaths. Although it's possible to get the same insight by reviewing the numbers in the data set, plotting the data makes it easier to gain insights about the data and to share those insights with others.

Functions for creating plots

Function	Description
`ggplot(data, mapping)`	Specifies the data and the *aesthetic mappings* for a plot.
`aes(x, y, color, fill)`	Returns the aesthetic mappings for a plot with the specified x and y axes, the specified color, and the specified fill color. All of the parameters are optional, so you only need to specify the ones you need for your plot.
`geom_line()`	Draws a line plot like the one shown in this figure.

How to create a line plot

```
ggplot(mortality_long, aes(x = Year, y = DeathsPer100K, color = AgeGroup)) +
  geom_line()
```

The plot displayed in the Plots tab

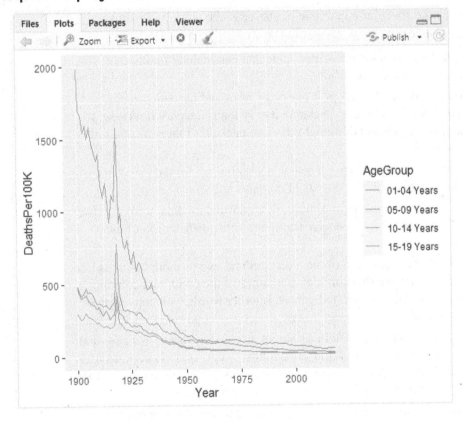

Description

- To create any plot, you start by calling the ggplot() function. Then, you use the aes() function to specify the *aesthectic mappings* for the plot, and you use the addition operator (+) to add calls to other plotting functions such as the geom_line() function.
- The ggplot(), aes(), and geom_line() functions are in the ggplot2 package.

Figure 3-11   How to create a line plot

Perspective

The simple but complete analysis presented in this chapter is designed to give you a taste of how data analysis works. In addition, it's designed to introduce you to some of the most useful R packages for working with data analysis

If you don't understand all of the code presented in this chapter, don't worry! The skills presented in this chapter are explained in more detail in the following chapters.

Summary

- The *working directory* is the directory that R uses when it reads or writes files. You can use R code or the RStudio menus to set the working directory.

- A *relative path* specifies a path that's relative to the working directory.

- An *absolute path* specifies the entire path to the file.

- In R, a *package* contains prewritten code that can include functions, sample data, and documentation.

- The *tidyverse* is a collection of R packages designed for data science. When you load the tidyverse package, it only loads the *core tidyverse*. If necessary, you can load additional tidyverse packages that aren't part of the core tidyverse.

- A *tibble* is special type of data frame that's optimized for data analysis. Like a data frame, a tibble consists of *columns* and *rows*.

- The *five-number summary* (minimum, first quartile, median, third quartile, and maximum) provides a concise summary of the distribution of the values in a column.

- When you *melt* the data in a tibble, you combine two or more columns into just two columns: one that contains the names of the combined columns and one that contains the related values. In other words, you convert *wide data* to *long data*.

- *Tidy data* stores each *variable* in a column, each *observation* in a row, and each cell only contains one value. By contrast, *messy data* sometimes uses multiple columns to store the same variable or uses column names to store variables. This often results in multiple observations per row.

- When summarizing grouped data, it's common to use the *pipe operator* (%>%) to pipe the data into the group_by() and summarize() functions.

- To create any plot, you start by calling the ggplot() function. Then, you use the aes() function to specify the *aesthetic mappings* for the plot, and you use the addition operator (+) to add calls to other plotting functions such as the geom_line() function.

Exercise 3-1   Review and edit the child mortality analysis

In this exercise, you'll run the code in the child mortality script. This script includes all of the statements for the analysis presented in this chapter. As you run each statement, be sure that you understand what it does even if you're not yet comfortable coding it yourself.

This exercise replicates something that's likely to come up whenever you work in a team: being given an R script for an analysis, confirming you get the same results as your team member, and performing a few extra calculations or creating a new visualization.

View the data in the spreadsheet (optional)

1. If you have spreadsheet software such as Excel installed on your computer, use it to open this Excel file:

   `Documents/murach_r/data/child_mortality_rates.xls`

2. Note that the spreadsheet contains five columns of data, one for the year and four for the age groups. Then, scroll down to view all of the data.

Open RStudio and use it to set the working directory

3. Start RStudio.

4. Open the script named exercise_3-1 that's in this directory:

   `Documents/murach_r/exercises/ch03`

   Note that the comments make it easier for you to follow what the script does even if you don't yet understand all of the code.

5. Set the working directory to the same directory that contains the script. To do that, refer to figure 3-2. You can use the setwd() function, or you can select the appropriate item from the RStudio menus.

6. In the Console pane, enter the getwd() function to check the working directory.

Run the statements in the script

7. Confirm you have correctly installed the required packages and set the working directory by running the first four statements in the script. These statements load the packages you need, read the child mortality data, and display it.

   Note: If you get a warning message that says, "Error: `path` does not exist", you need to make sure that the Excel file exists and that the working directory is set to the same directory as the script.

8. Run the rest of the script statement by statement. For in-depth explanations of how each statement works, you can refer to the appropriate figure from this chapter.

   Tip: While in the Console pane, you can use the up arrow to display and re-run statements. If you do that, remember that each statement runs using the variables currently in memory.

9. After you run the saveRDS() function, check to make sure the mortality_long.rds file has been saved to the current working directory.

10. Note that when statements are coded on multiple lines, each line typically ends with commas or operators like the addition (+) or pipe (%>%) operators.

Write and run some statements of your own

11. At the end of the script add a statement that uses the head() function to view only the first 3 rows of the mortality_data tibble. Then, add a statement that uses tail() to display the last 8 rows of the mortality_long tibble.

12. Display the summary data for the fifth column (15-19 years) of the mortality_data tibble.

13. Rename the four age group columns in mortality_data with the following names Years1_4, Years5_9, Years10_14, and Years15_19. This makes it easier to perform the calculation in the next step.

14. Use the mutate() function to add a new column to mortality_data named AllAges that adds the rate for each age group.

15. Create a line plot of the AllAges column. The aes() function should look like this:

```
aes(x = Year, y = AllAges)
```

Section 2

The essential skills for data analysis

Section 1 introduced some concepts that apply to data analysis and presented some basic skills for working with RStudio and R. In addition, it showed how to use the tidyverse package to perform a simple but complete data analysis. Now, section 2 presents the essential skills that you'll need to create professional data analyses. Many of these skills use functions that are available from the tidyverse package.

Chapter 4

How to visualize data

This chapter shows how to visualize data using the ggplot2 package. The ggplot2 package is the most common package for creating plots, graphs, and similar visualizations in R. It can handle almost all of your data visualization needs.

How to get some data to plot

Before you can plot data, you need to get some data to plot. That's why this chapter starts by introducing you to the datasets package.

How to use the datasets package

The datasets package allows you to access many sample data sets. These data sets are smaller and less complex than most real-world data sets, but they are useful for learning how to create visualizations. That's because they provide easy access to different types of data and most of them have already been cleaned.

The first example in figure 4-1 shows how to load the packages for this chapter. First, it loads the tidyverse package introduced in the previous chapter that includes the ggplot2 package that's used for data visualization. Then, it loads the datasets package.

Once you've loaded the datasets package, you can list all of the available data sets by calling the data() function. Unlike most other functions, this function displays its results in the Source pane, as shown in the second example. Then, you can scroll to view the entire list of data sets.

Unfortunately, the list of data sets doesn't give you much information about the kind of data that each data set contains. To learn more about each data set, you can view its data as shown in the third and fourth examples. Or, you can view its documentation as described at the end of this chapter.

To view the data for a data set, you can enter the name of the data set on the console. In this figure, the third example views all rows of the data set named iris. Then, the fourth example views the first six rows of a data set named ChickWeight. To do that, the fourth example passes the ChickWeight data frame to the head() function to select only the first six rows. This is sometimes helpful if the data set is long and you don't want to view all rows of the data set.

This chapter uses data from the iris and ChickWeight data sets. For now, all you need to know about the iris data set is that it contains the length and width measurements for the sepals and petals for three species of iris flowers. Meanwhile, the ChickWeight data set contains the weight (in grams) of 50 chicks for the first 21 days of their lives, where each chick is on one of four possible diets. As you progress through this chapter, you'll use data visualization to learn more about both of these data sets.

How to load the packages for this chapter

```
library("tidyverse")    # the core tidyverse packages (tibble, ggplot2, etc.)
library("datasets")     # the sample data sets (iris, ChickWeight, etc.)
```

How to list all available data sets

```
data()
```

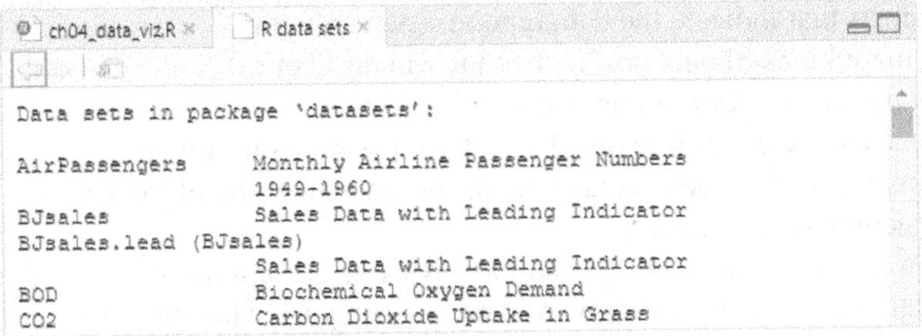

```
 📁 ch04_data_viz.R ✕      □ R data sets ✕                                    ━□

  [ ]   🔳

  Data sets in package 'datasets':

  AirPassengers       Monthly Airline Passenger Numbers
                      1949-1960
  BJsales             Sales Data with Leading Indicator
  BJsales.lead (BJsales)
                      Sales Data with Leading Indicator
  BOD                 Biochemical Oxygen Demand
  CO2                 Carbon Dioxide Uptake in Grass
```

How to view the data set named iris

```
iris
  Sepal.Length Sepal.Width Petal.Length Petal.Width Species
1          5.1         3.5          1.4         0.2  setosa
2          4.9         3.0          1.4         0.2  setosa
3          4.7         3.2          1.3         0.2  setosa
4          4.6         3.1          1.5         0.2  setosa
5          5.0         3.6          1.4         0.2  setosa
...
```

How to view the first six lines of the data set named ChickWeight

```
head(ChickWeight)
  weight Time Chick Diet
1     42    0     1    1
2     51    2     1    1
3     59    4     1    1
4     64    6     1    1
5     76    8     1    1
6     93   10     1    1
```

Description

- The datasets package provides sample data sets that you can use to practice your data analysis skills.
- Once you load the datasets package, you can use the data() function to list all data sets in the package.
- To view the entire data frame for a sample data set, enter the name of the data set.
- To select the first six rows of a data frame, pass the data frame to the head() function.
- The iris data set contains the length and width measurements for the sepals and petals for three species of iris flowers.
- The ChickWeight data set contains the weight (in grams) of 50 chicks for the first 21 days of their lives. Each chick is on one of four possible diets.

Figure 4-1   How to use the datasets package

How to get the irises data

To make it easier to view and work with the iris data set, the first example in figure 4-2 uses the as_tibble() function to convert the iris data set to a tibble named irises. Then, it displays that tibble on the console. This makes it easy to see that the data set contains 150 rows and 5 columns. In addition, it provides labels in angle brackets that indicate the data type for each column.

There are two main kinds of data you'll encounter in data sets: discrete data and continuous data. *Discrete data* can have one value from a set of values. For example, a value that can be either true or false is discrete. In this figure, the species column contains discrete data because each row can store one of three values: setosa, versicolor, or virginica.

Discrete data in tibbles can also be, but isn't always, *categorical data*. Categorical data can be used to assign a row of data to a category when plotting. Categorial variables in a tibble can be stored using the *factor type*, a special data type that's used to group rows into categories. As it turns out, the Species column in the irises tibble uses the factor data type. To indicate this, the output for the tibble displays <fct> at the top of this column.

In contrast to categorical data, *continuous data* can be any value within a range of numbers. In this figure, the first four columns contain continuous data. The values for these columns range between 0 and 8, but a range can be as long or short as needed. As it turns out, the first four columns of the irises data set use the double data type to store continuous data. To indicate this, the output for the tibble displays <dbl> at the top of these columns.

The function for converting a data frame to a tibble

Function	Description
`as_tibble(x)`	Converts the specified argument to a tibble. This function is typically used to convert data frames but also works with some other data objects such as lists, matrices, and tables.

How to convert the iris data frame to a tibble

```
irises <- as_tibble(iris)
irises
# A tibble: 150 x 5
   Sepal.Length Sepal.Width Petal.Length Petal.Width Species
          <dbl>       <dbl>        <dbl>       <dbl> <fct>
 1          5.1         3.5          1.4         0.2 setosa
 2          4.9         3            1.4         0.2 setosa
 3          4.7         3.2          1.3         0.2 setosa
 4          4.6         3.1          1.5         0.2 setosa
 5          5            3.6          1.4         0.2 setosa
 6          5.4         3.9          1.7         0.4 setosa
 7          4.6         3.4          1.4         0.3 setosa
 8          5           3.4          1.5         0.2 setosa
 9          4.4         2.9          1.4         0.2 setosa
10          4.9         3.1          1.5         0.1 setosa
# ... with 140 more rows
```

An iris sepal and petal

Description

- *Sepals* and *petals* are parts of a flower.
- *Discrete data* is data that can have one of a set of values. In the irises data set, the species column contains discrete data. Discrete data is often called *categorical data* because it can be used to assign a row to a category.
- *Continuous data* is data that can be any value within a range. In the irises data set, the length and width columns contain continuous data.
- In a tibble, <fct> indicates that the column stores objects of the *factor type*. This data type stores an object that assigns a row to a category.
- In a tibble, <dbl> indicates that the column stores values of the double type. This data type is often used to store a continuous range of numbers such as measurements.

Figure 4-2   How to get the irises data

How to get the chicks data

To make it easier to view and work with the ChickWeight data set, the first example in figure 4-3 uses the as_tibble() function to convert the ChickWeight data frame to a tibble named chicks. Then, it displays that tibble on the console. This makes it easy to see that the data set contains 578 rows and 4 columns. In addition, it provides labels in angle brackets that indicate the data type for each column.

When examining the data set, you might notice that the first column name uses all lowercase letters while the other column names use title case. To use a consistent naming convention for all columns, the second example uses the rename() function to rename the first column from "weight" to "Weight".

When using rename(), be sure to code the new column name first, on the left side of the equals operator. You can rename as many columns as you like at once by separating each new/old name pair with a comma. If a column name contains spaces, you must enclose that name in quotation marks (""). Otherwise, you don't need to enclose column names in quotation marks.

After you rename a column, you can use the names() function to view all of the column names for the tibble as shown in the third example. This shows that the names() function described in chapter 2 works the same for a tibble as it does for a vector or a data frame.

The output for the chicks tibble uses <ord> to show that the Chick column uses the *ordered factor type*. Like a factor, an ordered factor contains one of a set number of values. However, an ordered factor stores its values in an order that you can specify. It's typical to store the values in numerical (1, 2, 3, etc.) or alphabetical (a, b, c, etc.) order, but you can specify any order you want.

Meanwhile, the Diet column uses the factor data type. However, if you later find that you want or need to specify an order for the diets, you could convert this column to the ordered type.

A function for working with column names

Function	Description
`rename(data, newname = oldname, ...)`	Rename one or more columns in the data object.

How to convert the ChickWeight data frame to a tibble

```
chicks <- as_tibble(ChickWeight)
chicks
# A tibble: 578 x 4
   weight  Time Chick Diet
    <dbl> <dbl> <ord> <fct>
1      42     0 1     1
2      51     2 1     1
3      59     4 1     1
4      64     6 1     1
5      76     8 1     1
6      93    10 1     1
7     106    12 1     1
8     125    14 1     1
9     149    16 1     1
10    171    18 1     1
# ... with 568 more rows
```

How to fix the capitalization for the first column

```
chicks <- rename(chicks, Weight = weight)
```

How check the column names

```
names(chicks)
[1] "Weight" "Time"   "Chick"  "Diet"
```

Description

- The *ordered factor type* (or just *ordered type*) is a special version of the factor type where each value is stored in a specified order that you can control.
- In a tibble, <ord> indicates that the column stores an ordered factor.
- The rename() function is part of the core tidyverse package.
- The names() function is part of the R language.

Figure 4-3   How to get the chicks data

How to select rows based on a condition

You can select rows from a data set based on a condition by using the filter() function. This function takes two arguments, the name of a data object and a condition. Then, it generates a data object of the same type containing only the rows that meet the condition. For example, passing a data frame to the filter() function returns a data frame, but passing a tibble to the filter() function returns a tibble.

To create a condition for the filter() function, you can use the relational and logical operators described in chapter 2. For instance, the first example in figure 4-4 uses the equality operator (==) to select all rows where the value in the Diet column equals 2. In other words, it selects all chicks on diet 2. Since this example passes the tibble named chicks, the filter() function returns a tibble for the selected chicks.

The second example uses the greater than operator (>) to select all rows in the irises tibble where the value in the Sepal.Length column is greater than 5. In other words, it selects all irises where the sepal length greater than 5. Again, since this example passes a tibble, the filter() function returns a tibble.

If the condition you specify isn't true for any rows, the filter() function returns a data set with zero rows. In most cases, that's not what you want. Fortunately, it's easy enough to modify your condition and run the code again until it works the way you want.

A function for selecting rows from a tibble

Function	Description
`filter(data, condition)`	Selects the rows of the data object that meet the condition and returns a data object of the same type.

The rows for the chicks on diet 2

```
filter(chicks, Diet == 2)
# A tibble: 120 x 4
   Weight  Time Chick Diet
    <dbl> <dbl> <dbl> <fct>
1      40     0    21 2
2      50     2    21 2
3      62     4    21 2
...
```

The rows of the iris data set that have a sepal length greater than 5

```
filter(irises, Sepal.Length > 5)
# A tibble: 118 x 5
   Sepal.Length Sepal.Width Petal.Length Petal.Width Species
          <dbl>       <dbl>        <dbl>       <dbl> <fct>
1           5.1         3.5          1.4         0.2 setosa
2           5.4         3.9          1.7         0.4 setosa
3           5.4         3.7          1.5         0.2 setosa
...
```

Description

- The filter() function is part of the core tidyverse package.

Figure 4-4   How to select rows based on a condition

An introduction to ggplot2

Now that you know how to get some data, you're almost ready to start using the ggplot2 package to create some plots! But first, this chapter shows how to use two functions to create an empty base plot. Then, it introduces some functions that you can use to add common plots to a base plot.

How to create a base plot

With the ggplot2 package, every plot starts with a call to the ggplot() function. The first table in figure 4-5 summarizes the two most common parameters of the ggplot() function, data and mapping. As you would expect, you can pass a data argument to set the data for the plot. This data argument can be any type of data frame, including a tibble. In this figure, the code example uses the tibble named chicks that was created earlier in this chapter.

The mapping argument allows you to use the aes() function (short for *aesthetics*) to specify some aspects of the plot. The second table summarizes five of the most common parameters of this function. However, the way these parameters work varies depending on the type of plot, and you don't need to pass all of them to every plot.

When using the aes() function, you typically specify columns for the x and y axes as well as a column for the color or fill mapping. In some situations, you may even use both color and fill. In other situations, you may want to use the size argument to improve your visualizations.

The code example shows two techniques for creating a blank plot for the chicks tibble. The first technique uses positional arguments, and the second uses named arguments. Both techniques are easy to read and work equally well. As a result, you can use whichever technique you prefer. However, it's a good practice be consistent within an analysis.

Both of the ggplot() functions in this figure use the aes() function to set the x axis to the Time column and the y axis to the Weight column. In addition, they set the color to the Chick column. As a result, the base plot that's generated by these functions has x and y axes with appropriate ranges and labels. However, it doesn't use color anywhere. That's because these function calls only set the data and aesthetics for the plot, not anything else. To actually display a plot, you need to add one of the functions shown in the next figure to the ggplot() function.

The first two parameters of the ggplot() function

Parameter	Description
data	Sets the data frame to use for the plot.
mapping	Sets the aesthetic mappings for the plot.

Common parameters of the aes() function

Parameter	Description
x	Sets the column for the x axis.
y	Sets the column for the y axis.
color	Sets the color for the lines or points.
fill	Sets the fill color for the shapes.
size	Sets the size for the lines or points.

How to call the ggplot() and aes() functions

With positional arguments
```
ggplot(chicks, aes(x = Time, y = Weight, color = Chick))
```

With named arguments
```
ggplot(data = chicks,
       mapping = aes(x = Time, y = Weight, color = Chick))
```

The base (empty) plot

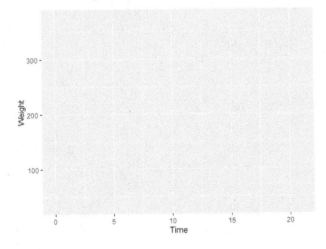

Description

- Every plot that you create using the ggplot2 package starts with a call to the ggplot() function. The data and mapping parameters are common to all plot types.
- The ggplot() function creates the base (empty) plot and some aesthetic mappings.
- The aes() function, short for *aesthetics*, allows you to set some aspects of the plot.

Figure 4-5   How to create a base plot

Functions for common plot types

The table in figure 4-6 summarizes eight functions in the ggplot2 package that you can use to display eight types of plots that are commonly used in data analysis. In addition, it shows examples of the first three types of plots. As you progress through this chapter, you'll learn how to create all eight of these types of plots, and you'll learn more about how each type of plot works.

Functions for common plot types

Function	Description
geom_line()	Creates a line plot.
geom_point()	Creates a scatter plot.
geom_bar()	Creates a bar plot.
geom_boxplot()	Creates a box plot.
geom_histogram()	Creates a histogram.
geom_density()	Creates a KDE plot.
stat_ecdf()	Creates an ECDF plot.
geom_density_2d()	Creates a 2D KDE plot.

A line plot

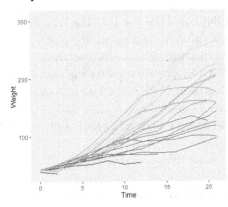

A scatter plot

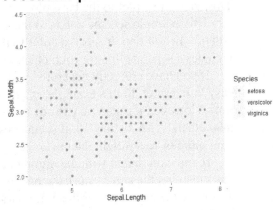

A bar plot

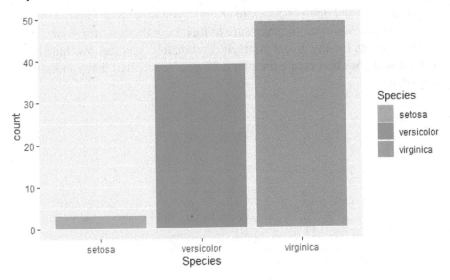

Figure 4-6   Functions for common plot types

How to create relational plots

Relational plots show the relationship between two or more variables. Line plots (which you probably called line graphs in school) and scatter plots are typical examples. *Line plots* work well for showing how variables change over time, and *scatter plots* work well for showing the relationship between two variables.

How to create a line plot

The example in figure 4-7 shows how to add a line plot to a base plot. Here, the ggplot() function creates the base plot as described in figure 4-5. Then, this example uses the addition operator (+) to add a call to the geom_line() function. This adds a line plot to the base plot.

This example sets the x axis to the Time column and the y axis to the Weight column. In addition, it sets the color to the Chick column. As a result, the plot includes a separate line for each chick and it uses a separate color for each line.

If you have correctly generated your line plot, the data should be clear and easy to interpret. In this example, it's obvious that each chick gained weight over time. However, it's also easy to see that some chicks gained much more weight than others. Also, by day 10, it seems possible to predict whether a chick will gain more or less weight than average.

If you experiment with the arguments in this example, you'll notice that changing the arguments can change the plot dramatically. For example, removing the argument for the color parameter results in a plot that's difficult to interpret.

The ggplot() function does its best with the arguments you give it, but it can't read your mind and doesn't generate errors if you mix up or forget arguments. As a result, you need to make sure to look over the code for your plots and visualizations for any errors yourself. Fortunately, you can continue to edit and run the functions that create the plot until you're satisfied that you have the plot you want.

How to create a line plot for all chicks

```
ggplot(chicks, aes(x = Time, y = Weight, color = Chick)) +
  geom_line()
```

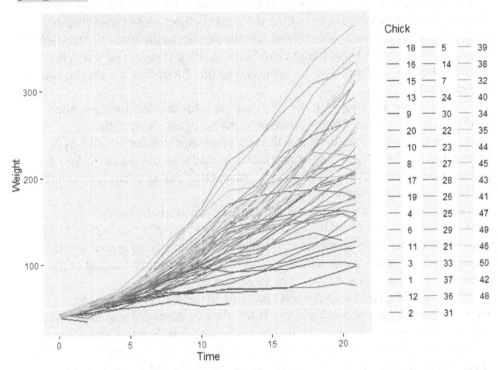

Description

- To add a plot to the base plot that's created by the ggplot() function, code the addition operator (+) and a call to the function that displays the type of plot that you want.

- A *line plot* works well for showing how data changes over time.

Figure 4-7   How to create a line plot

How to create a scatter plot

The first example in figure 4-8 uses the geom_point() function to create a scatter plot that shows the relationship between two columns in the irises tibble. The code for this plot begins by calling the ggplot() function to create a base plot. In this case, the base plot sets the data for the plot to the irises tibble, it sets the x and y axes to the Sepal.Length and Sepal.Width columns, and it sets the color for the dots on the plot to the Species column. These dots are also known as *data points*.

After creating the base plot, the first example adds the dots for the scatter plot by adding the geom_point() function to the base plot. Within the geom_point() function, the code sets the size parameter to a literal value of 3. This makes the dots slightly larger and easier to read than the default value of 1. This is often helpful if you're planning to present this plot to others, or even if you just don't want to squint.

Notice that the size parameter is not passed as an aesthetic for the geom_point() function (even though it could be). The general rule is if you want to use a parameter to set an aspect of the plot like the size or color to a single value, you don't pass it as an aesthetic. And if you want to set an aspect of the plot to use a variable, you do pass it as an aesthetic.

The second example uses the aes() function in ggplot() to set the size parameter to the result of a calculation. It sets the size of each dot to the value of the Sepal.Length column multiplied by the value of the Sepal.Width column. This calculation makes the size of the dots roughly correspond with the size of the sepals. In other words, if a sepal has a long length and width, the plot displays a larger dot for it. On the other hand, if a sepal has a short length and width, the plot displays a smaller dot for it.

How to create a scatter plot

```
ggplot(irises, aes(x = Sepal.Length, y = Sepal.Width, color = Species)) +
  geom_point(size = 3)
```

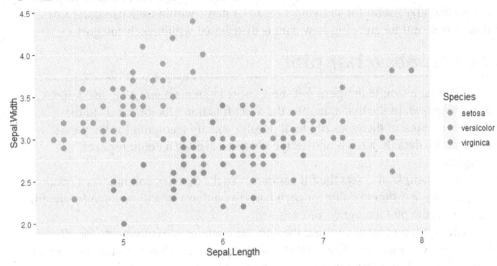

How to change the size of the dots on the scatter plot

```
ggplot(irises, aes(x = Sepal.Length, y = Sepal.Width,
       color = Species, size = Sepal.Length * Sepal.Width)) +
  geom_point()
```

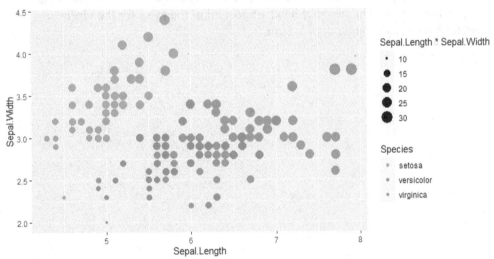

Description

- A *scatter plot* works well for showing the relationship between two variables.
- To set the size of all points to a literal value (like 1.5 or 3), set the size parameter outside of the aes() function.
- To set the size of the points based on the values in a column, set the size parameter within the aes() function.

Figure 4-8   How to create a scatter plot

How to create categorical plots

Categorical plots are used to compare different categories of data. *Bar plots* are particularly useful for showing a count of items within each category. *Box plots* work well for showing how data is distributed within each category.

How to create a bar plot

The first example in figure 4-9 shows how to create a *bar plot*, also known as a *bar graph*. In the first example, the aes() function sets the x axis to the Species column of the irises tibble. It doesn't specify a column for the y axis because, by default, a bar plot uses the y axis to display a count for each category.

This example also sets the fill parameter to the Species column. As a result, the plot uses a different color for each bar. If you don't specify a value for the fill parameter, the plot uses gray for each bar.

The resulting plot shows that the data set contains measurements for 50 flowers for each species. This confirms that each species has the same number of rows. However, it doesn't make for an interesting plot.

To create a more interesting bar plot, the second example sets the data for the plot to the data that's returned by the filter() function. As a result, the data for this plot only includes rows for irises with a sepal length greater than 5.5 cm. The resulting plot shows that the setosa species only has a few rows with sepals longer than 5.5 cm. On the other hand, almost all of the rows for the virginica species have sepals longer than 5.5 cm.

How to create a bar plot

```
ggplot(irises, aes(x = Species, fill = Species)) +
  geom_bar()
```

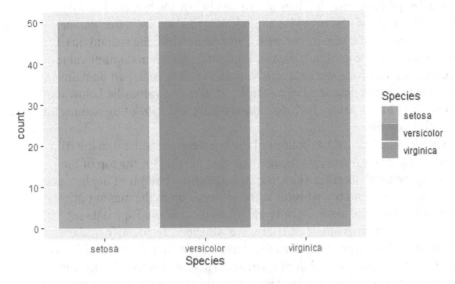

How to create a bar plot for selected rows

```
ggplot(filter(irises, Sepal.Length > 5.5), aes(x = Species, fill = Species)) +
  geom_bar()
```

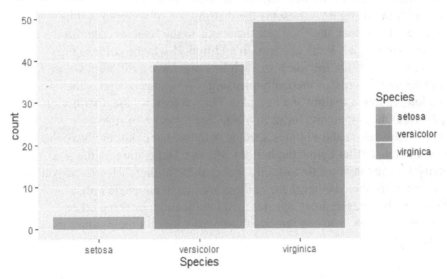

Description

- A *bar plot* shows the count of the rows in a category.
- You can use the filter() function to select certain rows from the data set.
- When creating a bar plot, the y axis displays a count of rows by default. As a result, you don't need to specify the y axis.

Figure 4-9   How to create a bar plot

How to create a box plot

Figure 4-10 shows how to create a *box plot*. A box plot is also called a *box and whisker plot* because the lines above and below the box are known as whiskers. This type of plot provides a quick way to visualize a *five-number summary* that includes the minimum value, the first quartile, the second quartile (commonly known as the median), the third quartile, and the maximum value.

These five values divide the data into four parts where each part contains 25% of the values in the data set. For example, 25% of the values lie below the first quartile, 50% of the values lie below the median, and 75% of the values lie below the third quartile.

In a box plot, the line for the bottom of the box represents the first quartile, the line through the box represents the median, and the line for the top of the box represents the third quartile. Then, the whisker at the bottom of the box goes down to the minimum, and the whisker at the top goes up to the maximum.

In addition, a box plot automatically identifies the *outliers* for a data set and represents them as dots above or below the whiskers. To identify outliers, the geom_boxplot() function calculates the *interquartile range* (*IQR*), which is the distance between the first and third quartiles. Then, it checks whether any values are more than 1.5 times the IRQ above the third quartile or more than 1.5 times the IRQ below the first quartile. If so, the geom_box() plot identifies these values as outliers and displays them as dots. Because of this, the top and bottom whiskers are the largest and smallest values that aren't outliers.

The box plot in this figure represents the distribution of all sepal widths for each species. To do that, the code sets the x axis to the Species column and the y axis to the Sepal.Width column. In addition, it sets the color to the Species column. As a result, the lines for each box plot have a different color. This isn't necessary, but it's aesthetically pleasing. Then, the code adds the geom_boxplot() function to draw the box plot. This plot shows the distribution of the sepal width measurements for each of the three species of irises.

In this plot, the setosa and viginica species both have one outlier above the top whisker and one outlier below the bottom whisker. Depending on the goal of your analysis, you may be able to safely ignore these outliers. However, you might also want to probe whether these outliers are due to a genetic mutation, lack of water, faulty measurement, species misclassification, or some other factor. Then, if necessary, you can remove these outliers or fix their values as described in chapter 6.

How to create a box plot

```
ggplot(irises, aes(x = Species, y = Sepal.Width, color = Species)) +
  geom_boxplot()
```

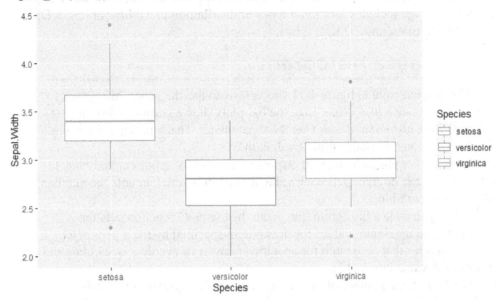

How to interpret a box plot

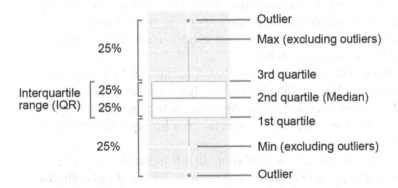

Description

- A *box plot* (also known as a *box and whisker plot*) provides a way to visualize the *five-number summary* (the minimum, the first quartile, the median, the third quartile, and the maximum) for the data.

- *Outliers* are values that fall outside the expected range of values.

- The geom_boxplot() function calculates outliers as being more than 1.5 times the *interquartile range (IRQ)* above the third quartile or below the first quartile.

Figure 4-10   How to create a box plot

How to create distribution plots

A *distribution plot* shows how the values for a column are distributed. The ggplot2 package includes four main types of distribution plots: histograms, KDE plots, ECDF plots, and 2D KDE plots.

How to create a histogram

The first example in figure 4-11 shows how to use the geom_histogram() function to create a *histogram*. Like the bar plots shown earlier in this chapter, the code for a histogram doesn't set the y parameter. That's because the y axis displays the count for each group by default.

Unlike a bar plot, a histogram displays the count for groups called *bins*. In other words, a histogram puts each value into a bin before it counts the number of values in each bin.

When you code a histogram, the geom_histogram() function gets the minimum and maximum values for the column specified by the x parameter. Then, it divides that range into the specified number of evenly spaced bins, and it puts each value into its bin.

By default, the number of bins is 30. As a result, if you don't specify a value for the bins parameter, the geom_histogram() function creates 30 bins and displays a warning on the console as shown in the first and second examples. In that case, you may want to adjust the number of bins to better suit your analysis as shown in the third example. Alternately, you can get rid of the warning message by setting the bins parameter to 30. This tells the geom_histogram() function that you really want to use 30 bins.

In the third example, the code sets the number of bins to 20. This makes the plot easier to understand by creating larger bins. It shows that setosa irises have shorter sepals than virginica and versicolor irises, with virginica having the longest. In addition, for all species, it shows that the bins for sepal lengths longer than 7 cm contain the fewest values.

The calculation for each bin can be confusing. In this figure, for example, the minimum sepal length is 4.3 cm and the maximum is 7.9 cm. As a result, the second example calculates the number of bins by subtracting 4.3 from 7.9 and dividing by 20. This yields bins that are .18 cm wide. In other words, the sepal lengths in bin 1 range from 4.3 to 4.48, the lengths in bin 2 range from 4.48 to 4.66, the lengths in bin 3 range from 4.66 to 4.84, and so on.

How to create a histogram

```
ggplot(irises, aes(x = Sepal.Length, fill = Species)) +
  geom_histogram()
```

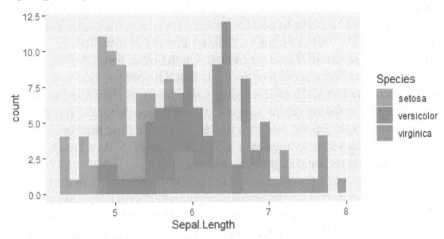

A warning when you don't specify the bins argument

```
`stat_bin()` using `bins = 30`. Pick better value with `binwidth`.
```

How to set the number of bins

```
ggplot(irises, aes(x = Sepal.Length, fill = Species)) +
  geom_histogram(bins = 20)
```

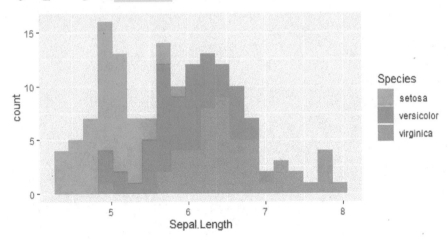

Description

- A *histogram* divides continuous data into groups called *bins* and displays the count for each bin.
- To specify the number of bins for a histogram, set the bins parameter of the geom_histogram() function. If you don't specify a number of bins, this function uses the default value of 30 and displays a warning.

Figure 4-11　How to create a histogram

How to create a KDE plot

A *KDE* (*Kernel Density Estimate*) *plot* is effectively a histogram that has been smoothed into a curve. Instead of dividing the data into bins and plotting the count, a KDE plot uses a curve to plot the probability of a value occurring. For instance, the first KDE plot in figure 4-12 shows that the probability of a setosa having a sepal length of 7 is close to zero. On the other hand, the probability of a setosa having a sepal length of 5 is much higher.

By default, each line in a KDE plot is drawn individually as shown in the first example. Here, the color argument for ggplot() sets the color for the lines. Then, the size parameter for geom_density() makes the lines a little thicker and easier to see. That's because these lines have a default value of .5. As a result, setting the size parameter to 1 makes the lines twice as thick.

If you want, you can stack the lines of a KDE plot to more closely resemble a histogram. To do that, you can set the position argument of geom_density() to "stack" as shown by the second example. When you do that, you typically pass a fill argument to ggplot() to fill in the area for each category. Using fill works well for a stacked KDE plot since the lines don't overlap. However, it doesn't work well for a regular KDE plot since it obscures the lines where they overlap.

How to create a KDE plot

```
ggplot(irises, aes(x = Sepal.Length, color = Species)) +
  geom_density(size = 1)
```

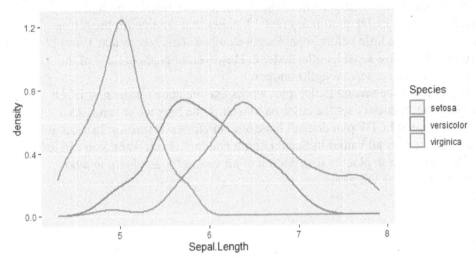

How to create a stacked KDE plot

```
ggplot(irises, aes(x = Sepal.Length, color = Species, fill = Species)) +
  geom_density(position = "stack")
```

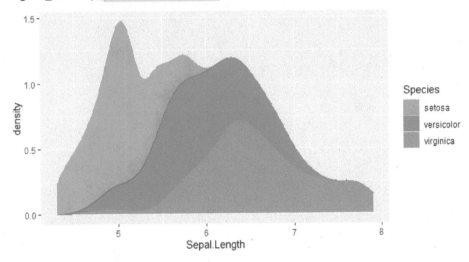

Description

- A *KDE* (*Kernel Density Estimate*) *plot* is a histogram smoothed into a curve.
- To change the thickness of the line in any plot, you can set the size parameter to a literal value like 1, 2, or 3.
- To stack the curves in a KDE plot, you can set the position parameter of the geom_density() function to "stack".

Figure 4-12 How to create a KDE plot

How to create an ECDF plot

In contrast to a KDE plot, an *ECDF (Empirical Cumulative Distribution Function) plot* shows what percent of the data falls at or below a plotted x value. So, the first ECDF plot in figure 4-13 shows that the line for the setosa species reaches 1 a little before sepal length equals 6. This means that 100% of the setosa irises have sepal lengths under 6. However, only about 50% of the versicolor irises have sepal lengths under 6.

The code for the second ECDF plot works like the code for the first ECDF plot. However, it doesn't set the color parameter to the Species column. As a result, the second ECDF plot doesn't have one line for each species. Instead, it has a single line for all values in Sepal.Length column. If you want, you can use a similar technique to plot the distribution of all values for a column in a box plot, histogram, or KDE plot.

How to create an ECDF plot for multiple categories

```
ggplot(irises, aes(x = Sepal.Length, color = Species)) +
    stat_ecdf(size = 1)
```

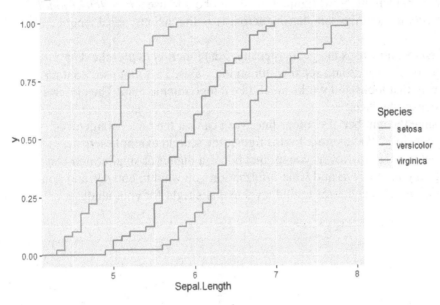

How to create an ECDF plot for all values in a column

```
ggplot(irises, aes(x = Sepal.Length)) +
    stat_ecdf(size = 1)
```

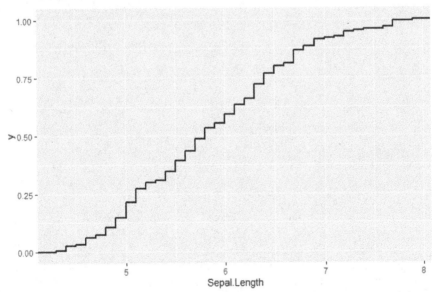

Description

- An *ECDF* (*Empirical Cumulative Distribution Function*) *plot* shows the percentage of values at or below a given value.

Figure 4-13   How to create an ECDF plot

How to create a 2D KDE plot

The plots in the previous figures examined the density of data across one dimension, sepal length. Now, figure 4-14 shows how to use a *two-dimensional (2D) KDE plot* to examine data density across two dimensions, sepal length and width.

The first example uses the geom_density_2d() function to plot the data for sepal length on the x axis and sepal width on the y axis. This plot uses contour lines in a way that looks and works much like a topographic map. The closer the lines, the denser the data.

To change the number of contour lines, you can set the bins parameter of the geom_density_2d() function. In this figure, the second example sets the bins parameter to 15. However, you might choose a bigger or smaller number depending on your analysis and what information you want to convey, and you can typically use trial and error to find a value that's right for your analysis.

How to create a 2D KDE plot

```
ggplot(irises, aes(x = Sepal.Length, y = Sepal.Width, color = Species)) +
    geom_density_2d(size = 1)
```

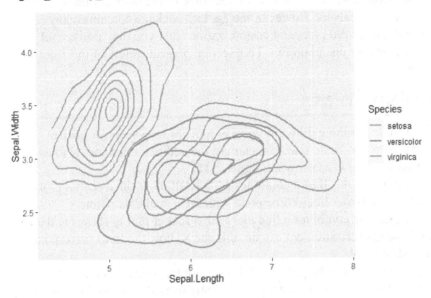

How to set the number of bins

```
ggplot(irises, aes(x = Sepal.Length, y = Sepal.Width, color = Species)) +
    geom_density_2d(size = 1, bins = 15)
```

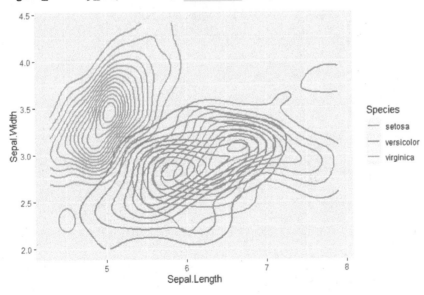

Description

- The geom_density_2d() function creates a two-dimensional (2D) KDE plot.
- To change the number of contour lines for the plot, you can set the bins parameter of the geom_density_2d() function.

Figure 4-14   How to create a 2D KDE plot

More skills for working with plots

So far, this chapter has shown how to create some of the most common types of plots for data analysis. However, the ggplot2 package contains many more options for specialized plots and customizations that you may find useful, aesthetically pleasing, or just plain fun. This section presents a few of the more popular options.

How to combine plots

You may want to combine different types of plots to save space or present more detail in your plots. To do that, you can use the addition operator (+) to add another function to ggplot() as shown in figure 4-15.

The first example in this figure combines a 2D KDE plot with a scatter plot. This shows how the contour lines correspond with the actual data points.

The second example combines a line plot with a scatter plot to show the data points for each weight measurement. Again, this shows how the lines correspond with the actual data points.

Not all types of plots can be combined. For example, you can't combine a plot that uses both x and y values and a plot that only uses an x value. So, you can't combine a scatter plot with a histogram. Furthermore, combining plots can make your plots more difficult to read. So, before you attempt to combine two plots, you should carefully consider if separate plots would be a better way to visualize the data.

How to combine a 2D KDE plot with a scatter plot

```
ggplot(irises, aes(x = Sepal.Length, y = Sepal.Width, color = Species)) +
    geom_density_2d(bins = 15, size = 1) +
    geom_point(size = 3)
```

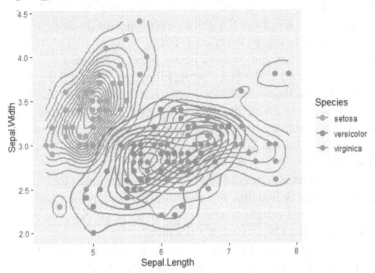

How to combine a line plot with a scatter plot

```
ggplot(data = chicks, aes(x = Time, y = Weight, color = Chick)) +
    geom_line() +
    geom_point()
```

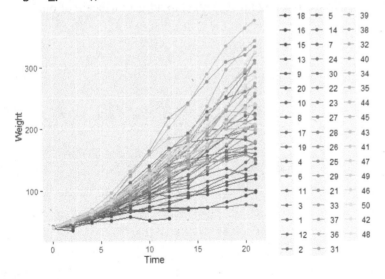

Description

- You can combine plots by using the addition operator (+).
- You can combine as many plots as you want, but they must use the same values for the x and y axes.

Figure 4-15   How to combine plots

How to create a grid of plots

In some cases, you may want to split a single plot into a grid of plots (also called *subplots*). For example, you might want to create a grid of line plots for the chicks data set where each plot corresponds to the diet. That way, it's easier to see how each diet impacted the weight of the chicks over time. To do that, you can use the facet_wap() function as shown in figure 4-16.

This example begins by using the ggplot() and geom_line() functions to create a line plot. Then, it adds a call to the facet_wrap() function. Within the function, the facets argument uses the vars() function to specify the column of categorical data (Diet) that's used to divide the plots into a grid. Next, the nrow (number of rows) and ncol (number of columns) arguments specify that the grid should have two rows and two columns. You don't need to specify both of these arguments, but you do need to specify at least one.

The resulting grid of plots makes it possible for you to view the line plots for each type of diet and make quick insights. For example, the grid in this figure tells you the chicks on diet 4 gained weight consistently, but those chicks didn't gain as much weight as some of the chicks on the other diets.

By default, the scales argument is set to "fixed", which specifies that the axes should use the same scale for each subplot. Since this makes it easy to compare the data from different plots, that's usually what you want. As a result, you only need to specify the scales argument if you don't want to use the same scale for each subplot. In that case, you can set the scales argument to "free" to allow facet_wrap() to adjust the scale of both axes. Or, you can set the scales argument to "x_free" or "y_free" to allow facet_wrap() to adjust the scale of one axis or the other.

Common facet_wrap() parameters

Parameter	Description
facets	Specifies the categorical variable. It's typically used with the vars() function.
nrow, ncol	Specifies the number of rows and columns in the grid. You must specify at least one or the other. Alternately, you can specify both.
scales	Specifies whether the x and y axes should be the same for all plots. By default, this is set to "fixed". This keeps the axes the same for all plots. However, setting this parameter to "free", "x_free", or "y_free" allows the x and/or y axes to vary for each grid.

How to create a grid of plots

```
ggplot(chicks, aes(x = Time, y = Weight, color = Chick)) +
  geom_line() +
  facet_wrap(facets = vars(Diet), nrow = 2, ncol = 2)
```

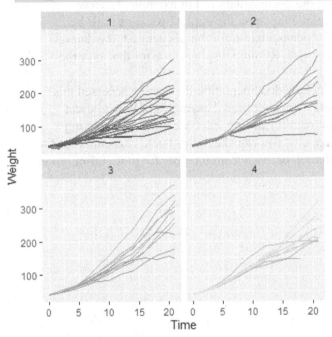

Description

- The plots displayed by the facet_wrap() function can be used to display several plots together in a grid. These plots are sometimes referred to as *subplots*.

Figure 4-16   How to create a grid of plots

How to view documentation

When you're working on an analysis, you may want to view the *documentation* for a function, package, or a data set in the datasets package. Fortunately, RStudio makes this easy to do.

To view documentation, you type a question mark (?) at the prompt in the Console pane. Then, you enter the name of the function, package, or data set as shown by the first example in figure 4-17. When you do, RStudio displays the documentation in the Help tab.

In this figure, the Help tab displays the documentation for the ggplot() function. This documentation begins by displaying information about the data and mapping parameters described earlier in this chapter. In addition, it contains a lot of other information for more complicated uses, and you can scroll down to view it.

Before you can use RStudio to view the documentation, you must load the package that contains the documentation. For example, to view the documentation for the ggplot() function, the ggplot2 package must be loaded. Similarly, to view the documentation for the iris data set, the datasets package must be loaded. Otherwise, RStudio displays an error that says the documentation can't be found.

While most R packages include documentation that can be accessed this way, some don't. Also, you may want additional examples of how to use a function, especially when you're still learning and unfamiliar with parsing documentation. When this happens, you can check this book's appendix for the function name, search the internet, or ask a team member for more information.

Statements for viewing documentation

```
? ggplot       # for a function
? tidyverse    # for a package
? iris         # for a data set in the datasets package
```

The Console pane and the Help tab for a function

Description

- You can use RStudio to view the documentation for a function, a package, or a data set in the datasets package.

- To view documentation, go to the Console pane, type a question mark (?) at the prompt, and enter the name of the function, package, or data set. Then, RStudio displays the documentation in the Help tab.

- Before you can use RStudio to view the documentation, the package that contains the documentation must be installed and loaded.

Figure 4-17   How to view documentation

How to save a plot

Sometimes you may want to add a plot to an existing document or presentation. RStudio provides an easy way to save a plot as an image file or a PDF file. It also provides an easy way to copy a plot to the clipboard so you can paste it into a document or presentation.

To save a plot, you can start by viewing the plot in the Plots tab. Then, you can click the Export button and select the item you want from the resulting menu as shown in figure 4-18.

When you select an item from the Export menu, RStudio displays a dialog that allows you to change the resolution of the plot and to preview the final output. If you're saving the plot to a file, this dialog allows you to specify the filename and directory for the file. In addition, if you're saving the plot as an image, this dialog lets you select an image format such as PNG, JPEG, or TIFF.

The Plots tab with its Export menu displayed

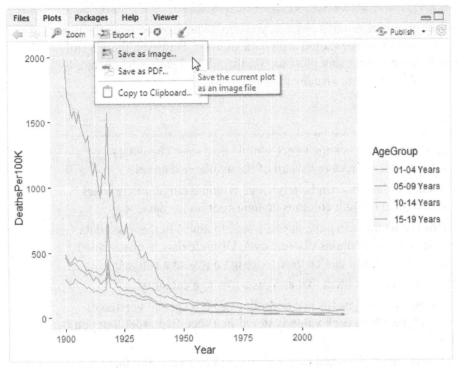

Description

- To save a plot, click the Export button in the Plots tab and select the item you want from the resulting menu. Then, use the resulting dialog to specify the options for the file.
- For an image file, you can specify the filename, the directory, an image format such as PNG or JPEG, and the resolution.
- For a PDF file, you can specify the filename, the directory, and the resolution.
- For the clipboard, you can specify the resolution. Then, you can paste the file into a document or presentation.

Figure 4-18   How to save a plot

Perspective

Now that you've completed this chapter, you should be able to create the types of plots that you need to get started with data analysis. However, there's still much more to learn about using plots to visualize data. That's why chapter 8 presents more skills for data visualization.

Summary

- You can use the datasets package to get sample data sets. The data() function of the datasets package lists all of the available data sets.

- *Continuous data* is data that can be any value within a range. In the irises data set, the length and width columns contain continuous data.

- *Discrete data* is data that can have one of a set of values. In the irises data set, the species column contains discrete data. Discrete data is often called *categorical data* because it can be used to assign a row to a category.

- The *factor type* stores an object that assigns a row to a category.

- The *ordered factor data type* (or just *ordered type*) is a special version of the factor data type where each value is stored in a specified order that you can control.

- The ggplot2 package is the standard for plotting in R. Every plot that you create with ggplot2 starts with a call to the ggplot() function.

- The aes() function, short for *aesthetics*, allows you to set aspects of the plot such as the column to use for the x and y axes.

- *Relational plots* such as line plots and scatter plots show the relationship between two or more variables.

- A *line plot* works well for showing how variables change over time.

- A *scatter plot* works well for showing the relationship between two variables.

- *Categorical plots* such as bar plots and box plots are used to compare different categories of data.

- A *bar plot* works well for showing a count of items within each category.

- A *box plot* (also known as a *box and whisker plot*) provides a way to visualize the *five-number summary* (minimum value, first quartile, median, third quartile, and maximum value) for a range of values.

- *Outliers* are values that fall outside the expected range of values.

- To identify outliers, the geom_boxplot() function makes a calculation that uses the *interquartile range (IQR)*, which is the distance between the first and third quartiles.

- A *distribution plot* shows how the values for a column are distributed. The ggplot2 package includes four main types of distribution plots: histograms, KDE plots, ECDF plots, and 2D KDE plots.

- A *histogram* divides continuous data into groups called *bins* and plots the count for each bin.

- A *KDE (Kernel Density Estimate) plot* is a histogram smoothed into a curve.

- An *ECDF (Empirical Cumulative Distribution Function) plot* shows the percentage of values at or below a given value.

- A *two-dimensional (2D) KDE plot* examines data density across two columns.

- You can use the facet_wrap() function to display several plots together in a grid. These plots are sometimes referred to as *subplots*.

- You can combine plots by adding multiple plot types to one call to the ggplot() function.

- You can use the ? operator to view the *documentation* for a function, data set, or package.

- RStudio provides functionality for saving a plot or copying it to your clipboard.

Exercise 4-1   Edit some existing plots

In this exercise, you'll run a script that creates the plots shown in this chapter. Then, you'll edit them to see how different arguments affect the plots.

Open and run an existing script

1. Open the R script named exercise_4-1.R located in this folder:
   ```
   Documents/murach_r/exercises/ch04
   ```

2. Step through the script statement by statement using Ctrl+Enter (Windows) or Command+Enter (macOS). Note that it creates the plots used in the figures for this chapter.

Edit some of the plots

3. Find the statement for the first scatter plot and change the x and y arguments to be for petals instead of sepals. Your new code should look like this:
   ```
   aes(x = Petal.Length, y = Petal.Width, color = Species)
   ```
   You now have a whole new insight into the differences between the three species of irises.

4. Find the statement for displaying a bar plot with a subset of the data. Edit the ggplot() function to change the subset of data being graphed to show sepal lengths less or equal to 5.5, and delete the fill argument. Note how this changes the plot.

5. Find the statement for displaying a histogram using a specified number of bins. Try changing the number of bins to 5 and generate the new plot. Does the smaller number of bins help or hinder visualizing the data? Try additional numbers of bins like 15, 50, and 100 to see how the histogram is affected.

6. Find the statement for combining a 2d density plot with a scatter plot. Edit the data to use the filter() function to only display versicolor irises. Remember that the equals operator uses two equals signs (==). Your function should look like this:
   ```
   filter(irises, Species == "versicolor")
   ```

7. Find the statement for creating a two-by-two grid of plots using the facet_wrap() function. Edit the statements to generate a grid with one column and four rows instead.

8. Experiment with changing the arguments for any of the plots. Try small or large numbers to see the effects, and use the filter() function to create plots of interesting data subsets.

Exercise 4-2   Create your own plots

In this exercise, you'll create your own plots using a data set from the datasets package. This is effectively a mini analysis.

Create a script and read data into a tibble

1. Clear the Environment and Console panes by clicking their respective broom icons. The one for the Console pane can be hard to see, but it's in the upper right and in grey.

2. Create a new script and name it exercise_4-2.R.

3. In the new script, add statements that load the tidyverse and datasets packages. Then, run those statements.

4. Create a new tibble named swchars from the starwars data set.

5. Display the contents of the tibble to the console. Note that it contains a list of 87 characters from the first seven Star Wars movies along with various data about them like height, hair color, gender, species, etc.

Create some plots

6. Create a bar plot that displays the count of characters of each species. Note that unlike the irises tibble, the starwars data set does not capitalize column names. The resulting plot may be hard to read, but it should make it clear that humans are overwhelmingly the most common species in the Star Wars films.

7. Create another bar plot that displays the count of characters by gender. Note how this helps you understand the data set.

8. Create a scatter plot with character height on the x axis, mass on the y axis, and the species as the color of each point. The console will warn you that 28 rows were not plotted due to lack of data. In a real-world analysis, this should cause you to reevaluate your data set and consider if you need to update or clean it.

9. Note that this scatter plot contains one significant outlier with a mass over 1000 kg. Because of this outlier, it's hard to interpret the rest of the data. Also, because there are so many species, the color of each datapoint doesn't help you interpret the data.

10. Create another scatter plot that uses the filter() function to restrict mass to under 1000 kg and uses gender to set the color of each data point. Additionally, use the size argument to make the points larger and easier to see.

11. Note that this scatter plot has several obvious insights. For example, height and mass have a linear relationship, and feminine characters are clustered in the middle of the plot, unlikely to be extremely short, tall, light, or heavy.

12. Use the facet_wrap() function to generate a grid of scatter plots that map height to mass for characters with a mass less than 1000 kg. Display one subplot for each gender with three columns of subplots, and set the color of the data points based on the species. This should display three subplots (feminine, masculine, and NA).

Chapter 5

How to get data

This chapter shows you how to get the data that you're going to analyze. That includes finding the data on a website as well as reading it into a tibble. For some readers, this chapter may present more than you need to know. That's because it not only shows you how to get data from simple files like CSV or Excel files but also how to get data from zip files, databases, and JSON files. If that's more than you need, you can read just the parts that apply to the types of data that you want to get.

Basic skills for getting data

This chapter starts by presenting some sources for finding the data you need for an analysis. Then, it shows you how to read data from CSV and Excel files. Next, it shows how to download data from the internet and how to unzip compressed files.

How to find the data you want

Figure 5-1 begins by summarizing some common data sources. To start, the data you need for an analysis may be available from spreadsheets or databases maintained by the organization that you're doing the analysis for. In that case, the organization can give you access to the data that you need.

Often, though, you need to find the data that you're going to analyze from third-party websites. That's why this figure lists three websites that can help you find the data that you're looking for.

This figure also lists the websites that provide the data for the analyses presented in this book. These are the websites for the Centers for Disease Control, the U.S. Forest Service, the NBA (National Basketball Association), and FiveThirtyEight, a well-known website for American political polling.

When you look for data on third-party websites like these, remember that your analysis will only be as good as the data you build it on. Carefully consider your sources and remember that data can be biased via selective inclusions or omissions. Sometimes data may be from an unrepresentative sample or even be outright wrong.

The download page for the Childhood Mortality data

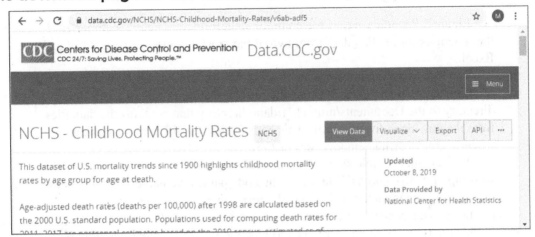

Common sources of data

Internal data sets and databases

This can include everything from departmental spreadsheets to any of the databases used by a corporation.

Third-party websites

Many websites let you download data for your own analysis.

Good places to look for external datasets

Google Dataset Search: https://toolbox.google.com/datasetsearch

Kaggle: https://www.kaggle.com/datasets

Registry of Open Data on AWS: https://registry.opendata.aws

The websites for the analyses presented in this book

- **Child Mortality**: Centers for Disease Control (data.cdc.gov)
- **Wildfires**: U.S. Forest Service (fs.fed.us)
- **Basketball Shots**: NBA Stats (stats.nba.com)
- **Polling**: FiveThirtyEight (fivethirtyeight.com)

Description

- Not all data sources contain accurate, unbiased data. Your analysis will only be as good as the data you're using, so be careful to use trusted data sets.

Figure 5-1   How to find the data you want

How to read data from CSV and Excel files

Figure 5-2 summarizes some common data formats. Then, it summarizes the functions for reading data from two of the most common formats, CSV and Excel.

The first example in this figure shows how to set the working directory for this chapter. To do that, you can use the setwd() function to set the working directory to the Documents/murach_r/data directory that contains the data files from the download for this book. This code is slightly different depending on whether you're using Windows or macOS.

The second example shows how to read data from a CSV file located in the working directory into a tibble. Here, the code passes the name of the CSV file to the read_csv() function and stores the result in a variable named mortality_data. Since the read_csv() function is in the readr package that's part of the core tidyverse, you don't need to load the readr package separately if you've already loaded the tidyverse.

The third example works the same as the first example, except that it specifies the full path to the CSV file. You can do this if you don't want to change the working directory.

The fourth example reads the CSV file directly from a website. This may be preferable if the data is constantly being updated, and you want to make sure your analysis is always using the most current data.

However, there are several disadvantages to this approach. First, reading data from the internet is slower than reading data that has already been downloaded to your computer. Second, you might not want the data for your analysis to be changing as you work on it. As a result, it often makes sense to download the data for your analysis before you begin working on it.

The fifth example shows how to read data from an Excel file. Here, the code passes two arguments to the read_excel() function. The first argument specifies the name of the Excel file. The second argument specifies the sheet (or page) of the Excel file. This is necessary because a single Excel file can contain multiple spreadsheets.

In the fifth example, the code sets the sheet parameter to 1 to indicate that the function should read the first sheet in the Excel file. However, if you want to read data from the third sheet, you could pass a value of 3. Or, if you want to specify the spreadsheet by name, not number, you can code the name of the spreadsheet like this:

```
sheet = "Mortality Wide"
```

Unlike the read_csv() function, the read_excel() function is in the readxl package that's *not* part of the core tidyverse. As a result, if you want to use the read_excel() function, you need to load the readxl package separately even if you've already loaded the tidyverse.

Common data formats

Type	Extension	Description	Contents
CSV	.csv	Comma-separated values	One table
TSV	.tsv	Tab-separated values	One table
Excel	.xlsx, .xls	Excel spreadsheet	One or more sheets
JSON	.json	JavaScript Object Notation	Nested data
XML	.xml	Extensible Markup Language	Nested data
Zip	.zip	Archive format	One or more files
Stata	.dta	Stata statistical package	Complex data
SAS	.sd7, .sd6	SAS statistical package	Complex data
SPSS	.sav	SPSS statistical package	Complex data

Common functions for reading data into a tibble

Function	Description
read_csv(file)	Reads the specified CSV file and returns a tibble.
read_excel(file, sheet)	Reads the specified sheet from the specified Excel file and returns a tibble. To specify the sheet, you can use a number or a name.

How to set the working directory for this chapter

```
setwd("~/murach_r/data")              # for Windows
setwd("~/Documents/murach_r/data")  # for macOS
```

How to read a CSV file from the working directory

```
mortality_data <- read_csv("child_mortality_rates.csv")
```

How to read a CSV file from a path

```
mortality_data <-  read_csv("~/murach_r/data/child_mortality_rates.csv")
```

How to read a CSV file from a website

```
mortality_data <- read_csv(
  "https://data.cdc.gov/api/views/v6ab-adf5/rows.csv")
```

How to read a sheet from an Excel file

```
library("readxl")       # read_excel() not part of core tidyverse
mortality_data <- read_excel("child_mortality_rates.xlsx", sheet = 1)
```

Description

- A *CSV (comma-separated values)* file uses commas to separate the values for its columns and stores each row on a new line.
- The read_csv() function is in the readr package that's part of the core tidyverse.
- The read_excel() function is in the readxl package that's *not* part of the core tidyverse.

Figure 5-2   How to read data from CSV and Excel files

How to download data

When you get data from the internet, it often makes sense to download the file that contains the data and save it to your computer before you read it into a tibble. To do that, you can use the download.file() function presented in figure 5-3. This function takes two arguments. The first specifies the URL for the source file on the internet, and the second specifies the filename for the destination file on your computer.

By default, the download.file() function saves the destination file in the current working directory. As a result, you typically want to set the working directory before calling this function. That way, you can save it to the directory of your choice.

The first example shows how to download a CSV file from a website. Here, the first statement defines a URL that points to a CSV file that's available from the FiveThirtyEight website. Note that this URL is too long to fit on one line in this book, so it's been split into two lines in the example. However, for this code to work, the URL must be coded on a single line in the code editor.

Then, the second statement defines a name of "polls.csv" for the destination file. Finally, the third statement passes the URL and the destination filename to the download.file() function. As a result, this code downloads the specified CSV file and stores it with the specified filename in the current working directory.

If you download a file in a format like CSV or Excel, you can open it in Excel or another spreadsheet program. Then, you can use your spreadsheet program to examine the data set and learn more about what kind of data it contains.

Once you download your data set, you can read it into a tibble as shown in the previous figure. For instance, the second example shows how to use the read_csv() function to read the CSV file downloaded by the first example. To do that, the code passes the dest_file variable to the read_csv() function. That way, you can be sure that both examples are working with the same destination file.

A function for downloading data

Function	Description
download.file(url, destfile)	Downloads the data at the URL and saves it to the destination file.

How to download data and save it in a file

```
url <- "http://projects.fivethirtyeight.com/general-model/
        president_general_polls_2016.csv"
dest_file <- "polls.csv"
download.file(url, dest_file)
```

The CSV file after it has been opened in a spreadsheet program

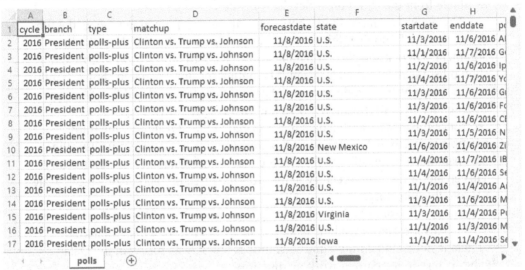

How to read the downloaded file into a tibble

```
polls <- read_csv(dest_file)
```

Description

- If you download a delimited file such as a CSV, TSV, or Excel file, you can open it in a spreadsheet program to view the data before you read it into a tibble.

Figure 5-3   How to download data

How to work with a zip file

In some cases, the data you want may be stored in a *zip file*, which is a file that consists of one or more files in a compressed format so they require less storage space. In that case, you would download the zip file, save it on your computer, *unzip* the files, and read the file that contains the data into a tibble. Figure 5-4 shows how to do this.

The first example shows how to download a zip file and save it to your computer using the download.file() function from the previous figure. Here, the URL for the zip file is too long to fit on one line, so it has been wrapped onto a second line in this book. However, for the code to work, this URL must be coded on a single line. This URL specifies a zip file that contains wildfire data that's available from the U.S. Forest Service website.

The second example shows how to use the unzip() function to view the files in the zip file. To do that, this code passes the name of the zip file as the first argument. Then, it sets the list parameter to TRUE to indicate that the function should list the files, not unzip them. The output for this statement shows that this zip file contains two subdirectories (Data and Supplements) and five files (one SQLite file, an XML file, a PDF file, and two HTML files).

Since this book uses the data that's stored in the SQLite file, the third example shows how to extract just this file. To do that, this example uses the files parameter to specify the SQLite file that's stored in the Data subdirectory. This creates a Data subdirectory in the working directory and unzips the SQLite file into this directory.

The fourth example shows how to rename the SQLite file to a shorter name. This makes it easier to read the database name and to remember what kind of data it stores. In addition, this code moves the SQLite file from the Data subdirectory to the current working directory.

The unzip() function

Function	Description
unzip()	Extracts the specified files from the zip file and saves them in the working directory.

Parameter	Description
zipfile	The path and filename for the zip file.
files	The files from the zip file that should be extracted.
list	If TRUE, lists the files stored in the zip file instead of extracting them.

How to download the fires zip file

```
fires_url <- "https://www.fs.usda.gov/rds/archive/products/RDS-2013-
    0009.4/RDS-2013-0009.4_SQLITE.zip"

download.file(fires_url, "fires.zip")
```

How to view the files in the fires zip file

```
unzip("fires.zip", list = TRUE)
                          Name       Length                 Date
1        Data/FPA_FOD_20170508.sqlite 795785216 2015-05-17 16:12:00
2                             Data/           0 2015-05-24 12:15:00
3          _metadata_RDS-2013-0009.4.xml    51400 2015-08-09 12:22:00
4 Supplements/FPA_FOD_Source_List.pdf  109336 2015-05-09 14:20:00
5                       Supplements/       0 2015-05-09 14:20:00
6     _fileindex_RDS-2013-0009.4.html    4398 2015-05-22 12:44:00
7      _metadata_RDS-2013-0009.4.html   89005 2015-08-09 12:22:00
```

How to extract a file from the fires zip file

```
unzip("fires.zip", files = "Data/FPA_FOD_20170508.sqlite")
```

How to rename the SQLite file and move it to the working directory

```
file.rename(c("Data/FPA_FOD_20170508.sqlite"),
            c("fires.sqlite"))
```

Description

- If the data that you download is in a zip file, you need to unzip the file before you can read it.
- Unzipping a file that's stored in a subdirectory of the zip file creates the subdirectory in the working directory.

Figure 5-4   How to work with a zip file

How to read data from a database

A database stores its data in one or more tables. Multiple tables in a database are typically related to one another. To read data from a database, you can run a *SQL* (*Structured Query Language*) statement known as a *query* to select the data you want to read. In conversation, SQL is sometimes pronounced as "sequel", and it's sometimes spelled out as S-Q-L.

The next three figures show how to work with a SQLite database, which is a popular, open-source, relational database that can be distributed as a file and embedded into programs. These examples show how to work with SQLite databases because you don't need to set up a database server to work with them.

The concepts presented here are similar for other databases such as MySQL, SQL Server, and Oracle. However, since those databases run on servers, you need to specify more information to connect to them, such as a URL, username, and password. If you need to connect to such a database, you'll typically be able to get the connection details from team members or an administrator.

How to connect to a database

To read data from a database, you need to connect to the database as shown in figure 5-5. To start, you need to load the packages for working with a database. That's why the first example shows how load the packages for working with a SQLite database. Here, the DBI package contains some functions for working with any type of database. Then, the RSQLite package contains the SQLite() function that returns a *database driver* for working with a SQLite database specifically.

The second example begins by defining the name of the database file. Then, it connects to the database by calling the dbConnect() function and passing it the database driver and a path to the database. This assigns the *connection object* that's returned from the function to a variable named con.

How to list the tables in a database

The third example shows how to list the tables in the database. To do that, it passes the connection object to the dbListTables() function. The resulting vector shows the names of all tables in the database. This includes the Fires table.

How to list the columns of a table

The fourth example uses the dbListFields() function to list the names of the columns of the Fires table. This is important because you typically need to know the names of the columns of a table before you can code a query that reads data from that table. As it turns out, the Fires table contains 39 columns, and you probably don't want to read all of them into a tibble.

The fifth example in figure 5-5 shows how to use the dbDisconnect() function to close the database connection. If you don't close the database

The functions for connecting to a database

Function	Description
dbConnect(drv, path)	Returns a connection object using the specified driver (drv) to connect to the database at the specified path.
SQLite()	Returns a driver for the SQLite database.
dbDisconnect(conn)	Disconnects from the database and frees resources used by the connection.

The functions for examining the contents of a database

Function	Description
dbListTables(conn)	Lists the tables in the connected database.
dbListFields(conn, name)	Lists the files of the specified table (name).

How to load the packages for working with a SQLite database

```
library(DBI)
library(RSQLite)
```

How to connect to a database

```
db_file <- "fires.sqlite"
con <- dbConnect(SQLite(), db_file)
```

How to list the tables in the database

```
dbListTables(con)
[1] "ElementaryGeometries"      "Fires"
[3] "KNN"                       "NWCG_UnitIDActive_20170109"
[5] "SpatialIndex"              "geom_cols_ref_sys"
[7] "geometry_columns"          "geometry_columns_auth"
...
```

How to list the columns in a table

```
dbListFields(con, "Fires")
[1] "OBJECTID"                  "FOD_ID"
[3] "FPA_ID"                    "SOURCE_SYSTEM_TYPE"
[5] "SOURCE_SYSTEM"             "NWCG_REPORTING_AGENCY"
[7] "NWCG_REPORTING_UNIT_ID"    "NWCG_REPORTING_UNIT_NAME"
[9] "SOURCE_REPORTING_UNIT"     "SOURCE_REPORTING_UNIT_NAME"
[11] "LOCAL_FIRE_REPORT_ID"     "LOCAL_INCIDENT_ID"
[13] "FIRE_CODE"                "FIRE_NAME"
...
```

How to close the connection and free its resources

```
dbDisconnect(con)
```

Description

- Once you've connected to the database you can use the dbListTables() function to list the tables in the database and the dbListFields() function to list the columns in a table.

Figure 5-5   How to connect to a database

connection, the code in this chapter still runs correctly. However, it's a good practice to close the database connection after you're done with it because it frees the system resources used by the connection.

How to code a query

Now that you know how to list the names of the tables and columns in a database, you're ready to code a query that gets the data from a single table. To do that, you can use the SELECT statement that's summarized in figure 5-6. This statement consists of four clauses: SELECT, FROM, WHERE, and ORDER BY. The keywords for these clauses aren't case sensitive, but this book capitalizes them to show that they're part of SQL.

When coding a query, the SELECT and FROM clauses are typically required, but the WHERE and ORDER BY clauses are optional. That's because you use the SELECT clause to specify the columns to select and the FROM clause to specify the table that stores the columns. In this figure, all of the examples select columns from the Fires table.

If you want to select all columns, you can use the * wildcard as shown in the second example. However, if a table has many columns, you typically only want to select some of them for your analysis. To do that, you can specify a comma-separated list of column names as shown in the third example.

If a column stores values for dates or times, you typically need to use the DATETIME() function to convert the values to a string that represents the value. When you do that, you typically want to use the AS keyword to provide a new name for the resulting column as shown by the fourth example. In this case, the code just provides the same name as the original column, which is often what you want.

The optional WHERE clause filters the rows in the result set. To do that, it uses relational operators with column names. For instance, the fifth example selects rows where the fire_size column is greater than 100 acres. This reduces the number of rows in the data set from roughly 1.8 million to around 50 thousand. This dramatically reduces the amount of time that it takes the query to run. As a result, if you know in advance that you want to filter the rows in your data set, you can use a WHERE clause to efficiently do that.

The optional ORDER BY clause sorts the result set by the specified column. By default, the ORDER BY clause sorts the column in ascending sequence from smallest to largest. However, if you want to sort in descending sequence from largest to smallest, you can add DESC keyword after the column name as shown in the sixth example.

As you review this figure, you should know that it only presents some basic skills for coding a query that selects data from a single table. For many analyses, that's all you need to do. However, a SELECT statement can also join data from multiple tables and provide many other advanced functions. For more information about coding queries, you can search online or get a book dedicated to teaching SQL such as *Murach's MySQL*.

The basic syntax for a SELECT statement

```
SELECT columns
FROM table
WHERE boolean_expression
ORDER BY column ASC|DESC
```

How to select all columns and rows

```
SELECT * FROM Fires
```

How to select specified columns

```
SELECT fire_name, fire_state, fire_size
FROM Fires
```

How to select a column that stores date/time values

```
SELECT fire_name, fire_state, fire_size,
       DATETIME(discovery_date) AS discovery_date
FROM Fires
```

How to select specified rows

```
SELECT fire_name, fire_state, fire_size
FROM Fires
WHERE fire_size > 100
```

How to sort by a specified column

```
SELECT fire_name, fire_state, fire_size
FROM Fires
ORDER BY fire_size DESC
```

Description

- *SQL* (*Structured Query Language*) provides statements that you can use to work with a relational database.

- You can use the SELECT statement to code a *query* that selects data from a database.

- In a query, the SELECT and FROM clauses are typically required, but the WHERE and ORDER BY clauses are optional.

- In the SELECT clause, you can use the * wildcard to select all columns.

- In the SELECT clause, you can use the DATETIME() function to convert date/time values to a string that represents the date/time value.

- In the SELECT clause, you can use the AS keyword to specify a name for the column.

- When you use the ORDER BY statement, the column is sorted in ascending order by default. If you want to sort in descending order, you can code the DESC keyword after the column name.

Figure 5-6   How to code a query

How to use a query to read data

Now that you understand how to code a query, you're ready to use a query to read data from a database. The first example in figure 5-7 shows how to code a query in R. To do that, the first statement stores the query as a string. Here, the query selects five columns from the Fires table, and it doesn't filter any rows or sort the result set.

The second statement sends the query to the database. To do that, it passes two arguments to the dbSendQuery() function. The first argument is the connection object created earlier in this chapter, and the second argument is the query defined by the first statement.

The third statement gets the results from the response object that's returned by the second statement. To do that, the third statement passes the response object to the dbFetch() function. This returns a data frame for the query, but the code uses the as_tibble() function to convert the data frame to a tibble.

The resulting tibble shows that the data set contains five columns and over 1.8 million rows. In addition, it shows that the first four column names use uppercase letters, and the fifth column name uses lowercase letters. That's because the Fires table uses uppercase for its column names, so its columns are returned in uppercase unless you use the AS keyword in the query to specify another name for the column. Then, the column uses the case that you specify. In this figure, the query uses the AS keyword to specify lowercase for the fifth column. However, if you want all of the columns to use uppercase, you could use the AS keyword to specify uppercase like this:

```
DATETIME(discovery_date) AS DISCOVERY_DATE
```

When you're done reading the data, it's a good practice to call the dbClearResult() function. This clears the result set and frees the system resources used by it.

The functions for querying a database

Function	Description
dbSendQuery(conn, statement)	Runs the query on the database specified by the connection object and returns a response object.
dbFetch(res)	Returns the results from the response object as a data frame.
dbClearResult(res)	Clears the result set and frees resources used by it.

How to execute a SQL query

```
fires_sql <- "
  SELECT fire_name, fire_size, state, fire_year,
         DATETIME(discovery_date) AS discovery_date
  FROM Fires
"
response <- dbSendQuery(con, fires_sql)
fires <- as_tibble(dbFetch(response))
```

The resulting tibble

```
fires
# A tibble: 1,880,465 x 5
   FIRE_NAME   FIRE_SIZE STATE FIRE_YEAR discovery_date
   <chr>           <dbl> <chr>     <int> <chr>
 1 FOUNTAIN         0.1  CA         2005 2005-02-02 00:00:00
 2 PIGEON           0.25 CA         2004 2004-05-12 00:00:00
 3 SLACK            0.1  CA         2004 2004-05-31 00:00:00
 4 DEER             0.1  CA         2004 2004-06-28 00:00:00
 5 STEVENOT         0.1  CA         2004 2004-06-28 00:00:00
 6 HIDDEN           0.1  CA         2004 2004-06-30 00:00:00
 7 FORK             0.1  CA         2004 2004-07-01 00:00:00
 8 SLATE            0.8  CA         2005 2005-03-08 00:00:00
 9 SHASTA           1    CA         2005 2005-03-15 00:00:00
10 TANGLEFOOT       0.1  CA         2004 2004-07-01 00:00:00
# ... with 1,880,455 more rows
```

How to clear the response and free its resources

```
dbClearResult(response)
```

Description

- Because dbFetch() returns a data frame, you usually want to convert the result to a tibble prior to analyzing the result.
- It's good practice to use dbClearResult() after you're done with a connection.

Figure 5-7   How to use a query to read data

How to read data from a JSON file

A *JSON (JavaScript Object Notation)* file stores data as a list of *key-value pairs*. You can think of a *key* as being like a variable name or column name. Each key has a *value*. That value can be a number, a string, a vector of values, or a list of more key-value pairs. When using JSON files for data analysis, you usually want to use data that is nested in these lists within lists.

How to read a JSON file into a list

The first example in figure 5-8 shows how to download a JSON file and save it on your computer in a file named shots.json. This works the same as downloading any other type of file.

The second example shows how to load the RJSONIO package. This package contains functions such as the fromJSON() function that make it possible for R to use JSON files for input and output (IO).

The third example shows how to read the JSON file into a variable. In other words, it shows how to use the JSON file as input. To do that, it uses the fromJSON() function to read the JSON from the shots.json file into the variable named json_data. Since JSON data isn't usually tabular data that fits into a data frame, the fromJSON() function returns the JSON data as a list. As a result, before you can use this data in an analysis, you typically need to do some processing to load the data in the list into a data frame.

If you read the JSON data into a list and store that list in a variable, you can use RStudio to view the data. To do that, click on the variable name in the Environment pane. This displays the list data in the Source pane as shown in this figure. Here, the top level of the JSON data contains three keys: resource, parameters, and resultSets. The latter two are both lists that can be expanded by clicking on the triangles next to their names.

A function for reading JSON data

Function	Description
fromJSON(txt)	Reads the data from the JSON file (txt) into a list.

How to download a JSON file

```
json_url <- "https://www.murach.com/python_analysis/shots.json"
download.file(url = json_url, destfile = "shots.json")
```

How to load a package for reading JSON files

```
library("RJSONIO")
```

How to read JSON data into a list

```
json_data <- fromJSON("shots.json")
```

The root level of the list when viewed in RStudio

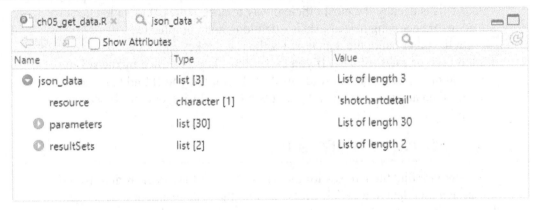

Description

- A *JSON (JavaScript Object Notation)* file stores data as a list of *key-value pairs*. A *key* is equivalent to a variable name or column name. Each key has a value. A *value* can be a number, string, vector of values, or list of more key-value pairs.

- After you use the fromJSON() function to read a JSON file into a list variable, you can view the imported JSON data in RStudio. To do that, click the variable name in the Environment pane.

Figure 5-8   How to read a JSON file into a list

How to get the index for a list

Once you've displayed the list for the JSON data in RStudio, you can explore the data by clicking on the arrows to expand or collapse each level. For instance, the first example in figure 5-9 shows the resultSets list after it has been expanded. This shows that there are two result sets with indexes of 1 and 2.

Here, the first result set has a key named headers. The data type for this key indicates that it stores a vector of 24 strings. In addition, RStudio displays the first few items of this vector, so you can see that they contain strings like "GRID_TYPE", "GAME_ID", and so on. If you study these strings, you'll find that they provide the names for the columns that are stored in the rowSet list.

To extract this vector from the JSON data, you need to know its index. To get that index, you can click on its key. When you do, RStudio displays the index for that key in the lower left corner of the Source pane. Then, you can copy the index from the RStudio pane and paste it into your script to access the specified value. This is usually the most efficient way to access the data in a list because the code for accessing nested lists is often long and complicated.

On the other hand, the data type for the rowSet key indicates that it contains 11,846 lists (or rows) of data. If you expand this key as shown by the second screen in this figure, you can study its data. If you do, you'll find that each row contains data about a shot taken by Stephen Curry during one of his seasons in the NBA.

How to get data from a list

After you find the indexes for the data you want to extract from a list, you can assign that data to a variable as shown by the code example in this figure. Here, the first statement assigns the vector stored at the index for the headers key to a variable named column_names. Then, the second statement assigns the list of lists stored at the index for the rowSet key to a variable named rows.

How to get the index for the column names

How to get the index for the rows

How to get data from a list

```
column_names <- json_data[["resultSets"]][[1]][["headers"]]
rows <- json_data[["resultSets"]][[1]][["rowSet"]]
```

Description

- After you display the JSON data in RStudio, you can explore it by clicking the triangles to expand and collapse each nested list.

- To view the index for a value, click on the value. When you do, RStudio displays the index on the bottom left corner of the Source pane.

- To copy the index for a value into your code, you can select the index from the bottom left corner of the Source pane, copy the text to the clipboard, and paste it into your source code file.

Figure 5-9   How to get data from a list

How to build a tibble from the data in a list

If you can get the data for the column names and rows of a data set, you can build a tibble from that data as shown by figure 5-10. Here, the code starts by creating an empty data frame named shots. To do that, it calls the data.frame() function.

After creating the data frame, this code loops through each row in the rows variable that was created in the previous figure. Since the rows variable is a list that stores one list for each row, this loops through each of the 11,846 lists (or rows) in the rows variable. Within the loop, the code uses the rbind() function to add the row to the end of the data frame.

After adding the rows to the data frame, the code sets the column names for the data frame. To do that, it sets the vector returned by the names() function to the vector of names stored in the column_names variable.

Finally, with the data frame complete, this code converts the data frame to a tibble and stores it in a variable named shots. The result is a tibble with 11,846 rows and 24 columns that stores the data for the Basketball Shots analysis presented in chapter 11.

How to create a tibble for the data

```
# create an empty data frame
shots <- data.frame()

# loop through each row and add it to data frame
for (row in rows) {
  shots <- rbind(shots, row)
}

# set the column names
names(shots) <- column_names

# convert data frame to tibble
shots <- as_tibble(shots)
```

The shots tibble when viewed in RStudio

	GRID_TYPE	GAME_ID	GAME_EVENT_ID	PLAYER_ID	PLAYER_NAME	TEAM_ID	TEAM_NAME	PERIOD	MINUT
1	Shot Chart Detail	0020900015	4	201939	Stephen Curry	1610612744	Golden State Warriors	1	
2	Shot Chart Detail	0020900015	17	201939	Stephen Curry	1610612744	Golden State Warriors	1	
3	Shot Chart Detail	0020900015	53	201939	Stephen Curry	1610612744	Golden State Warriors	1	
4	Shot Chart Detail	0020900015	141	201939	Stephen Curry	1610612744	Golden State Warriors	2	
5	Shot Chart Detail	0020900015	249	201939	Stephen Curry	1610612744	Golden State Warriors	2	
6	Shot Chart Detail	0020900015	277	201939	Stephen Curry	1610612744	Golden State Warriors	2	
7	Shot Chart Detail	0020900015	413	201939	Stephen Curry	1610612744	Golden State Warriors	4	
8	Shot Chart Detail	0020900015	453	201939	Stephen Curry	1610612744	Golden State Warriors	4	
9	Shot Chart Detail	0020900015	487	201939	Stephen Curry	1610612744	Golden State Warriors	4	
10	Shot Chart Detail	0020900015	490	201939	Stephen Curry	1610612744	Golden State Warriors	4	

Showing 1 to 11 of 11,846 entries, 24 total columns

Description

- You can use the data.frame() function to create a new data frame.
- You can use a for loop to loop through each item in a list.
- You can use the rbind() function to bind an item to the end of a data frame.
- You can use the names() function to set the column names for a data frame to a vector of names.
- You can use the as_tibble() function to convert a data frame to a tibble.

Figure 5-10   How to build a tibble from the data in a list

Perspective

Now that you've completed this chapter, you should have the skills you need for reading data from CSV, Excel, and JSON files into a tibble. In addition, you should be able to read data from a database into a tibble.

Unfortunately, reading data is often the easy part of getting the data. The hard part is finding the data that you want to analyze and figuring out which columns and rows you need for your analysis. Although there's no one right way to find and access data, this chapter has given you a few ideas for how to get started.

Summary

- A *CSV* (*comma-separated values*) file uses commas to separate the values for its columns and stores each row on a new line.
- A *zip file* contains one or more other files in compressed format.
- A relational database stores its data in one or more *tables* where multiple tables are typically related to one another.
- *SQL* (*Structured Query Language*) provides statements that you can use to work with a relational database.
- You can use the SELECT statement to code a *query* that selects data from a database.
- A *database driver* makes it possible to use R functions to work with a database.
- To connect to a database, you need to create a *connection object*.
- A *JSON (JavaScript Object Notation)* file stores data as a list of *key-value pairs*. A *key* is equivalent to a variable or column name. Each key has a value. A *value* can be a number, string, vector of values, or a list of more key-value pairs.

Exercise 5-1   Run the example script and modify it

In this exercise, you'll run a script that contains the examples presented in this chapter. In addition, you'll modify some statements in this script and add some statements too.

Run and edit the example script of plots

1. Open the R script named exercise_5.1.R located in this folder:
    `Documents/murach_r/exercises/ch05`

2. Run the script statement by statement by clicking on the Run button or by using the keystroke shortcut.

3. After you read the polls.csv file into the polls tibble, click on its name in the Environment pane to view it. Then, review the data for this data set, and click on some column names to sort the data by those columns.

4. When you list the column names for the Fires table of the SQLite database, review the names and consider what type of data each column might store.

5. When you read the data from the SQLite database into the fires tibble, note how long it takes to read the data. Then, use RStudio to view the tibble and sort by the FIRE_SIZE column. Note that many of the fires are very small.

6. Modify the SELECT statement so it only selects fires that are larger than 100 acres. Note how this reduces the number of rows in the data set and helps the query run more quickly. If you get an error, read the error message closely and double-check your SELECT statement to make sure you coded it correctly.

7. Modify the SELECT statement so it uses uppercase for the name of the DISCOVERY_DATE column and includes the LATITUDE and LONGITUDE columns.

8. After you read data from the shots.json file into the shots tibble, click on its name in the Environment pane to view its data. Then, click on the json_data variable and explore its data. Note that the first result set is named Shot_Chart_Detail, and that there's a second result set named LeagueAverages.

9. Add code to the end of the script that creates a tibble that stores the data for the LeagueAverages result set. To do that, you can copy the code for the shots tibble and modify it so it stores the data in a tibble named league_avgs.

Exercise 5-2   Create your own tibbles from downloaded data

In this exercise, you'll create your own tibbles from CSV, SQLite, and JSON files downloaded from the internet.

Create a tibble from a CSV file downloaded from the internet

1. Open the script named exercise_5.1.R so you can use it as reference.
2. Create a new script named exercise_5-2.R to store your code.
3. Use Google Dataset Search, Kaggle, or the Registry of Open Data on AWS to find a data set available as a CSV file. Ideally, choose a data set that interests you.
4. Download the CSV file for your chosen data set and save it in your working directory.
5. If your CSV file is in a zip file, unzip it and save the CSV file in your working directory.
6. Use the read_csv() function to create a tibble from the CSV file and store it with a variable name that describes the data set.
7. Explore your tibble by clicking on the variable name in the Environment pane. Then, click on the column names to sort your tibble in various ways.
8. If you like, create some visualizations that help you understand this data set better.

Create a tibble from a SQLite file downloaded from the internet

9. Use Google Dataset Search, Kaggle, or the Registry of Open Data on AWS to find a data set available as a SQLite file.

    Tip: Depending on which search option you choose, you can use a filter or search term to limit your results to SQLite files.
10. Save the SQLite file for your chosen data set to your working directory. If your SQLite file is in a zip file, unzip it and save the contents to your working directory.
11. Connect to your database and display a list of tables in the database. Don't be surprised if your database only has one table since that's common.
12. Choose a table from the list and display a list of its columns.
13. Code a query that selects at least two columns from your chosen table. Then, send that query to the database, create a tibble from the response that's returned, and store that tibble in a variable. If you encounter errors, modify your code and try again until it works correctly.
14. When you have successfully created your tibble, click on the name in the Environment pane to view its data.
15. Free the resources for the response and connection objects.

Create a tibble from a JSON file downloaded from the internet

16. Use Google Dataset Search, Kaggle, or the Registry of Open Data on AWS to find a data set available as a JSON file. Again, this exercise will be more interesting if you pick a data set you personally find intriguing.

17. Save the JSON file for your chosen data set to your working directory. If the JSON file is stored in a zip file, unzip it.

18. Read the data from the JSON file into a variable named json_data. Don't forget to load the RJSONIO package first.

19. In the Environment pane, click the json_data variable and explore it. If your JSON file has more than one level, this variable should be a list of lists. If your data set has only one level, this variable might be a single list or a vector.

20. If your json_data variable has more than one level, use RStudio to explore it.

21. Create a tibble from the data that's available from the json_data variable. If your data set doesn't include column names, you can code a vector that contains your own column names.

22. If the row names are important, you can use cbind() and the names() function to add them as a column, like this:

```
my_data_frame <- cbind(my_data_frame, names(rows))
```

23. If your json data only has one value per key, you can skip using rbind() and cbind() and simply create a data frame using the names and rows as vectors.

```
my_data_frame <- data.frame(Keys = names(rows), Values = rows)
```

24. Use RStudio to display and view the tibble that stores data from your JSON file.

Chapter 6

How to clean data

Unfortunately, most data sets aren't perfect. They may have missing data, obvious errors, incorrect types, and more. All of these issues need to be dealt with before you can begin your analysis. This is called cleaning the data.

Needing to clean a data set is common whether you get your data from a third-party source or from your organization's internal databases and spreadsheets. In fact, it's common for data cleaning to take a full quarter of the time spent working on a typical data analysis project.

The consequences of skipping data cleaning can be dire. If you don't clean your data, the results of your analysis may be misleading or inaccurate. That's why this chapter presents the skills that you need to clean data.

Introduction to cleaning data

When you clean the data for an analysis, you should have a general idea of how to get started and what to look for.

A general plan for cleaning data

Figure 6-1 presents a basic workflow for cleaning the data in a data set. Of course, you won't have to do all of these tasks for every data set, but this provides a good guide for most data sets.

Before you start cleaning your data, you should make sure that you've set the goals for your analysis. That will help you decide which rows and columns are okay to *drop* (that is, remove) and otherwise guide you during the cleaning process. Above all, be sure that you don't drop rows or columns that may be needed for your analysis.

After you've set the goals for your analysis, you should identify any problems that need to be fixed. If there's any documentation for your data set, start by reading it. For instance, the zip file for the wildfires data contains an HTML file that contains documentation that reveals that the data has been collected from multiple reporting sources, which lets you know that you should be on the lookout for duplicate data.

After you've reviewed the documentation, you can examine the data as shown in the next few figures. At that point, you can plan the cleaning that needs to be done.

To simplify the data, you usually start by dropping unneeded rows and columns. That may include rows and columns that don't present useful information, duplicate rows, and rows that aren't within the scope of your analysis. After that, you may want to rename columns so it's easier to tell what they contain.

After you've simplified the data, you're ready to fix any data problems. That includes finding and fixing missing data, correcting data types, and handling outliers, which are values that lie far outside of the normal range of values for a column.

A general plan for cleaning a data set

Set the goals
1. Set the goals for your analysis.

Identify problems
2. Review any documentation that's available to help you understand the data set, such as where it came from and what each column represents.
3. Read the data into a tibble and examine it.
4. Identify what cleaning needs to be done, if any.

Simplify the data
5. Drop (remove) duplicate rows.
6. Drop rows and columns that you're certain aren't needed for the analysis.
7. Rename columns so the names are easier to understand.

Fix data problems
8. Find and fix missing values.
9. Fix data type problems, like dates or numbers that are imported as strings.
10. Find and fix outliers.

Documentation that identifies some files in the fires zip file

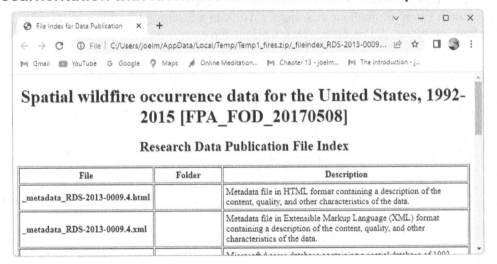

Description
- Most data sets aren't clean when you import them. To get accurate results, you need to clean them before you analyze them.
- Data cleaning commonly can take as much as 25% of the total time for an analysis.
- When you *drop* (remove) rows or columns, you need to be sure that you aren't dropping data that will affect your analysis.

Figure 6-1   A general plan for cleaning data

How to display column names and data types

After you review the documentation for a data set and read the data into a tibble, you need to examine the data. If the tibble has many columns, you can view a list of the column names and their data types by passing the tibble to the str() function as shown in figure 6-2. Note that here str stands for structure, not string.

The examples in this figure set the strict.width argument of the str() function to "cut". This aligns the column names and displays one line per column. If you don't set this parameter, rows that are too long to fit on one line will wrap to the next line. This makes the output more difficult to read.

The first example shows the names and data types for the seven columns of fires tibble. Here, chr means the character type, int means the integer type, and num means the double type. Therefore, these columns store strings, integers, and decimal numbers. In addition, the output shows the first few values for each column. This is often helpful in understanding the data.

The second example shows the names and data types for the columns of the polls tibble. It's been truncated for display in this figure, but if you try this example in RStudio, you'll see all 27 columns listed.

The third example uses *piping* to accomplish the same task as the first example. This technique is often helpful for writing code in a more readable way, and it's used throughout this chapter. You can use piping like this whenever the first parameter of a function accepts a data object.

A function for displaying the structure of an object

Function	Description
str(object)	Displays the structure of an object. For a tibble, this includes the column names and data types.

Parameter	Description
strict.width	Specify if the width argument should be followed strictly. Value can be no, wrap, or cut (one line per column).

The column names and data types of the fires tibble

```
str(fires, strict.width = "cut")
tibble [49,797 x 6] (S3: tbl_df/tbl/data.frame)
 $ FIRE_NAME     : chr [1:49797] "POWER" "FREDS" "AUSTIN CREEK" "THOMPSON BUTT"..
 $ FIRE_SIZE     : num [1:49797] 16823 7700 125 119 119 ...
 $ STATE         : chr [1:49797] "CA" "CA" "NC" "SD" ...
 $ FIRE_YEAR     : int [1:49797] 2004 2004 2005 2005 2005 2005 2005 2005 2005 2..
 $ DISCOVERY_DATE: chr [1:49797] "2004-10-06 00:00:00" "2004-10-13 00:00:00" ""..
 $ CONTAIN_DATE  : chr [1:49797] "2004-10-21 00:00:00" "2004-10-17 00:00:00" ""..
```

The column names and data types of the polls tibble

```
str(polls, strict.width = "cut")
tibble [12,624 x 27] (S3: tbl_df/tbl/data.frame)
 $ cycle            : int [1:12624] 2016 2016 2016 2016 2016 2016 2016 2016 201..
 $ branch           : chr [1:12624] "President" "President" "President" "Presi"..
 $ type             : chr [1:12624] "polls-plus" "polls-plus" "polls-plus" "po"..
 $ matchup          : chr [1:12624] "Clinton vs. Trump vs. Johnson" "Clinton v"..
 $ forecastdate     : chr [1:12624] "11/8/16" "11/8/16" "11/8/16" "11/8/16" ...
 $ state            : chr [1:12624] "U.S." "U.S." "U.S." "U.S." ...
 $ startdate        : chr [1:12624] "11/3/2016" "11/1/2016" "11/2/2016" "11/4/"..
 $ enddate          : chr [1:12624] "11/6/2016" "11/7/2016" "11/6/2016" "11/7/"..
 $ pollster         : chr [1:12624] "ABC News/Washington Post" "Google Consume"..
 $ grade            : chr [1:12624] "A+" "B" "A-" "B" ...
 $ samplesize       : int [1:12624] 2220 26574 2195 3677 16639 1295 1426 1282 8..
 $ population        : chr [1:12624] "lv" "lv" "lv" "lv" ...
 $ poll_wt          : num [1:12624] 8.72 7.63 6.42 6.09 5.32 ...
 $ rawpoll_clinton  : num [1:12624] 47 38 42 45 47 ...
 $ rawpoll_trump    : num [1:12624] 43 35.7 39 41 43 ...
 $ rawpoll_johnson  : num [1:12624] 4 5.46 6 5 3 3 5 6 6 7.1 ...
 $ rawpoll_mcmullin : num [1:12624] NA NA NA NA NA NA NA NA NA NA ...
 ...
```

How to use piping with the str() function

```
fires %>% str(strict.width = "cut")
```

Description

- Displaying the column names and data types for a data set can help you understand the data.
- You can use *piping* whenever the first parameter of a function accepts a data object.

Figure 6-2   How to display column names and data types

How to examine the unique values for a single column

When you're examining a new data set, you often want to look at the unique values for a column. To do that, you can use the unique() and length() functions presented in figure 6-3.

The first example in this figure shows how to get the number of rows for a column. In particular, it shows how to get the number of rows in the type column of the polls tibble. Here, the code accesses the column by coding the name of the tibble followed by a dollar sign ($) and the name of the column. This returns a vector that stores the values for the column. Then, it passes this vector to the length() function. This shows that the type column has 12,624 rows, just like all of the columns in the polls data set.

The second example shows how to use the unique() function to get the unique values for a column. This works like the first example, but it displays a list of the three unique values in the type column. In other words, even though this column has 12,624 values, it only has 3 unique values.

The third example shows how to get a count of the unique values for a column. To do that, this code nests a call to the unique() function within a call to the length() function. The output confirms that there are three unique values in the type column.

When you nest two or more functions as in the third example, you can sometimes improve the readability of your code by using piping instead of nesting. That's why the fourth example shows how to use piping to get the same result as the third example.

To pipe, the code begins with the data for the operation, which is the type column. Then, the first pipe operator (%>%) *pipes* that data into the unique() function, and the second pipe operator pipes the vector that's returned by the unique() function into the length() function.

This book uses both nested functions and piping in its examples. It attempts to use whichever technique results in code that's easier to read and understand. In general, if there are multiple levels of nesting, it often makes sense to use piping.

Some functions for examining data

Function	Description
length(x)	Returns the number of rows in a column.
unique(x)	Returns a list of the unique values in a column.

How to get the number of rows in a column

```
length(polls$type)
[1] 12624
```

How to get the unique values of a column

```
unique(polls$type)
[1] "polls-plus" "now-cast"    "polls-only"
```

How to get a count of the unique values of a column

```
length(unique(polls$type))
[1] 3
```

How to use piping instead of nested functions

```
polls$type %>% unique() %>% length()
[1] 3
```

Description

- In a tibble, each column is a vector of values. As a result, you can use the length() and unique() functions on columns.

- If there are multiple levels of nested functions, piping can make your code more readable.

Figure 6-3   How to examine the unique values for a single column

How to display the unique values for all columns

Although you can examine the unique values one column at a time as shown in the previous figure, it usually makes sense to examine the unique values for every column in a data set at once as shown in figure 6-4. Here, the code starts by using the apply() function to apply the unique() function to every column in the polls data set. This returns a list that contains each column name and a list of its unique values. Then, the code pipes the list into the str() function with the strict.width parameter set to "cut". As a result, this displays each column with its values on a single line.

The output displayed by the example is both informative and readable. You can examine this output to draw conclusions based on the values of the data. For instance, the type column in the output has just three values: "polls-plus", "now-cast", and "polls-only". The "now-cast" values identify the actual values of the votes that were cast in the poll while the other two values represent calculated values that were based on the actual values. As a result, you may want to drop the "polls-plus" and "polls-only" rows and base your analysis on just the "now-cast" rows.

A function for applying another function to all columns

Function	Description
apply(X, MARGIN, FUN)	Applies the specified function (FUN) to the specified data object (X). If the MARGIN parameter is set to 2, this function applies the specified function to each column. If MARGIN is set to 1, it applies the specified function to each row.

How to display the unique values for all columns

```
polls %>% apply(MARGIN = 2, FUN = unique) %>%
  str(strict.width = "cut")
List of 27
 $ cycle            : chr "2016"
 $ branch           : chr "President"
 $ type             : chr [1:3] "polls-plus" "now-cast" "polls-only"
 $ matchup          : chr "Clinton vs. Trump vs. Johnson"
 $ forecastdate     : chr "11/8/16"
 $ state            : chr [1:57] "U.S." "New Mexico" "Virginia" ...
 $ startdate        : chr [1:352] "11/3/2016" "11/1/2016" "11/2/2016" ...
 $ enddate          : chr [1:345] "11/6/2016" "11/7/2016" "11/5/2016" ...
 $ pollster         : chr [1:196] "ABC News/Washington Post" "Googl"..
 $ grade            : chr [1:11] "A+" "B" "A-" ...
 $ samplesize       : chr [1:1767] " 2220" "26574" " 2195" ...
 $ population       : chr [1:4] "lv" "rv" "a" ...
 $ poll_wt          : chr [1:4399] "8.720654e+00" "7.628472e+00" ..
 $ rawpoll_clinton  : chr [1:1312] "47.00" "38.03" "42.00" ...
 $ rawpoll_trump    : chr [1:1385] "43.00" "35.69" "39.00" ...
 $ rawpoll_johnson  : chr [1:585] " 4.00" " 5.46" " 6.00" ...
 $ rawpoll_mcmullin : chr [1:17] NA "24.00" "27.60" ...
 $ adjpoll_clinton  : chr [1:12569] "45.20163" "43.34557" "42.02638" ...
 $ adjpoll_trump    : chr [1:12582] "41.724300" "41.214390" "38.816200" ...
 $ adjpoll_johnson  : chr [1:6630] " 4.6262210" " 5.1757920" " 6.8447340" ...
 $ adjpoll_mcmullin : chr [1:58] NA "24.00000" "27.70142" ...
 $ multiversions    : chr [1:2] "" "*"
 $ url              : chr [1:1305] "https://www.washingtonpost.com/news/t" ..
 $ poll_id          : chr [1:4208] "48630" "48847" "48922" ...
 $ question_id      : chr [1:4208] "76192" "76443" "76636" ...
 $ createddate      : chr [1:222] "11/7/16" "11/8/16" "11/6/16" ...
 $ timestamp        : chr [1:3] "09:35:33  8 Nov 2016" "09:24:53  8 Nov" ..
```

Description

- Examining the unique values in a column can help you understand the data and identify problems that need to be fixed.

Figure 6-4   How to display the unique values for all columns

How to count the unique values for all columns

Although the output from the previous figure includes the count of unique values for each column, it's mixed in with other data. Instead, you may want to only display the count of unique values for each column as shown in figure 6-5. To do that, the example in this figure begins by using the apply() function to apply the unique() function to every column in the polls data set, just like the example in this previous data set. This returns a list of unique values for each column.

The lapply() function works like apply() except that it works with lists. It does not need an argument to specify rows or columns since lists do not have them. In the example, lapply() applies the length() function to the list that's returned by the apply() function. Note that this code must use lapply() because it's applying the function to a list, not a data frame.

Finally, this code pipes the list with lengths into the str() function to make the output more readable. Since this list isn't wide enough for a line to wrap to the next line, there's no need to set the strict.width parameter to "cut", but you could if you wanted to.

Once you have the output, you can analyze it to help determine how to clean the data. The first two columns in this example, cycle and branch, have only one unique value each. Columns that only have one value can usually be dropped because they don't contribute information to the data set.

The third column, type, has only three unique values. Columns with a small number of unique values usually represent groups or categories and can be converted to the factor type to make it easier to create visualizations. If you look at the rawpoll_ columns, you can see that there are not many polls for Johnson or McMullin compared with Clinton and Trump. For this reason, you may want to drop these columns to focus on the columns with more data. Whether you do or not depends on the goals you set for your analysis.

The last group of columns that you should consider are the poll_id and question_id columns. Each of these columns has exactly one third as many unique values as there are rows in the data set. As a result, you might guess that there is a relationship between them and the columns with three unique values, such as the type and the timestamp columns.

A function for applying another function and returning a list

Function	Description
`lapply(X, FUN)`	Applies the specified function (FUN) to a list or vector.

How to get a count of unique values for all columns

```
polls %>% apply(MARGIN = 2, FUN = unique) %>%
  apply(FUN = length) %>%
  str()
List of 27
 $ cycle             : int 1
 $ branch            : int 1
 $ type              : int 3
 $ matchup           : int 1
 $ forecastdate      : int 1
 $ state             : int 57
 $ startdate         : int 352
 $ enddate           : int 345
 $ pollster          : int 196
 $ grade             : int 11
 $ samplesize        : int 1767
 $ population        : int 4
 $ poll_wt           : int 4399
 $ rawpoll_clinton   : int 1312
 $ rawpoll_trump     : int 1385
 $ rawpoll_johnson   : int 585
 $ rawpoll_mcmullin  : int 17
 $ adjpoll_clinton   : int 12569
 $ adjpoll_trump     : int 12582
 $ adjpoll_johnson   : int 6630
 $ adjpoll_mcmullin  : int 58
 $ multiversions     : int 2
 $ url               : int 1305
 $ poll_id           : int 4208
 $ question_id       : int 4208
 $ createddate       : int 222
```

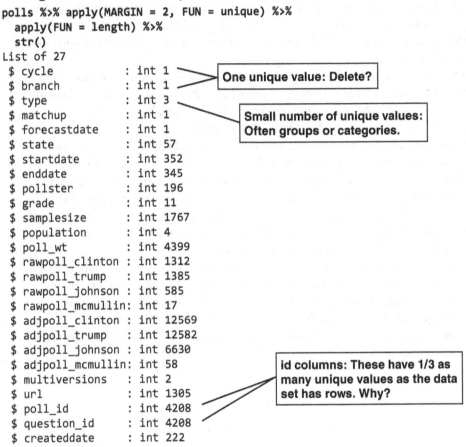

One unique value: Delete?

Small number of unique values: Often groups or categories.

id columns: These have 1/3 as many unique values as the data set has rows. Why?

Description

- Only one unique value in a column usually means that column isn't relevant to your analysis.

- A limited number of unique values indicates values that may be useful as categories and can be converted to the factor data type.

- The lapply() function works like apply(), but for lists or vectors rather than data objects.

Figure 6-5   How to count the unique values for all columns

How to display the value counts

Figure 6-6 presents the table() function. It returns each unique value in a column as well as a count of the number of times each non-null value occurs. This data is sometimes known as the *value counts* for a column. Value counts are useful for learning about your data set.

The first example shows the first 20 values for the state column of the polls data set. It can be inferred that this column contains the names of each state in the United States. In addition, it contains a value of "U.S." for national polls that include all states.

The second example shows how to use the table() function to get the value counts for the state column. This shows that the polls were not conducted evenly for each state. For example, there were only 114 polls conducted in Alaska but there were 237 conducted in Arizona (a swing state).

Sometimes, it may be more helpful to view this data as a percentage. That's why the third example shows how to calculate this. Here, the code calls the table() function for the state column and divides the count for each value by the sum of the counts for all values. Then, the code multiplies the result by 100 and rounds the results to two decimal places. This displays the value for each state as a percentage of the total. For example, it's now easy to see polls for California make up 1.69% of all of the polls in this data set.

The fourth example shows how to get the value counts for combinations of two variables. To do that, you use the bracket selector to specify two column names when you pass the data to the table() function. Then, the table() function returns the counts for each pairing of the columns. In this example, there were 9 polls that were unrated for California, 36 that were rated "A", 42 that were rated "A-", and so on.

A function for returning value counts

Function	Description
table(x)	Returns the count for each unique value in a column, or a table that shows the count for each unique combination of two or more columns.

The first 20 values in the state column

```
polls$state %>% head(20)
 [1] "U.S."         "U.S."         "U.S."         "U.S."
 [5] "U.S."         "U.S."         "U.S."         "U.S."
 [9] "New Mexico"   "U.S."         "U.S."         "U.S."
[13] "U.S."         "Virginia"     "U.S."         "Iowa"
[17] "U.S."         "Wisconsin"    "North Carolina" "Georgia"
```

How to display the count for each value in the state column

```
polls$state %>% table()
```

Alabama	Alaska	Arizona	Arkansas
129	114	237	135
California	Colorado	Connecticut	Delaware
213	240	132	120

...

How to display the counts as a percentage of the total

```
(table(polls$state) / sum(table(polls$state)) * 100) %>% round(2)
```

Alabama	Alaska	Arizona	Arkansas
1.02	0.90	1.88	1.07
California	Colorado	Connecticut	Delaware
1.69	1.90	1.05	0.95

...

How to display the count for each unique combination

```
table(polls[c("state","grade")])
```

	grade									
state	A	A-	A+	B	B-	B+	C	C-	C+	D
Alabama	6	0	42	0	45	0	0	0	36	0	0
Alaska	3	0	0	0	57	3	6	9	36	0	0
Arizona	42	9	45	3	63	9	9	0	39	18	0
Arkansas	0	0	42	0	48	0	0	0	36	9	0
California	9	36	42	9	66	0	9	0	42	0	0
Colorado	18	9	57	6	69	21	9	12	39	0	0
Connecticut	0	0	45	0	51	0	0	0	36	0	0

...

Description

- You can use the table() function to get the *value counts* for a single column or to show a table of value counts for two columns.

Figure 6-6   How to display the value counts

How to sort the data

Figure 6-7 starts by presenting the arrange() function that you can use to sort the rows in a data frame by one or more columns. The first argument specifies the data set to be sorted. Then, the function accepts one or more columns to sort the data by.

The first example in this figure shows how to sort the fires data by the FIRE_SIZE column. By default, the arrange() function sorts the data from lowest to highest. This resulting tibble shows that the lowest fire size is 100 and that the first three rows have this size.

However, if you want to sort from highest to lowest, you can pass the column name to the desc() function as shown in the second example. The resulting tibble shows the largest fire in the first row, the second largest fire in the second row, and so on.

The third example shows how to sort by several columns. To do that, you can code multiple columns separating each column name with a comma. Then, the arrange() function begins by sorting by the first column. For any ties in the first column, it uses the second column. In this figure, the resulting data set shows that the rows are sorted by state first, then by year, and finally by size in descending order. As a result, the top of the resulting tibble displays the largest fires in Alaska in 1992.

When you use the group_by() function to group your data, you often want to sort the data by each group. To do that, you can use the .by_group parameter as shown in the fourth example. This works much like the third example, except that it creates a group for the STATE column, which allows you to calculate summary data for the states.

A function for sorting data

Function	Description
arrange(.data, ...)	Sorts the rows of a data frame by the specified column or columns. To sort in descending order, enclose the column in the desc() function.

Parameter	Description
.by_group	Specifies whether to sort by groups. Default is FALSE.

How to sort the data by a column from smallest to largest

```
fires %>% arrange(FIRE_SIZE)
# A tibble: 49,797 x 5
   FIRE_NAME          FIRE_SIZE STATE FIRE_YEAR DISCOVERY_DATE
   <chr>                  <dbl> <chr>     <int> <chr>
 1 NA                      100. GA         2007 2007-02-13 00:00:00
 2 NADINE NAFF             100. WA         2006 2006-06-09 00:00:00
 3 NA                      100. SC         2002 2002-03-01 00:00:00
...
```

How to sort the data from largest to smallest

```
fires %>% arrange(desc(FIRE_SIZE))
# A tibble: 49,797 x 5
   FIRE_NAME          FIRE_SIZE STATE FIRE_YEAR DISCOVERY_DATE
   <chr>                  <dbl> <chr>     <int> <chr>
 1 INOWAK                606945 AK         1997 1997-06-25 00:00:00
 2 LONG DRAW            558198. OR         2012 2012-07-08 00:00:00
 3 WALLOW                538049 AZ         2011 2011-05-29 00:00:00
...
```

How to sort by multiple columns

```
fires %>% arrange(STATE, FIRE_YEAR, desc(FIRE_SIZE))
# A tibble: 49,797 x 5
   FIRE_NAME  FIRE_SIZE STATE FIRE_YEAR DISCOVERY_DATE
   <chr>          <dbl> <chr>     <int> <chr>
 1 213225         48087 AK         1992 1992-06-11 00:00:00
 2 232414         35090 AK         1992 1992-07-19 00:00:00
 3 BTT 3 48       25600 AK         1992 1992-06-23 00:00:00
...
```

How to sort by multiple columns with groups

```
fires %>% group_by(STATE) %>%
    arrange(FIRE_YEAR, desc(FIRE_SIZE), .by_group = TRUE)
# A tibble: 49,797 x 5
# Groups:   STATE [50]
   FIRE_NAME  FIRE_SIZE STATE FIRE_YEAR DISCOVERY_DATE
   <chr>          <dbl> <chr>     <int> <chr>
 1 213225         48087 AK         1992 1992-06-11 00:00:00
 2 232414         35090 AK         1992 1992-07-19 00:00:00
 3 BTT 3 48       25600 AK         1992 1992-06-23 00:00:00
...
```

Figure 6-7   How to sort the data

How to simplify a data set

After you're done examining the data in a data set, you're ready to clean it. One good way to start is to drop the rows and columns that you don't need. After that, you may want to rename the remaining columns so the names are easier to remember, type, or understand.

How to filter and drop rows

In most cases, you drop rows based on conditions. To do that, you can use the filter() function to select the rows you want to keep. Any rows that aren't selected are dropped. The filter() function takes two arguments: a data object like a tibble and a condition to use for the filter. You can combine as many conditions as you like by using the logical operators. Some useful ones are listed in the second table of this figure for reference.

The first example in figure 6-8 shows how to use a simple condition to filter the polls data to only show rows where the state value is not "U.S." Then, the second example shows how to combine multiple conditions. In this case, the code uses the AND operator (&) to specify that only rows that both have a value of "California" for state and a poll grade of "A" should be selected.

The third example shows how to filter data with the %in% operator. Here, the code selects only rows where the state is California, Oregon, or Washington.

In any data set, it's possible to have duplicate rows. That's especially true if the data comes from several sources, as with the fires data set. In that case, you typically want to remove all but one of the rows in each set of duplicates. To do that, you can use the duplicated() function to find duplicate rows as shown in the fourth example. The tibble that's returned shows that there are 441 rows that have the exact same information as another row in the data set.

The fifth example drops these duplicate rows. To do that, it uses the NOT operator (!) to select all rows that are not duplicates. Then, it assigns the tibble that's returned by the filter() function to the same fires variable that was passed to the filter() function. This overwrites the fires variable and therefore removes the duplicate rows from the tibble.

Although the fifth example works, it's a little long and hard to read. To make your code short and easy to read, you can use the unique() function to drop duplicate rows as shown by the sixth example.

Two functions for selecting rows

Function	Description
`filter(.data, ...)`	Returns the rows of the data object that match the given condition.
`unique(data)`	Drops duplicate rows and returns the unique rows.

Some logical operators and functions

Operator/Function	Name	Description
`&`	AND	Returns true if both the first and the second conditions are true.
`|`	OR	Returns true if either the first or the second condition is true.
`!`	NOT	Reverses the logical value.
`%in%`	IN	Returns true if the value on the left is in the vector on the right.
`xor(x, y)`	XOR	Returns true if only one of the conditions is true, but not both.
`duplicated(x)`		Returns true if the value is a duplicate in the specified data set.

How to filter for one condition

```
polls %>% filter(state != "U.S.")
```

How to filter for multiple conditions

```
polls %>% filter(state == "California" & grade == "A")
```

How to filter for a value in a vector of values

```
polls %>% filter(state %in% c("California", "Oregon","Washington"))
```

How to view duplicate rows

```
fires %>% filter(duplicated(fires))
# A tibble: 441 x 6
  FIRE_NAME FIRE_SIZE STATE FIRE_YEAR DISCOVERY_DATE       CONTAIN_DATE
  <chr>         <dbl> <chr>     <int> <chr>                <chr>
1 NA             1500 CO         2005 2005-07-24 00:00:00  NA
2 NA              250 FL         2004 2004-06-14 00:00:00  NA
3 NA              125 KY         2003 2003-03-28 00:00:00  NA
...
```

How to drop duplicate rows

```
fires <- fires %>% filter(!duplicated(fires))
```

Another way to drop duplicate rows

```
fires <- fires %>% unique()
```

Description

- If a tibble contains rows that you don't need for your analysis, you can drop them.
- You can combine multiple conditions using the AND, OR, and XOR logical operators.

Figure 6-8   How to filter and drop rows

How to drop columns

Before you drop columns, you need to identify the columns to drop. To do that, you can use similar techniques to those for identifying rows to drop.

The first example shows that the polls data set contains several columns with only one value (cycle, branch, matchup, and forecastdate). Since these columns don't provide information that you can analyze, dropping them helps to clean your data set.

You can select columns by calling the select() function and providing either the indexes or names of the columns. The second example in figure 6-9 shows several ways to select columns by index. Here, the first three statements select columns with the specified indexes. This is useful if you want to drop all other columns. Then, the fourth and fifth statements show how to use a minus (-) to select all columns other than the specified columns. This is useful if you just want to drop the specified columns.

The third example shows how to select columns by name. The statements in this example select the same columns as the statements in the second example. However, they use names, not indexes, to specify the columns. The advantage of this approach is that it works even if the indexes of the columns change, which happens whenever you drop a column. For instance, if you drop the fourth column, it changes the indexes for all columns after the fourth column.

When using names to specify columns, you don't need to code quotations around the names. However, when specifying a vector of column names, you can include quotations around the names if you want, which is sometimes helpful.

The second and third examples show how to select columns, but they don't actually drop the columns. It's a good idea to confirm that you have selected the correct columns before you drop them. Then, to actually drop the columns, you need to overwrite the variable that stores the tibble for the analysis as shown in the fourth and fifth examples. Here, both examples drop the same seven columns.

A function for selecting columns

Function	Description
select(.data, ...)	Selects the specified column or columns from the specified data set. You can specify columns by index, name, or range. You can select all columns except the specified columns by preceding the specification with a minus sign.

The unique polls data for the first six columns

```
..$ cycle        : int 1
..$ branch       : int 1
..$ type         : int 3
..$ matchup      : int 1
..$ forecastdate : int 1
..$ state        : int 57
```

One unique value: Delete?

Small number of unique values: Often groups or categories.

How to use indexes to select columns

```
polls %>% select(1)              # col with specified index
polls %>% select(1:5)            # all cols in specified range
polls %>% select(1, 2, 4)        # each col with a specified index
polls %>% select(-1, -2, -4)     # all cols except specified indexes
polls %>% select(-c(1, 2, 4))    # another way to code previous
```

How to use names to select columns

```
polls %>% select(cycle)                       # specified column name
polls %>% select(cycle:forecastdate)          # all cols in specified range
polls %>% select(cycle, branch, matchup)      # each col with a specified name
polls %>% select(-cycle, -branch, -matchup)   # all cols except specified names
polls %>% select(-c(cycle, branch, matchup))  # more ways to code previous
polls %>% select(-c("cycle", "branch", "matchup"))
```

How to drop columns with specified indexes

```
polls <- polls %>% select(-1 ,-2, -4, -5, -23, -24, -25)
```

How to drop the columns with the specified names

```
polls <- polls %>% select(-cycle, -branch, -matchup, -forecastdate, -url,
                          -poll_id,-question_id)
```

Description

- Most data sets have more columns than you need for your analysis, so it makes sense to simplify the data set and drop columns you don't need.
- Columns that have only one value in the entire column are usually not needed for an analysis, so they can typically be dropped.
- When you drop columns, the indexes of the other columns will change. Consider using names to be sure you're dropping the correct columns.
- Column names in the select() function can be in quotes or not.

Figure 6-9   How to drop columns

How to rename columns

When you get a new data set, it's often hard to tell what the columns in a data set contain based on their column names. This is one reason studying the documentation or the data itself prior to starting an analysis is important.

After you figure out what each column contains, you may want to change the column names so they're more descriptive and easier to remember. That's always okay if you're working on your own, but if you're working in a group, you may need to get everyone in your group to agree on the changes. Once you decide on the name changes that you want to make, you can use any of the functions shown in figure 6-10 to rename the columns.

The first example shows how to use the rename() function to rename a single column in the fires data. To do that, you can pipe the data set into the rename() function. Then, within the rename() function, you can code the name of the new column, an equals sign, and the name of the old column. In this figure, for instance, the code renames the FIRE_SIZE column to SIZE. The assignment order seems backwards at first, so make sure to code the new column name first, followed by the old column name.

The second example uses the rename() function to shorten the names of two more columns. This works like the first example, except that you code a comma between each pair of new and old names.

To rename all columns by applying a function to them, you can use the rename_with() function as shown in the third example. This function is commonly used to apply the toupper(), tolower(), and str_to_title() functions. In this figure, for instance, the rename_with() function applies the tolower() function to all column names in the fires data set to make the column names lowercase.

The fourth example shows how to use the names() function to rename all columns in a tibble. In this case, the code uses title case for each of the column names. To do that, the code assigns a vector of column names to the column names returned by the names() function. The disadvantage of this technique is that it requires you to code each column name, even the ones that don't need to be changed. So, it can become tedious for data sets that have a large number of columns. However, it often works well for data sets that have a small number of columns.

Functions for renaming columns

Function	Description
rename(.data, ...)	Renames one or more columns and returns a tibble.
rename_with(.data, .fn)	Renames the columns by applying the specified function and returns a tibble.
names(x)	Returns a vector of column names for the specified data set.

How to rename a single column

```
fires <- fires %>% rename(SIZE = FIRE_SIZE)
fires
# A tibble: 49,356 x 6
   FIRE_NAME        SIZE STATE FIRE_YEAR DISCOVERY_DATE      CONTAIN_DATE
   <chr>           <dbl> <chr>     <int> <chr>               <chr>
 1 POWER           16823 CA         2004 2004-10-06 00:00:00 2004-10-21 00:00:00
 2 FREDS            7700 CA         2004 2004-10-13 00:00:00 2004-10-17 00:00:00
...
```

How to rename multiple columns

```
fires <- fires %>% rename(YEAR = FIRE_YEAR, NAME = FIRE_NAME)
fires
# A tibble: 49,356 x 6
   NAME             SIZE STATE  YEAR DISCOVERY_DATE      CONTAIN_DATE
   <chr>           <dbl> <chr> <int> <chr>               <chr>
 1 POWER           16823 CA     2004 2004-10-06 00:00:00 2004-10-21 00:00:00
 2 FREDS            7700 CA     2004 2004-10-13 00:00:00 2004-10-17 00:00:00
...
```

How to rename columns with a function

```
fires <- fires %>% rename_with(tolower)
fires
# A tibble: 49,797 x 6
   name             size state  year discovery_date      contain_date
   <chr>           <dbl> <chr> <int> <chr>               <chr>
 1 POWER           16823 CA     2004 2004-10-06 00:00:00 2004-10-21 00:00:00
 2 FREDS            7700 CA     2004 2004-10-13 00:00:00 2004-10-17 00:00:00
...
```

How to rename all columns

```
names(fires) <- c("Name","Size","State","Year","DiscoveryDate","ContainDate")
fires
# A tibble: 49,356 x 6
   Name             Size State  Year DiscoveryDate       ContainDate
   <chr>           <dbl> <chr> <int> <chr>               <chr>
 1 POWER           16823 CA     2004 2004-10-06 00:00:00 2004-10-21 00:00:00
 2 FREDS            7700 CA     2004 2004-10-13 00:00:00 2004-10-17 00:00:00
...
```

Description

- When using rename(), the new column name goes on the left, not the right.

Figure 6-10   How to rename columns

How to work with missing values

When you read data into a tibble, R may use an *NA value* to indicate that a value is not available or missing. An NA value doesn't have a data type. It's used only to mark a value as not available or missing.

Values are often missing for a reason. For example, in the fires data set, some fire names are missing because the fire was too small or in too remote of a region to name. Similarly, in the polls data set, some values are missing because there wasn't a poll conducted for a particular candidate. However, values are sometimes missing due to errors or other problems.

When your tibble contains missing values, you need to decide whether to keep them or fix them. In general, there are two ways to fix a missing value. You can drop the row that contains the missing value, or you can replace the missing value with the correct value or a value that makes sense for your analysis.

For instance, in some cases, it might make sense to replace a missing numeric value with 0 or -1. However, in other cases, doing that might skew your data set. So, before you fix any missing values, you should carefully think about why the data might be missing and consider the best strategy for handling the missing value.

How to find missing values

To start, you need to find any missing values in your data. To do that, you can use the complete.cases() function as shown in the first example in figure 6-11. Here, the code uses the NOT operator (!) before the complete.cases() function to return all rows in the fires data set that have missing values.

If you only care about missing values from a single column, you can pass that column to complete.cases() as shown in the second example. Here, the code only displays rows that have missing values in the Name column of the fires data set.

The third example shows how to use the is.na() function to accomplish the same task as the second example. This approach yields code that's shorter and easier to understand. As a result, it typically makes sense to use is.na() if you want to select all rows where a single column has a missing value and to use complete.cases() to select all rows where any column in the row contains a missing value.

The fourth example sorts the result set from the third example by the Size column in descending order. This shows that some fires with missing names still have large fire sizes. For example, the first fire in the sorted result set has a size of 71,760 acres. As a result, you probably don't want to drop these rows from your analysis.

Functions for working with missing values

Function	Description
complete.cases(...)	Returns a vector of Boolean values that indicates whether the row contains an NA value. If you pass a tibble, it checks whether the row contains an NA value in any of its columns. If you pass a column, it checks whether the value for the column is an NA value.
is.na(col)	Returns a vector of Boolean values that indicates whether each value in the specified column is an NA value.
nrow(data)	Returns the total number of rows in a data set.
replace_na(col, val)	Replaces all NA values in the column with the specified value.
ifelse(test, yes, no)	Returns the yes parameter if the test condition is TRUE or the no parameter if the test condition is FALSE.

How to select all rows that have missing values

```
fires %>% filter(!complete.cases(fires))
# A tibble: 20,140 x 6
   Name              Size State  Year DiscoveryDate        ContainDate
   <chr>            <dbl> <chr> <int> <chr>                <chr>
 1 PEPPIN           64488 NM     2004 2004-05-15 00:00:00  NA
 2 ROCHELLE HILLS     230 WY     2005 2005-08-26 00:00:00  NA
...
```

How to select all rows that have missing names

```
fires %>% filter(!complete.cases(Name))
# A tibble: 11,916 x 6
   Name  Size State  Year DiscoveryDate        ContainDate
   <chr> <dbl> <chr> <int> <chr>               <chr>
 1 NA     150  MT     1992 1992-10-20 00:00:00 1992-10-23 00:00:00
 2 NA     222  ID     1992 1992-05-17 00:00:00 1992-05-20 00:00:00
...
```

Another way to select all rows that have missing names

```
fires %>% filter(is.na(Name))
```

How to sort the rows that have missing names

```
fires %>% filter(is.na(Name)) %>%
  arrange(desc(Size))
# A tibble: 11,916 x 6
   Name   Size State  Year DiscoveryDate        ContainDate
   <chr> <dbl> <chr> <int> <chr>                <chr>
 1 NA    71760 AK     1997 1997-06-24 00:00:00  NA
 2 NA    70000 OK     1996 1996-02-22 00:00:00  NA
...
```

Description

- R uses an *NA value* to indicate that a value is not available or missing.

Figure 6-11   How to work with missing values (part 1)

If you want to compare the count of missing values to the total number of rows, you can use the nrow() function to get the total number of rows. For instance, the first example in part 2 of figure 6-11 shows that the data set contains 49,797 rows. Then, you can use the sum() function to count the total number of rows that contain a missing value in any of their columns as shown by the second example. This shows that 20,140 rows have missing values.

Similarly, you can use the sum() function to count the number of rows that have a missing value in a specific column as shown by the third example. This shows that 11,916 rows have missing values in the Name column.

The second and third examples work because the complete.cases() and is.na() functions both return a vector of Boolean values that indicate whether the row contains a missing value. Since TRUE evaluates to 1 and FALSE evaluates to 0, you can use the sum() function to count the rows that have missing values.

When you're trying to understand the missing values in a new data set, it often makes sense to check every column for missing values as shown in the fourth example. This code uses the apply() function to apply the sum() and is.na() functions to every column in the fires data set. To do that, it uses a lambda expression, which you'll learn more about in the next chapter. For now, though, you can copy this code anytime you need it.

The result that's displayed by the fourth example shows the number of missing values for each column. This shows that the Name column contains 11,916 missing values, the ContainDate column contains 17,579 missing values, and the other columns don't contain any missing values.

How to fix missing values

For the fires data set, it doesn't make sense to drop any of the rows that contain missing values. However, if you want to drop rows that contain NA values, you can use statements like the ones shown in the fifth and sixth examples. These examples assign the tibble that's returned to a variable named fires_dropna. As a result, this tibble doesn't contain any NA values.

Similarly, it doesn't make sense to replace any of the missing values in the fires data set. However, if you wanted to replace these values, you could use statements like the ones the seventh and eight examples. The seventh example shows how to use the replace_na() function to replace all NA values in a column with a specified value. In this case, it replaces all NA values in the Name column with a value of "Unknown".

If you want to replace NA values based on a condition, you can use the ifelse() function instead of the replace_na() function. The eighth example shows how to use ifelse() to replace NA values. In this case, the code checks whether the Name column is missing and the Size column is smaller than 200 acres. If so, it replaces the missing value for the Name variable with a value of "Unnamed Small Fire". Otherwise, it leaves the NA value in the Name column for fires that are 200 or more acres.

How to count the number of rows in a tibble

```
nrow(fires)
[1] 49797
```

How to count the rows that contain one or more missing values

```
sum(!complete.cases(fires))
[1] 20140
```

How to count the rows that have missing names

```
sum(is.na(fires$Name))
[1] 11916
```

How to get a count of NA values for all columns

```
fires %>% apply(MARGIN = 2, function(col) sum(is.na(col)))
         Name        Size       State       Year DiscoveryDate  ContainDate
        11916           0           0          0             0        17579
```

How to drop all rows that have one or more missing values

```
fires_dropna <- fires %>% filter(complete.cases(fires))
```

How to drop all rows that have a missing name

```
fires_dropna <- fires %>% filter(complete.cases(Name))
```

How to replace missing values

```
fires_replna <- fires %>% mutate(Name = replace_na(Name, "Unknown"))
```

How to conditionally replace missing values

```
fires_replna <- fires %>% mutate(Name = ifelse(is.na(Name) & Size < 200,
                                 "Unnamed Small Fire", Name))
```

Description

- Since a Boolean value of TRUE evaluates to a numerical value of 1 and FALSE evaluates to 0, you can use the sum() function to count the number of rows that meet a condition.

- You can use the apply() function to apply the sum() and is.na() functions to all columns in a result set.

Figure 6-11   How to work with missing values (part 2)

How to work with data types

When you import data into a tibble, dates and numbers are sometimes imported as strings instead of date and numeric types. So, before you can use any of the functions that work with date and numeric data types, you need to convert the string types to the right data types.

How to select columns by data type

When you import data into a tibble, numbers and dates are sometimes imported with the character type instead of a numeric or date type. To determine if the columns have the correct data types, you can display the names and data types for each column using str() as described earlier in this chapter.

However, it's also often helpful to use the functions described in figure 6-12 to select columns by their data types. To do that, you can pass the data set to the select_if() function. Then, you can pass a function such as is.numeric() or is.character() to select the columns.

The first example uses the is.numeric() function to select all of the numeric columns in the fires data set. This shows that the fires data set has two numeric columns, and it shows that the Year column is of the integer type while the Size column is of the double data type, as you would expect.

The second example uses the is.character() function to select all columns of the character type. This shows that the fires data set has four columns that are stored with the character type. However, the DiscoveryDate and ContainDate columns should be stored as the Date type for easier analysis.

The third example shows that no columns are stored with the factor type. However, for the fires data set, you could convert the year column to the factor type since it's primarily used to categorize a fire.

The fourth example loads the lubridate package that contains the is.Date() and is.timepoint() functions, and the fifth example uses the is.timepoint() function to check if a column is stored with a date or time type. This shows that no columns are stored with the Date or timepoint types.

Functions for selecting columns by data type

Function	Description
select_if(.data, .predicate)	Selects all columns in the tibble that meet the specified condition (predicate) and returns a tibble.
is.numeric(x)	Returns TRUE if the specified column is any numeric type.
is.integer(x)	Returns TRUE if the specified column is the integer type.
is.double(x)	Returns TRUE if the specified column is the double type.
is.character(x)	Returns TRUE if the specified column is the character type.
is.factor(x)	Returns TRUE if the specified column is the factor type.
is.Date(x)	Returns TRUE if the specified column is the Date type.
is.timepoint(x)	Returns TRUE if the specified column is the Date or timepoint type.

How to select all numeric columns

```
fires %>% select_if(is.numeric)
# A tibble: 29,657 x 2
    Size  Year
   <dbl> <int>
 1 16823  2004
 2  7700  2004
...
```

How to select all character columns

```
fires %>% select_if(is.character)
# A tibble: 29,657 x 4
   Name       State DiscoveryDate        ContainDate
   <chr>      <chr> <chr>                <chr>
 1 POWER      CA    2004-10-06 00:00:00  2004-10-21 00:00:00
 2 FREDS      CA    2004-10-13 00:00:00  2004-10-17 00:00:00
...
```

How to select all factor columns

```
fires %>% select_if(is.factor)
# A tibble: 29,657 x 0
...
```

How to load the lubridate package

```
library("lubridate")
```

How to select all date and time columns

```
fires %>% select_if(is.timepoint)
# A tibble: 29,657 x 0
...
```

Description

- To use the is.Date() and is.timepoint() functions, you must load the lubridate package.

Figure 6-12   How to select columns by data type

How to convert strings to numbers

If you encounter a column that uses the character type to store data that's mostly numeric, it's often because the data set used character strings such as "#" or "*" to represent missing data. For example, in figure 6-13, the first example shows a column named Chick that stores two numbers and a string of "#". The output for this example shows that R converts all of the values to the character type even though only the third value is a string.

When this happens, you often want to convert the strings to a numeric type. To do that, you can use the mutate() function to convert the specified column to the specified numeric type as shown in the second example. Here, the example uses the as.integer() function to convert the specified column to the integer type. This truncates the decimal value of 2.5 to an integer value of 2, which might not be what you want.

In addition, it coerces any data that it can't convert to a number to an NA value. If the function coerces one or more values to NA, it displays a warning like the one shown in the second example. This strategy works well if it makes sense to allow NA values in the column or if you plan on replacing the NA values later.

The third example shows another strategy for removing the invalid values. With this strategy, you use the mutate() function to replace the invalid values in the column with valid values before you convert them to a number. To do that, this example begins with a call to the mutate() function that uses the sub() function to replace the value of "#" with a value of "0". Then, it pipes the resulting tibble into another mutate() function that uses the as.double() function to convert the column to the double type. This strategy works well if a number like 0 or a calculated average makes sense for missing values in your data set. In addition, since it uses the as.double() function, it doesn't truncate the second value.

Functions for changing a column's data type

Function	Description
`mutate(.data, ...)`	Modifies one or more columns in a tibble and returns a tibble. If the specified column exists, it modifies the column. Otherwise, it adds a new column.
`as.integer(x)`	Converts the specified column to the integer type.
`as.double(x)`	Converts the specified column to the double type.

A column that stores numeric data with the character type

```
ids <- c(1,2.5,"#")
chicks = as_tibble(data.frame(Chick = ids))
chicks
# A tibble: 3 x 1
  Chick
  <chr>
1 1
2 2.5
3 #
```

How to convert the column to the integer type

```
chicks %>% mutate(Chick = as.integer(Chick))
# A tibble: 3 x 1
  Chick
  <int>
1     1
2     2
3    NA
Warning message:
Problem while computing `Chick = as.integer(Chick)`.
i NAs introduced by coercion
```

How to substitute numbers to remove NA values

```
chicks %>% mutate(Chick = sub("#", "0", Chick)) %>%
          mutate(Chick = as.double(Chick))
# A tibble: 3 x 1
  Chick
  <dbl>
1   1
2   2.5
3   0
```

Description

- Sometimes a column that contains mostly numeric data has the character type. This often happens when a symbol such as a # or * denotes missing data.

- If the as.integer() and as.double() functions encounter a string value that they can't convert to the intended data type, they coerce it to an NA value and display a warning.

Figure 6-13   How to convert columns to the numeric types

How to convert strings to dates and times

Figure 6-14 shows how to convert strings that store information about dates and times to the Date and timepoint types. If you're only working with dates, not times, you can use the as.Date() function, which is part of base R, to convert strings to the Date type. However, if you want to use the datetime type to include a time component in your data, you need to use the as_datetime() function from the lubricate package. Then, you may want to use the as_date() function from the lubricate package. This function works like the as.Date() function, but it uses a name that's consistent with the as_datetime() function.

The first example shows the columns in the polls data that store strings for dates and times. The first two columns store dates with the full year, the third column stores a date with an abbreviated year, and the fourth column stores a date and a time.

When you convert a string column to a date, you often need to specify the format of the string to tell your conversion function how to extract the parts of the date. To do that, you can use the format specifiers shown in this figure.

The second example shows how to convert the createddate column from the character type to the Date type that's used for storing dates that don't have a time component. To do that, this code uses the mutate() function to pass the createddate column to the as.Date() function with a format specifier that matches the date format for this column, including using a lowercase y for the abbreviated year.

The third example works like the second example, except that it converts the timestamp column to the timepoint type. This uses the as_datetime() function instead of the as.Date() function. Note that the format string includes the same spaces as the column and uses capital Y for the full year.

The fourth example shows how to convert multiple columns that use the same format specifier. It starts by assigning the format specifier to a variable. Then, uses this variable instead of coding the format specifier twice within the mutate() function.

When you specify the format, remember that you need to include any other characters in the date such as slashes, dashes, and spaces. That's why the format specifiers shown in this figure include front slashes (/), colons (:), and spaces. Note that the format specifier for the timestamp column uses two spaces between the date and time but just one space between the day, month, and year.

By default, the as.Date() function uses a format specifier of "%Y-%m-%d". As a result, if you need to convert a string to a date that's in the YYYY-MM-DD format, you don't need to set the format parameter. Instead, you can just call the as.Date() function like this:

```
fires <- fires %>% mutate(DiscoveryDate = as.Date(DiscoveryDate),
                          ContainDate = as.Date(ContainDate))
```

Functions for converting strings to dates and times

Function	Description
`as.Date(x, format)`	Converts the specified column to the Date type based on the specified format string.
`as_date(x, format)`	Converts the specified column to the Date type based on the specified format string. Available from the lubridate package.
`as_datetime(x, format)`	Converts the specified column to the timepoint type based on the specified format string. Available from the lubridate package.

Some date/time format specifiers

Code	Description
`%d`	Numeric day (3, 17, 31).
`%a`	Abbreviated name of day (Mon, Wed, Fri).
`%m`	Numeric month (1, 6, 12).
`%b`	Abbreviated name of month (Jan, Feb, Mar).
`%y`	Abbreviated year (96, 02, 22).
`%Y`	Full year (1996, 2002, 2022).
`%H`	Hours
`%M`	Minutes
`%S`	Seconds

The polls columns to convert

```
polls %>% select(startdate, enddate, createddate, timestamp)
# A tibble: 12,624 x 4
  startdate enddate    createddate timestamp
  <chr>     <chr>      <chr>       <chr>
1 11/3/2016 11/6/2016  11/7/16     09:35:33  8 Nov 2016
2 11/1/2016 11/7/2016  11/7/16     09:35:33  8 Nov 2016
...
```

How to convert a single column to the Date type

```
polls <- polls %>%
  mutate(createddate = as_date(createddate, format = "%m/%d/%y"))
```

How to convert a single column to the timepoint type

```
polls <- polls %>%
  mutate(timestamp = as_datetime(timestamp, format = "%H:%M:%S  %d %B %Y"))
```

How to convert multiple columns that use the same format specifier

```
date_fmt <- "%m/%d/%Y"
polls <- polls %>%
  mutate(startdate = as_date(startdate, format = date_fmt),
         enddate   = as_date(enddate, format = date_fmt))
```

Figure 6-14   How to convert columns to the date/time types

How to work with the factor type

The *factor type* stores categorical data. In other words, it stores data that's used to put each row in a category. There are two types of factors: nominal and ordered (or ordinal). *Nominal factors* aren't ordered, and *ordered* or *ordinal factors* are ordered.

When working with categorical data, it's usually helpful to use the factor type. This makes some operations on the data work better, and it typically allows R to use less memory when storing the column.

The first example in figure 6-15 shows the mortality_long data set from chapter 2. Here, the AgeGroup column is stored with the character type, but it should be stored with the factor type since it stores a string that puts each row into one of four categories (1-4 Years, 5-9 Years, 10-14 Years, or 15-19 Years).

The second example uses the factor() function to convert the AgeGroup column from the character type to the factor type. Then, the third example views the data for the AgeGroup column. This shows that data has been grouped into four levels. However, these age groups aren't in order from youngest to oldest, which is probably what you want in this case.

The fourth example uses the factor() function to convert the AgeGroup column to an ordered factor. This example uses the levels parameter to specify a vector that supplies the specific order for the categories. In addition, it sets the ordered argument to TRUE to indicate that the factor is now ordered. In other words, the first level listed is before the second level, which comes before the third level, and so on as shown by the fifth example. In addition, setting ordered to TRUE makes the data type for the column the ordinal data type.

The sixth example shows how to create an ordered factor for the Year column of the fires data set. Here, the code specifies a range of numbers for the years. To specify this range, the code specifies the first year in the range, a colon, and the last year in the range. This orders the factors from 1992 to 2015 as shown by the seventh example.

A function for converting a column to the factor type

Function	Description
factor(x)	Converts the specified column (x) to the factor type.
Parameter	**Description**
levels	Uses a vector to specify the order of the category.
ordered	Specifies whether the factor is an ordinal factor. Default is FALSE.

The mortality_long tibble

```
mortality_long
# A tibble: 476 x 3
    Year AgeGroup  DeathRate
   <dbl> <chr>         <dbl>
 1  1900 1-4 Years     1984.
 2  1901 1-4 Years     1695
...
```

How to convert the AgeGroup column to the factor type

```
mortality_long <- mortality_long %>% mutate(AgeGroup = factor(AgeGroup))
```

How to view the levels for the AgeGroup column

```
mortality_long$AgeGroup
...
Levels: 1-4 Years 10-14 Years 15-19 Years 5-9 Years
```

How to specify the order for the levels in a factor

```
mortality_long <- mortality_long %>%
   mutate(AgeGroup = factor(AgeGroup, levels = c("1-4 Years","5-9 Years",
                                "10-14 Years","15-19 Years"),
                 ordered = TRUE))
```

The levels for the column after it has been reordered

```
Levels: 1-4 Years < 5-9 Years < 10-14 Years < 15-19 Years
```

How to use a range of numbers to specify the new order

```
fires <- fires %>%
   mutate(Year = factor(Year, levels = 1992:2015, ordered = TRUE))
```

The levels of the Year column of the fires data set

```
24 Levels: 1992 < 1993 < 1994 < 1995 < 1996 < 1997 < 1998 < 1999 < ... < 2015
```

Description

- The *factor type* works with categorical data.
- If the factor can be ordered, it can be stored as a special type of factor called an *ordinal factor* or *ordered factor*. Otherwise, it's stored as a *nominal factor* (or just *factor*).

Figure 6-15 How to work with the factor type

How to work with outliers

Outliers are values that are far outside the normal range of values in a column. For instance, the death rate for the year 1918 is an outlier for the mortality data because that was the year of the Spanish Flu pandemic. As a result, far more children died that year than in the preceding or the following year.

In general, there are two types of outliers: mistakes in the data, and true outliers like the death rates in the year 1918. Either way, the outliers can affect the results of your analysis. When you clean the data for an analysis you need to look for outliers and decide if you need to fix or otherwise account for them.

How to assess outliers

Figure 6-16 shows two ways to find outliers. The first way to find outliers is to use the summary() function to examine the quartiles as shown in the first example. Here, the summary() function displays the quartiles for the mortality data. If the minimum or maximum values are particularly low or high, you might suspect an outlier value is in the set.

Since the maximum value for the DeathRate column is far higher (almost 10x higher) than the value for the third quartile, it seems likely that there are outliers in this column. However, while there is an outlier in this data set, the maximum value isn't it. As you can see from looking at the plot, death rates dropped dramatically from 1900 onwards, and the maximum recorded rate is for the first year in the data set. Once again, this is an example of why it's important to think carefully about your data set when performing an analysis.

The second way to find outliers is to plot the data as shown in the second and third examples. To do that, you can use a box plot to automatically calculate and display the outliers. Alternately, you can create another type of visualization, like a line or scatter plot.

The second example presents two plots that display the outliers in the mortality data. By looking at the line plot, it's clear there's only one true outlier in the mortality data set, the year 1918. Meanwhile, the box plot shows many outliers for the first age group. This is because the box plot doesn't take the trend of the data into account when it calculates the outliers.

The third example presents plots of the irises data. Here, some of the outliers are easy to pick out in the scatter plot, such as the outlier on the low end for the setosa species (shown on the left). However, the rest of the outliers are harder to pick out unless they are computed numerically as shown by the box plot. Therefore, they may not be outliers at all.

When you consider how to address outliers, it's important to think about the possible reasons for the outliers. For example, in the mortality data or the fires data, the outliers (the highest death rates and the largest fires) are certainly not mistakes and instead are points of interest in the data. On the other hand, the outliers in the iris data might be data entry mistakes or exist due to another

Summary data for the mortality data set

```
summary(mortality_long)
 Year             AgeGroup        DeathRate
 Min.   :1900    1-4 Years  :119   Min.   :   11.40
 1st Qu.:1929    5-9 Years  :119   1st Qu.:   40.58
 Median :1959    10-14 Years:119   Median :   89.50
 Mean   :1959    15-19 Years:119   Mean   :  192.92
 3rd Qu.:1989                      3rd Qu.:  222.57
 Max.   :2018                      Max.   : 1983.80
```

Two plots that display outliers in the mortality data

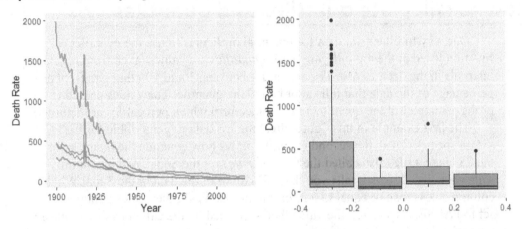

Two plots that display outliers in the irises data

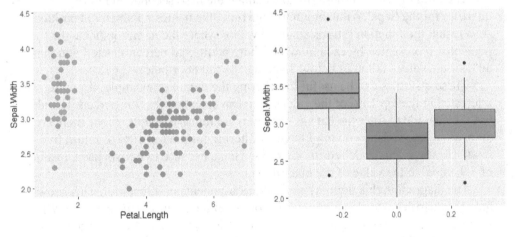

Figure 6-16   How to assess outliers

variable. For example, maybe the smallest setosa didn't get enough water or light.

Beyond considering the reasons for the outliers, you need to consider your goals for the analysis as well. For example, if you're writing a report on the economic impact of wildfires, it may make sense to only study the top 10% of fires by size because these fires are likely to have caused the most damage. On the other hand, if you are trying to classify the irises by species given only the numeric data, outliers can make this task much more difficult, so you may want to drop them entirely.

How to calculate quartiles and quantiles

One way to filter your data for potential outliers is to use the quantile() function listed at the top of figure 6-17. *Quantiles* are values taken from regular intervals in the data. *Percentiles* are values between 1 and 100 that represent a percentage of the data that falls at or below the quantile. The values passed to the probs parameter of the quantile() function control which percentiles are returned.

The first example in this figure filters the iris data to get a tibble for the setosa species. Then, the second example shows how generate the quantile for the 25th percentile (also called the first *quartile*) for the Sepal.Width column. This value is 3.2, which tells you that 25% of all of the data in the Sepal.Width column is less than or equal to 3.2. In this function call, the names parameter is set to FALSE. As a result, the value that's returned by the call doesn't include a name that indicates the percentile.

The third example shows how to calculate the first, second, and third quartiles for the Sepal.Width column. This time, the names parameter is omitted. As a result, the function returns a named vector where the names indicate the percentile represented by each number. These additional percentiles tell you that 50% of the values fall at or below 3.4 and 75% fall at or below 3.675.

These quantiles can be useful for filtering the data. For example, if you're interested in the top 10% of the values, you can calculate the 90th percentile and use the quantile to filter the data as shown in the fourth example. Since there are 50 rows of setosa data, you might expect filtering for the top 10% to return five rows. However, this code returns six rows. That's because two rows have a value of 3.9, equal to the value of the quantile.

The diagram in this figure reviews the components of a box plot. This shows how the quartiles work. In addition, it shows that the *interquartile range (IQR)* is the distance between the first and third quartiles. This is useful because a box plot uses the IQR to create a filter for its outliers.

A function for calculating quantiles

Function	Description
`quantile(x)`	Calculates the quantiles for the specified numeric vector.

Parameter	Description
`probs`	Specifies a probability percent value or vector of probability percent values.
`names`	Specifies whether to include names in the result that's returned. Default is TRUE.

How to get a tibble for the setosa species

```
setosa <- iris %>% filter(Species == "setosa") %>% as_tibble()
```

How to get the first quartile

```
quantile(setosa$Sepal.Width, probs = 0.25, names = FALSE)
[1] 3.2
```

How to get all three quartiles

```
quantile(setosa$Sepal.Width, probs = c(0.25, 0.5, 0.75))
  25%   50%   75%
3.200 3.400 3.675
```

How to use a quantile to select the rows in the top 10%

```
setosa %>% filter(Sepal.Width >= quantile(Sepal.Width, probs = 0.9))
# A tibble: 6 x 5
  Sepal.Length Sepal.Width Petal.Length Petal.Width Species
         <dbl>       <dbl>        <dbl>       <dbl> <fct>
1          5.4         3.9          1.7         0.4 setosa
2          5.8         4            1.2         0.2 setosa
3          5.7         4.4          1.5         0.4 setosa
4          5.4         3.9          1.3         0.4 setosa
5          5.2         4.1          1.5         0.1 setosa
6          5.5         4.2          1.4         0.2 setosa
```

The components of a box plot

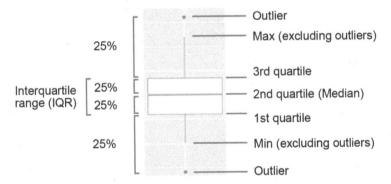

Figure 6-17   How to calculate quartiles and quantiles

How to calculate the fences for the box plot

The max and min values shown by the whiskers in a box plot are the highest and lowest values that are not outliers, not the limits themselves. These limits, also called *fences*, draw the line for what is and isn't considered an outlier. You can use the IQR, which is defined as the third quartile minus the first quartile, to calculate the value for the fences.

The first example in figure 6-18 shows how the use the IQR() function to get the IQR for the data. Alternately, you could calculate this value with the values returned from the quantile() function. However, using the IQR() function helps to keep your code clean.

The second example shows how to calculate the fences for the data. The lower fence is defined as Q1 − (1.5 * IQR) and the upper fence is defined as Q3 + (1.5 * IQR).

Once you've calculated the fences, you can add a horizontal line to the box plot for each fence as shown by the third example. To do that, this example uses the geom_hline() function to add a horizontal line at the specified value on the y axis. In addition, it sets the color and linetype parameters to draw the fences with a dashed red line.

How to fix the outliers

If you determine that an outlier is going to cause problems with your analysis, there are two ways to fix the outlier. First, you can drop the outlier. For instance, the fourth example shows how to drop the outliers for the setosa data by filtering for the data that lies between the fences. This provides a simple way to drop the outliers.

Second, you can replace the outlier with a value that doesn't skew the results of your analysis. For instance, you may want to replace the outlier with a minimum or maximum value that caps the values for a column. Or, you may want to replace the outlier with the mean value for the column. To do that, you can use the mutate() and ifelse() functions as shown in the fifth example.

A function for calculating the IQR

Function	Description
IQR(x)	Calculates the IQR (interquartile range) for the specified numeric column.

How to calculate the IQR for a column

```
IQR(setosa$Sepal.Width)
[1] 0.475
```

How to calculate the fences for a box plot

```
upper_fence <- quantile(setosa$Sepal.Width, 0.75) +
  (1.5 * IQR(setosa$Sepal.Width))
lower_fence <- quantile(setosa$Sepal.Width, 0.25) -
  (1.5 * IQR(setosa$Sepal.Width))
```

How to plot the fences for the box plot

```
ggplot(setosa) +
  geom_boxplot(aes(y = Sepal.Width), color = "blue") +
  geom_hline(aes(yintercept = upper_fence), color = "red", linetype = "dashed") +
  geom_hline(aes(yintercept = lower_fence), color = "red", linetype = "dashed")
```

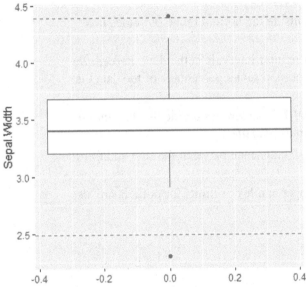

How to use the fences to drop the outliers

```
setosa <- setosa %>%
  filter(Sepal.Width > lower_fence & Sepal.Width < upper_fence)
```

How to use fences to replace the outliers

```
setosa %>% mutate(
  Sepal.Width = ifelse(Sepal.Width > lower_fence & Sepal.Width < upper_fence,
                       Sepal.Width, mean(Sepal.Width)))
```

Figure 6-18   How to use fences to fix outliers

Perspective

Now that you've finished this chapter, you should realize how time-consuming but important it is to clean the data for an analysis. In fact, if the data isn't clean, it's not only hard to analyze, but your results are likely to be inaccurate or misleading.

Summary

- The *value counts* for a column show each unique value as well as a count of the number of times each non-null value occurs.
- An *NA value* indicates that the value is not available or missing. An NA value doesn't have a data type. It's used only to mark a value as not available or missing.
- The *factor type* works with categorical data.
- If the factor can be ordered, it can be stored as a special type of factor called an *ordinal factor* or *ordered factor*. Otherwise, it's stored as a *nominal factor* (or just *factor*).
- *Outliers* are values that are far outside the normal range of values in a column.
- *Quantiles* are values taken from regular intervals in the data. *Percentiles* are values between 1 and 100 that represent a percentage of the data that falls at or below the quantile.
- The quantiles at the 25th, 50th, and 75th percentiles divide the data into four parts. As a result, they are known as *quartiles*.
- The *interquartile range* (*IQR*) is the distance between the first and third quartiles.
- The values that determine the upper and lower limits for outliers are also known as *fences*.

Exercise 6-1   Clean the Star Wars data set

In this exercise, you'll clean the Star Wars data set.

Load the Star Wars data set into a tibble and examine the data

1.   Create a new script in RStudio and name it exercise_6-1.R.

2.   Add statements to load the tidyverse and datasets packages.

3.   Convert the starwars data set to a tibble and assign it to a variable named swchars.

4.   Explore the swchars tibble by clicking on its name in the Environment pane.

5.   Note that there are a number of NA values, especially for characters from the seventh movie, The Force Awakens. Furthermore, the vehicles and starships columns both contain a number of empty lists.

6.   View the structure of the tibble. To do that, you can use the str() function. Note that while the first 11 columns display as you would expect, the lists contained in last 3 columns are displayed in their entirety. You can prevent the contents of the lists from displaying by setting the max.level parameter to limit the number of levels for nested lists like this:

```
str(swchars, strict.width = "cut", max.level = 2)
```

Whenever you encounter a function that doesn't work exactly how you want, remember that you can enter ? and the name of the function in the Console pane to view the documentation for the function and learn more about its parameters.

7.   Set a goal for your analysis. For this exercise, let's assume you want to explore any correlations between physical characteristics and species for Star Wars characters.

Simplify the data

8.   Check for duplicate rows. To do that, you can use the filter() and duplicated() functions. As it happens, there aren't any in this data set, but it's good to confirm.

9.   Drop columns that aren't needed for the analysis. To do that, use the select() function. Since the goal for this analysis uses physical characteristics, drop the columns for birth_year, homeworld, films, vehicles, and starships.

10.   Rename the columns to capitalize the first letter of each column. To do that, use the rename_with() function with the str_to_title() function.

11.   Rename the columns that still contain underscores so they use title case without any underscores. In other words, rename columns like Eye_color to EyeColor. To do that, use the rename() function.

Identify and handle NA values

12.   View the rows that contain a missing value in any column. To do that, use the complete.cases() function and the NOT operator (!). Note that the resulting tibble is nearly two fifths of the rows in the data set, and the missing values are in a variety of columns.

13. View the rows that have missing values for just the Height column. Note that the last row, for Captain Phasma, has no useful data in it.

14. Drop the row for Captain Phasma.

15. View the rows that have missing values for the Mass column. The result is 27 rows out of 86 – quite a lot. Let's say that the data available for characters' mass is too incomplete to properly analyze.

16. Drop the Mass column.

17. View the rows that have missing values in the HairColor column. Note that these are all characters without hair and that other characters without hair have the value "none".

18. Change the NA values in the HairColor column to "none". To do that, use the replace_na() function.

19. View the rows that have missing values in the Sex column. All three of these rows are missing not just a sex but also gender and species. Since they aren't useful for this analysis, drop them from the tibble.

20. Display all rows that have NA values. At this point, the only NA values still in the tibble should be five missing heights.

21. Display the unique values for the Species column. There are 37 different values, but many of them are only used once. That's not helpful for our analysis, but we don't want to drop these rows or the column.

22. Display a count of the unique values for the Species column to see if any values are used more than once or twice. The "Human" value has 35 rows and the "Droid" value has 6. Keep those two values, but replace all other values in the species column with a value of "Alien". To do that, use the ifelse() function.

23. Display the unique values for the Sex and Gender columns. To do that, use the unique() function. This shows that there are four unique values in the Sex column but only two for the Gender column.

24. Check whether the "female" and "male" values in the Sex column always correspond with "feminine" and "masculine" values in the Gender column. To do that, you can use a filter that combines and nests logical operators. Note that "female" and "male" are redundant with "feminine" and "masculine", and "none" is redundant with "Droid" in the Species column. That leaves one row with a value of "hermaphroditic", which can be considered an outlier.

25. Drop the Sex column since its information is redundant or an outlier.

26. View the unique values and their counts for the EyeColor column. There are a few common colors, like black, blue, and brown; some unusual colors, like hazel and red; and some combinations like "green, yellow". Consider how you might choose to handle this depending on your analysis goals.

Identify and fix wrong data types

27. View the number of unique values for each column of the cleaned tibble. This shows that the Gender and Species columns are both good candidates for being used as categories.

28. Convert the Gender and Species columns to the factor data type.

Identify and fix outliers

29. Use the summary() function for a quick overview of the tibble. Only one remaining column is numeric, so only the Height column will have calculated quartiles.

30. Create a box plot of character heights, grouped by species. This should show that there are no outliers for droids or humans, but many outliers for aliens. This makes sense, since we grouped together many disparate species together for the "Alien" category.

31. View the tibble in RStudio and sort by height. This shows that none of the height values are particularly unusual for the data set and it therefore makes sense to keep them.

Chapter 7

How to prepare data

This chapter shows how to prepare a data set for analysis. To do that, you may want to add or modify columns, group or bin data, apply functions, or reshape a data set.

How to add and modify columns

When you prepare your data, you often need to add columns or modify the data within one of the existing columns. You may also want to add a column that provides summary data for groups of rows.

The mutate() function provides a convenient way to add or modify columns in data objects such as tibbles. To use it, you specify the tibble to be modified, the name of the column you want to add or change, and the data that should be in that column. The name of the column and column data are also called a name-value pair.

In figure 7-1, the second and third examples show how to use the mutate() function to add a column. Then, the fourth example shows how to use the mutate() function to modify two columns with a single statement. To do that, you just separate the name-value pairs for each column with a comma.

How to work with date columns

When working with dates, it's common to want to extract a component of the date such as the month or year. If the date is stored in a column of the Date type, you can use the format() function to select components of the date. Once extracted, you can use the mutate() function to add them to the tibble as new columns or replace the data in the original column.

To illustrate, the first example in figure 7-1 shows the two columns in the fires tibble that are of the Date type. Then, the second example shows how to create a new column named Month that stores the month each fire was discovered. The value for the column comes from passing the DiscoveryDate column to the format() function and specifying that just the month (%m) should be returned. Additional format specifiers can be found in figure 6-14.

You can also perform calculations on dates. For instance, the third example shows how to subtract the DiscoveryDate column from the ContainDate column and add one to the result. This calculation returns a difftime object that represents the number of days between the two dates. Here, a fire that's contained on the same day that it's discovered is considered to be burning for one day.

The second and third examples cause the Month column to store its values as strings and the DaysBurning column to store its values as difftime objects. However, it makes more sense to convert the Month column to the factor type and the DaysBurning column to the integer type. To do that, the code in the fourth example uses the mutate() function to change the data type for both columns. Since the names of the columns already exist, the mutate() function doesn't add new columns. Instead, it modifies the existing columns.

The fifth example shows the date columns and the two new columns. This shows that the Month column now uses the factor type, and the DaysBurning column uses the integer type.

A function for adding and modifying columns

Function	Description
`mutate(.data, ...)`	Adds or modifies one or more columns to a data set and returns a data set of the same type. If the specified column exists, it modifies the column. Otherwise, it adds a new column.

A function for formatting dates

Function	Description
`format(x, format)`	Formats a date/time column according to the specified format.

The date columns in the fires data

```
fires %>% select(DiscoveryDate, ContainDate)
# A tibble: 238,955 x 2
   DiscoveryDate ContainDate
   <date>        <date>
 1 2004-10-06    2004-10-21
 2 2004-10-13    2004-10-17
 3 2005-02-12    2005-02-13
...
```

How to add a column named Month

```
fires <- fires %>% mutate(Month = format(DiscoveryDate, "%m"))
```

How to add a column named DaysBurning

```
fires <- fires %>% mutate(DaysBurning = ContainDate - DiscoveryDate + 1)
```

How to modify two columns with a single statement

```
fires <- fires %>% mutate(Month = as.factor(Month),
                          DaysBurning = as.integer(DaysBurning))
```

The original date columns and the two new columns

```
fires %>% select(DiscoveryDate, ContainDate, Month, DaysBurning)
# A tibble: 54,093 x 4
   DiscoveryDate ContainDate Month DaysBurning
   <date>        <date>      <fct>       <int>
 1 2004-10-06    2004-10-21  10             16
 2 2004-10-13    2004-10-17  10              5
 3 2005-02-12    2005-02-13  02              2
...
```

Description

- You can use arithmetic operators to perform calculations on Date objects, returning difftime objects that store a span of time. Then, you can convert the difftime objects to another data type like an integer.

Figure 7-1   How to work with date columns

How to use stringr to work with strings

Chapter 2 presented some of the functions from base R that you can use to work with strings such as paste(), grep(), toupper(), and tolower(). When you use R for data analysis, you're bound to come across these functions. However, the stringr package that's part of the core tidyverse provides string functions that often work better than the string functions provided by base R. Some of these functions are summarized at the top of figure 7-2, but the stringr package provides many more functions that make it easy to work with strings.

Each stringr function starts with the str_ prefix. This makes it clear that the function is in the stringr package and that it's for working with string data. In addition, each stringr function accepts the string that it's working with as its first argument. This makes it easy to use piping with stringr functions.

The first example shows how to use the str_c() function to combine multiple strings into a single string. This works much like the paste() function from base R, but the default separator is an empty string. As a result, if you want to use a different separator to separate strings, you need to specify it.

The second example shows how to use the str_detect() function to check if a string contains a substring. This function returns a Boolean value of TRUE if the specified pattern exists in the string or FALSE if it doesn't. By default, the str_detect() function is case-sensitive. As a result, a string of "cd-1" doesn't match a string of "CD-1". However, if you want the str_detect() function to perform a case-insensitive comparison, you can add (?i) to the beginning of the search pattern as shown in this example. This works for any stringr function that has a pattern argument.

The third example shows how to use the str_to_title() function to capitalize the first letter in every word of the string. Unfortunately, this function doesn't contain a parameter to specify the separator, so if you have a separator for your words that isn't a space, you'll need to replace it before calling this function.

The fourth example shows how to use piping to chain some stringr functions together to convert a column name from uppercase with underscores separating each word to title case. To do that, the code begins by piping the col_name string into the str_replace_all() function. This replaces each underscore in the string with a space. Then, it pipes the string that's returned into the str_to_title() function. This converts each word to title case. Finally, it pipes the string that's returned into the str_replace_all() function. This removes the spaces in the string by replacing each space with an empty string.

Although the examples in this figure show how to use stringr functions on a single string, you can also use them to process a vector of strings, like a column in a tibble. This makes it possible to use the code in the fourth example with the names() function to convert all of the names for a tibble to title case. In addition, it makes it possible to use stringr functions to modify all of the strings in a column as shown in the next figure.

Some functions from the stringr package

Function	Description
str_c(..., sep)	Combines multiple strings into a single string. Each combined string is separated by the specified separator string.
str_sub(string, start, stop)	Extracts part of a string from the start index to the stop index.
str_detect(string, pattern)	Returns a Boolean to indicate if a pattern is in the string.
str_replace(string, pattern, replacement)	Replaces the first found pattern with the replacement string.
str_replace_all(string, pattern, replacement)	Replaces all instances of the pattern with the replacement string.
str_to_lower(string)	Converts the string to lower case.
str_to_upper(string)	Converts the string to upper case.
str_to_title(string)	Converts the string to title case.

How to combine strings

```
first_name <- "Bob"
last_name <- 'Smith'
age <- 40
str_c(first_name, " ", last_name)          # returns "Bob Smith"
str_c(last_name, first_name, sep = ", ")    # returns "Smith, Bob"
str_c(first_name, " is ", age, " years old.") # returns "Bob is 40 years old."
```

How to check if a string contains a substring

```
state <- "Maine CD-1"
str_detect(state, "CD-")                    # returns TRUE
str_detect(state, "cd-")                    # returns FALSE
str_detect(state, "(?i)cd-")                # returns TRUE
```

How to convert a string to title case

```
fire_name <- "AUGUST COMPLEX"
str_to_title(fire_name)                     # returns "August Complex"
```

How to pipe a string into several functions

```
col_name <- "FIRE_DISCOVERY_DATE"
col_name %>% str_replace_all("_"," ") %>%
  str_to_title() %>%
  str_replace_all(" ", "")                  # returns "FireDiscoveryDate"
```

Description

- The stringr package is part of the core tidyverse and provides additional string functions beyond the ones provided by base R.
- You can prefix a pattern with (?i) to ignore case when searching.
- Each stringr function starts with str_ and accepts a string as its first argument. This makes it easy to identify stringr functions and to use piping with them.
- The stringr functions work on both single strings and vectors of strings.

Figure 7-2   How to use stringr functions to work with strings

How to work with string and numeric columns

The examples in figure 7-3 show how to use the mutate() function to create new columns from string and numeric columns. To start, the first example displays three columns of the fires data. Here, the FireName column stores strings and the Size and DaysBurning columns store numbers.

The second example shows how to use the mutate() function to work with the string columns. The code uses the str_to_title() function to convert the strings in the FireName column to title case. Then, it uses the str_c() function to create a new column named FullName that combines the fire name and year into a single descriptive string.

The third example shows how to use the mutate() function to work with numeric columns. Here, the code adds an AcresPerDay column and calculates its values by dividing the total number of acres burned by the number of days that the fire was burning. Then, it rounds the values to a single decimal place. As a result, this column represents the average number of acres that the fire burned per day.

Three columns in the fires data

```
fires %>% select(FireName, Size, DaysBurning)
# A tibble: 54,093 x 3
  FireName          Size DaysBurning
  <chr>            <dbl>       <int>
1 POWER            16823          16
2 FREDS             7700           5
3 AUSTIN CREEK       125           2
4 THOMPSON BUTTE     119           2
5 CHARLES DRAW       119           1
...
```

Modify the FireName column and add a column derived from it

```
fires <- fires %>% mutate(
  FireName = str_to_title(FireName),
  FullName = str_c("The ", FireName, " Fire (", Year, ")"))
```

Add a column using numeric calculations

```
fires <- fires %>% mutate(AcresPerDay = round(Size / DaysBurning, 1))
```

The new and modified columns

```
fires %>% select(FireName, FullName, AcresPerDay)
# A tibble: 54,093 x 3
  FireName       FullName                       AcresPerDay
  <chr>          <chr>                                <dbl>
1 Power          The Power Fire (2004)                1051.
2 Freds          The Freds Fire (2004)                 1540
3 Austin Creek   The Austin Creek Fire (2005)          62.5
4 Thompson Butte The Thompson Butte Fire (2005)        59.5
5 Charles Draw   The Charles Draw Fire (2005)           119
...
```

Description

- You can use any functions or operators that you would normally use with strings on string columns, such as the functions available from the stringr package.

- You can use any functions or operators that you would normally use with numeric variables on numeric columns.

Figure 7-3   How to work with string and numeric columns

How to group and bin data

When you analyze data, you often want to get summary statistics for different groups of data. For example, you may want to calculate summary statistics for each state in the fires data set. To do that, you can group the data and apply statistical functions to each group to summarize the data.

You may also want to create new groups by putting continuous data into groups known as *bins*, or *buckets*. This is known as *binning* data.

How to use statistical functions

The table in figure 7-4 shows some of the statistical functions available from R. These functions aren't a complete list, but they're a good place to start.

The first two functions are the mean() and median() functions. The mean() function calculates the average of all values, and the median() function returns the value that's the midpoint of all values so that half of the values are higher and half are lower.

The sd() function returns the *standard deviation* for a column or vector. This measures the amount of variation from the mean for the values in a column. A low standard deviation indicates that the values tend to be close to the mean. A high standard deviation indicates that the values are spread out over a wider range.

All of the functions presented in this figure accept at least two parameters. The first parameter is for the vector of values, such as a column from a data set. The second parameter, na.rm (NA remove), allows you to control how the function handles NA values. If a column doesn't contain any NA values, you don't need to set the second parameter. In that case, the function returns the result of the calculation as shown by the first example.

However, if you don't set the second parameter and a column contains any NA values, the statistical function returns an NA value. In the second example, the ContainDate column contains some NA values. As a result, the first statement, which does not set na.rm, returns an NA value. However, the second statement sets the rm.na parameter to TRUE. As a result, it removes all NA values before performing its calculation.

The third example shows how to use the sum() and is.na() functions to count the NA values for a column. Here, the first statement uses the is.na() function to return a vector of TRUE and FALSE values that indicate whether each value in a column is an NA value. Then, since TRUE evaluates to 1 and FALSE evaluates to 0, the sum() function calculates the count of NA values. The second statement works similarly, but it uses the NOT operator (!) to reverse the TRUE and FALSE values that are returned by the is.na() function.

Some statistical functions for working with vectors

Function	Returns
mean(x, na.rm)	Calculates the mean (average) value. If na.rm is set to TRUE, this function removes all NA values before calculating.
median(x, na.rm)	Median (middle) value.
sum(x, na.rm)	Sum of all values.
min(x, na.rm)	Minimum value.
max(x, na.rm)	Maximum value.
sd(x, na.rm)	Standard deviation.

When used with a column that doesn't contain any NA values

```
mean(fires$Size)
[1] 2435.754
median(fires$Size)
[1] 267
sum(fires$Size)
[1] 131757262
min(fires$Size)
[1] 100
max(fires$Size)
[1] 606945
sd(fires$Size)
[1] 14529.75
```

When used with a column that contains NA values

```
mean(fires$ContainDate)
[1] NA
mean(fires$ContainDate, na.rm = TRUE)
[1] "2005-01-11"
```

How to count the NA values for a column

```
sum(is.na(fires$ContainDate))    # count of NA values
[1] 20334
sum(!is.na(fires$ContainDate))   # count of other values
[1] 33759
```

Description

- R provides many functions that you can use to get statistics about a specified column.

- If the vector for a statistical function contains any NA values, the statistical function returns an NA value. To remove NA values from the vector before executing the function, set the na.rm parameter to TRUE.

- The *standard deviation* measures the amount of variation for the values in a column. A low standard deviation indicates that the values tend to be close to the mean. A high standard deviation indicates that the values are spread out over a wider range.

Figure 7-4 How to use statistical functions

How to summarize data

Figure 7-5 shows how to summarize, or aggregate, data. More specifically, it shows how to use the summarize() function to apply statistical functions to specified columns.

The first example uses summarize() with the mean() function on the Size column from the fires tibble. This returns a tibble that has one column named `mean(Size)` and one row that contains the mean fire size for the entire data set.

The second example uses summarize() to apply three statistical functions to the data set. It also specifies the name for each column in the resulting tibble.

The first column contains the mean size for all fires in the set, the second column contains the median size, and the third column contains a count of the number of fires. Note that the n() function doesn't require any arguments, but the mean() and median() functions require an argument that specifies the column name.

The fires data set doesn't have any groups so far. As a result, these examples apply the statistical functions to all rows in the data set. However, it's typically more useful to apply statistical functions to data that has groups as shown in the next figure.

A function for summarizing data

Function	Description
summarize(.data, colname = f(col), ...)	Applies the specified functions to each group in the data and returns a tibble that contains one column for each specified function.

A function that gets the count of rows

Function	Returns
n()	Count of rows in each group.

How to apply a function to a column with summarize()

```
fires %>% summarize(mean(Size))
# A tibble: 1 x 1
  `mean(Size)`
        <dbl>
1       2436.
```

How to apply multiple functions and name the resulting columns

```
fires %>% summarize(MeanSize = mean(Size),
                    MedianSize = median(Size),
                    Count = n())
# A tibble: 1 x 3
  MeanSize MedianSize Count
     <dbl>      <dbl> <int>
1    2436.        267 54093
```

Description

- If you don't provide a name for a column, summarize() generates one.
- The British spelling of summarise() works the same as summarize().

Figure 7-5   How to summarize data

How to group and summarize data

When you look for a variable to group your data by, you usually choose one that has a relatively small number of defined values. For the fires data set, you might want to group by the State column to compare fires in different states, or you might want to group by the Year or Month columns to compare fires in different years or different months of the year.

The first example in figure 7-6 shows how to group by a single column. To start, it pipes the fires data set into the group_by() function. Then, it sets the column to State. If you display a grouped tibble on the console, the data doesn't look different. However, there's a note at the top of the output that indicates how many groups the tibble contains.

The second example works like the first, but it groups by two columns, State and Year. When you group by more than one column, you create a number of groups equal to the number of unique values in the first column multiplied by the number of unique values in the second column. If there was a third column, you would multiply by that as well, and so on.

The third example combines the group_by() function with the summarize() function. Because the data set now contains groups, the statistical functions are applied separately to each group.

The tibble that's returned by the third example displays the mean size, median size, and count of the fires in each state. This provides some interesting insights about the differences between fires in various states. For example, the mean and median sizes of the fires in Alaska (AK) are much larger than for the other states. On the other hand, Alabama (AL) has many fires for its size but they tend to be small.

The fourth example shows how to combine group_by() and summarize() with multiple columns. This works much like the third example except that it groups the data by the State and Year columns. The resulting tibble provides a way to examine subgroups. In particular, it shows how to examine each year for each state. At this point, you could plot this data to analyze trends in the mean and median sizes of fires for each state.

If you assign grouped data to a variable, you may eventually want to ungroup the data. To do that, you can pass it to the ungroup() function as shown in the fifth example. This eliminates the groups but doesn't drop the columns that the data was grouped on.

Functions for working with the groups in a data set

Function	Description
group_by(.data, ...)	Groups the data by the specified column or columns.
ungroup()	Removes the groups from the data set.

How to group by a single column

```
fires %>% group_by(State)
# A tibble: 54,093 x 10
# Groups:   State [50]
...
```

How to group by multiple columns

```
fires %>% group_by(State, Year)
# A tibble: 54,093 x 10
# Groups:   State, Year [1,025]
...
```

How to group by a single column and summarize

```
fires %>% group_by(State) %>%
  summarize(MeanSize = mean(Size),
            MedianSize = median(Size),
            Count = n())
# A tibble: 50 x 4
   State MeanSize MedianSize Count
   <chr>    <dbl>      <dbl> <int>
 1 AK      17983.      1965  1790
 2 AL        246.       160  1292
...
```

How to group by multiple columns and summarize

```
fires %>% group_by(State, Year) %>%
  summarize(MeanSize = mean(Size),
            MedianSize = median(Size),
            Count = n())
# A tibble: 1,025 x 5
# Groups:   State [50]
   State  Year MeanSize MedianSize Count
   <chr> <int>    <dbl>      <dbl> <int>
 1 AK     1992   4413.        695    32
 2 AK     1993   7693.        970    89
...
```

How to ungroup

```
fires %>% ungroup()
```

Description

- The group_by() function can be used to get summary statistics for groups of data.
- You can group by one or more columns.

Figure 7-6   How to group and summarize data

Another way to group and summarize data

In addition to the technique shown in the previous figure, you can also group and summarize data by using the tapply() function to apply a function to a table or tibble. This function groups the data on the specified column before it applies the specified function as shown in figure 7-7.

The first example shows how to use the tapply() function to apply the mean() function to the fires tibble. Here, the code pipes the DaysBurning column in as the X argument. Then, it uses the INDEX argument to specify the column that should be used to group the data (State), and it uses the FUN parameter to specify the function that summarizes the data (mean). As a result, this example applies the mean() function to the DaysBurning column of the fires data set, and it groups the data set by the State column.

You can code any additional arguments for the function that you're applying by adding them after the FUN argument. In this example, the na.rm argument is passed for the mean() function after the FUN argument. Since this code sets the na.rm parameter to TRUE, the mean() function drops NA values before calculating the mean of the data. That's necessary here because the DaysBurning column contains NA values.

The output for this example shows that it calculates the mean number of days burning for each state. In addition, it shows that the results are returned as a named vector.

The second example shows how you can use the group_by() and summarize() functions to get the same data as the first example. This approach makes it clearer that the na.rm parameter applies to the mean() function. In addition, it returns a tibble, not a named vector.

A function for applying a function to a group

Function	Description
`tapply(X, INDEX, FUN)`	Applies the specified function (FUN) to the specified column (X) after grouping the data set by the specified column or columns (INDEX). Returns the resulting data as a named vector that uses the index columns for the names of each value.

Code that uses tapply() with mean()

```
fires$DaysBurning %>%
  tapply(INDEX = fires$State, FUN = mean, na.rm = TRUE)
       AK        AL        AR        AZ        CA        CO        CT        DE
43.682598  2.902724  2.983553 10.569061 11.463653 14.248908  1.333333  1.600000
       FL        GA        HI        IA        ID        IL        IN        KS
 6.572222  5.212329 42.156250  1.106796 18.765927  1.461538  2.428571  2.640732
...
```

Code that uses group_by() and summarize() to get the same data

```
fires %>% group_by(State) %>%
          summarize(MeanDays = mean(DaysBurning, na.rm = TRUE))
# A tibble: 50 x 2
   State MeanDays
   <chr>    <dbl>
 1 AK        43.7
 2 AL        2.90
 3 AR        2.98
 4 AZ        10.6
 5 CA        11.5
 6 CO        14.2
 7 CT        1.33
 8 DE        1.6
 9 FL        6.57
10 GA        5.21
# ... with 40 more rows
```

Description

- The tapply() function can be used to apply a function to a column in a table, tibble, or similar data object (remember t for table), grouped by values in another column.

- You can pass additional arguments to the function that you are applying by passing them after the FUN argument.

Figure 7-7   Another way to group and summarize data

How to rank rows

For some data sets, you may want to add a column that ranks the rows in the data set. To do that, you can use the functions summarized in figure 7-8. If you want to rank a column where there are unlikely to be any ties between rows, you can use the rank() function with its default values. However, if there are likely to be ties between rows, you need to determine how you want to handle the ties.

By default, the rank() function sets the ranking for the tied rows to the average ranking for the rows. For instance, the example in this figure shows that the states for rows 8, 9, and 10 have the same number of mean days burning. As a result, the rank() function assigns the average rank of 9 to rows 8, 9, and 10. Then, it assigns a rank of 11 to row 11.

If you don't want to use this method of resolving ties, you can set the ties.method parameter. For instance, the RankMin column sets the ties.method parameter to a value of "min". This assigns the minimum rank of 8 to rows 8, 9, and 10 and a rank of 11 to row 11.

Since setting the ties.method parameter to "first" is common, the dplyr package provides the row_number() function to make it easier to do this. This function resolves ties by letting the first row in the data set win the tie. As a result, when the data set is sorted by rank, it sets the rank equal to the row number.

Another way to resolve ties is to use the dense_rank() function. This function assigns the same ranking value for ties. However, it keeps the ranks consecutive. In other words, it doesn't skip any ranks after a tie. For instance, the example in this figure shows that the dense_rank() function assigns rows 8, 9, and 10 a rank of 8. Then, it assigns a rank of 9 to row 11.

By default, these functions rank the rows in ascending order from smallest to largest. However, you can use the desc() function to rank rows in descending order from largest to smallest. In this figure, for instance, all of the ranking columns use the desc() function to assign the highest rankings to the states with the largest number of mean days burning.

Some functions for ranking rows

Function	Description
rank(x, ties.method)	Returns a vector that ranks the specified vector. By default, if there is a tie, it sets the ranking for the tied rows to the average ranking for those rows. To change this, you can set the ties.method parameter to "first", "last", "random", "min", or "max".
row_number(x)	Works the same as the rank() method with its ties.method parameter set to "first".
dense_rank(x)	Returns a vector that ranks the specified vector. If there is a tie, it sets the ranking for all tied rows to the same rank and sets the next row to the next rank.

How to add ranking columns

```
fires %>% group_by(State) %>%
  summarize(MeanDays = mean(DaysBurning, na.rm = TRUE)) %>%
  mutate(MeanDays = as.integer(MeanDays),
         Rank = rank(desc(MeanDays)),
         RankMin = rank(desc(MeanDays), ties.method = "min"),
         RowNumber = row_number(desc(MeanDays)),
         DenseRank = dense_rank(desc(MeanDays))) %>%
  arrange(Rank) %>%
  head(11)
# A tibble: 11 x 6
   State MeanDays  Rank RankMin RowNumber DenseRank
   <chr>    <int> <dbl>   <int>     <int>     <int>
 1 AK          43     1       1         1         1
 2 HI          42     2       2         2         2
 3 WA          20     3       3         3         3
 4 ID          18     4       4         4         4
 5 MT          17     5       5         5         5
 6 NJ          16     6       6         6         6
 7 PA          15     7       7         7         7
 8 CO          14     9       8         8         8
 9 OR          14     9       8         9         8
10 WY          14     9       8        10         8
11 NM          12    11      11        11         9
```

Description

- The rank() function is available from base R, but the row_number() and dense_rank() functions are available from the dplyr package that's part of the core tidyverse.

- By default, the ranking functions rank the rows in ascending order from smallest to largest, but you can use the desc() function to rank rows in descending order from largest to smallest.

Figure 7-8   How to rank rows

How to add a cumulative sum

Figure 7-9 shows how to use the cumsum() function to add a *cumulative sum*, also known as a *running total*, that displays the total sum for a column as it grows with each row. To start, the first example shows the data for a tibble named fire_sizes. This tibble contains summary statistics about the size of the fires in each state for the years from 1992 to 1994.

The second example shows how to add a cumulative sum for all rows in the tibble. To do that, it uses the cumsum() function with the mutate() function to add a column named CumulativeSize that contains the cumulative sum for the Size column. The resulting tibble shows that the cumulative sum for the first row is the same as the first row, the second row is the first two rows added together, the third row is the first three rows added together, and so on. As a result, the cumulative sum for the last row in the result set should be the total acres burned for all rows in the data set.

The third example shows how to add a cumulative sum for the rows in each group in the tibble. This works like the second example, except that the group_by() function groups the tibble by state. As a result, the cumulative sum starts over with each new group. That's why the cumulative total starts over when the state switches from AK to AL.

A function for calculating the cumulative sum for a column

Function	Description
cumsum(x)	Returns a vector that contains the values for the cumulative sum of the specified vector.

The fire sizes summary data

```
fire_sizes
# A tibble: 119 x 3
  State  Year     Size
  <chr>  <int>    <dbl>
1 AK     1992  141207
2 AK     1993  684670.
3 AK     1994  260102.
4 AL     1992     611
5 AL     1993     785
...
```

How to add a cumulative sum for all rows

```
fire_sizes %>% mutate(CumulativeSize = cumsum(Size))
# A tibble: 119 x 4
  State  Year     Size CumulativeSize
  <chr>  <int>    <dbl>        <dbl>
1 AK     1992  141207       141207
2 AK     1993  684670.      825877.
3 AK     1994  260102.     1085978.
4 AL     1992     611      1086589.
5 AL     1993     785      1087374.
...
```

How to add a cumulative sum for each group

```
fire_sizes %>% group_by(State) %>%
  mutate(CumulativeSize = cumsum(Size))
# A tibble: 119 x 4
# Groups:   State [45]
  State  Year     Size CumulativeSize
  <chr>  <int>    <dbl>        <dbl>
1 AK     1992  141207       141207
2 AK     1993  684670.      825877.
3 AK     1994  260102.     1085978.
4 AL     1992     611          611
5 AL     1993     785         1396
...
```

Description

- A *cumulative sum*, also known as a *running total*, displays the total sum for a column as it grows with each row.
- The cumsum() function is available from base R.

Figure 7-9   How to add a cumulative sum

How to bin data

With R, you can use the cut() and the ntile() functions to put continuous data into *bins*, also known as *buckets*. A bin provides a way to group continuous data. Generally, each bin contains values between two defining values. In the first example in this figure, the Month value is converted to an integer. Then, the first bin contains values greater than zero and less than or equal to three, so 1, 2, or 3. This bin is then labeled "Q1". The second bin, Q2, contains values greater than three and less than or equal to six, so 4, 5, or 6. And so on for all four bins.

The cut() and ntile() functions are similar, but each function creates its bins in a different way. The cut() function creates bins based on values that you specify, while the ntile() function creates bins with approximately equal numbers of values based on the number of bins that you specify. Regardless of the function, binning data can help you summarize and plot the data.

The first example in figure 7-10 shows how to use the cut() function to create bins for the Month column. Here, the first statement passes the Month column to the cut() function, a vector of five break values (0, 3, 6, 9, 12), and a vector of four labels. This creates four bins for the data (1-3, 4-6, 7-9, and 10-12) and uses the corresponding labels (Q1, Q2, Q3, and Q4) as the values for the column. Since the break values specify the edges of the bins, break values are also sometimes referred to as *bin edges*.

The mutate() function specifies a name of Quarter for this column and adds it to the fires tibble. The second statement then selects the Month and Quarter columns to show the results of the binning.

The second example shows how to use the ntile() function to bin data. Instead of passing bin edges and labels to the function, ntile() only needs the column being binned and the number of bins to create. Then, it generates the bins on its own.

When the ntile() function generates the bins, it tries to keep the same number of values in each bin, and it labels each bin from 1 to the number of bins specified. For instance, the ntile() function in the second example bins the fires data by creating five bins of equal size. The smallest fifth of fires are assigned to bin 1 and the largest fifth of fires are assigned to bin 5, with these values recorded in the column SizeGroup. Then, the second statement in the example selects the two columns to show the results of the binning.

The third statement converts the SizeGroup column to a table to show the number of rows for each bin. In this case, the resulting table shows that the first three bins have one more row than the last two bins. That's because it isn't always possible to divide the rows into bins of equal sizes. When that happens, the ntile() function divides them as evenly as possible, which is usually what you want.

Some functions for binning data

Function	Description
cut(x, breaks, labels)	Puts the data into bins based on the values specified by the breaks argument. The labels argument can be used to specify a label for each bin. Returns a vector that stores factor objects.
ntile(x, n)	Puts the data into the specified number of bins where each bin has roughly the same number of rows. Returns a vector of integer values.

How to create bins by specifying values

```
# Create a new column named Quarter and use it to label each row with a bin
fires <- fires %>% mutate(Quarter = cut(as.integer(Month),
                                        breaks = c(0,3,6,9,12),
                                        labels = c("Q1","Q2","Q3","Q4")))

# Display the Month column and newly created Quarter column
fires %>% select(Month, Quarter)
# A tibble: 54,093 x 2
   Month Quarter
   <fct> <fct>
 1 10    Q4
 2 10    Q4
 3 02    Q1
 4 07    Q3
...
```

How to create bins with equal numbers of values

```
# Create a new column named SizeGroup and use it to assign a bin to each row
fires <- fires %>% mutate(SizeGroup = ntile(Size, n = 5))

# Display the Size column and newly created SizeGroup column
fires %>% select(Size, SizeGroup)
# A tibble: 54,093 x 2
    Size SizeGroup
   <dbl>     <int>
 1 16823         5
 2  7700         5
 3   125         1
 4   119         1
...

# The number of values in each bin
table(fires$SizeGroup)

    1     2     3     4     5
10819 10819 10819 10818 10818
```

Description

- You can use the cut() and ntile() functions to put continuous data into categories known as *bins*, or *buckets*.

Figure 7-10   How to bin data

How to apply functions and lambda expressions

Sometimes, you want to use a specific function to add or modify a column, but the function doesn't exist or you don't know where to find it. In that case, you can define your own function.

How to define functions that operate on rows

Creating a function intended specifically to operate on a row of your data set works much like creating a regular function. However, when you pass a row, you use double-bracket notation ([[ ]]) to access the values in each column by specifying the column name or index.

Figure 7-11 starts by defining a function that operates on a row of the fires tibble. This function replaces the NA values in the ContainDate column with the values of the DiscoveryDate column. However, if a valid date already exists in the ContainDate column, this code keeps that date. To do that, this example creates a function named get_contain_date() that accepts a row from the fires data set.

Within its braces, this custom function contains an if/else statement that checks whether the ContainDate column stores an NA value. To do that, it uses the column name in double brackets and quotes. If the ContainDate column stores an NA value, the function returns the date that's stored by the DiscoveryDate column. Otherwise, it returns the value in the ContainDate column.

The second example tests our new function with the first row of the fires data set. Here, the code uses the head() function to get the first row and pipe it into the get_contain_date() function.

The third example shows how to apply the get_contain_date() function to all rows in the fires data set. To do that, the code sets the X parameter of the apply() function to the fires data set, the MARGIN parameter to 1 so the function is applied to rows (not columns), and the FUN parameter to the get_contain_date() function. Running this statement with mutate() modifies the ContainDate column by converting all NA values to valid dates.

Unfortunately, running the code in the third example changes the data type for the ContainDate column to the character type. To fix this, you can use the mutate() function with the as.Date() function as shown in the fourth example.

Finally, the fifth example shows the first few rows of the DiscoveryDate and ContainDate columns after the get_contain_date() and as.Date() functions have been applied. This shows that both columns are stored with the Date type and the ContainDate column no longer has NA values.

The example in this figure is a simple example that's designed to show how to define a function that operates on multiple columns in a row. However, the fires data set contains some large fires that have NA values in the ContainDate

A function for applying functions to rows or columns

Function	Description
`apply(X, MARGIN, FUN)`	Applies the specified function (FUN) to the specified data object (X). If MARGIN is set to 2, it applies the specified function to each column. If MARGIN is set to 1, it applies the specified function to each row. Returns a list of columns with the value returned by the function.

How to define a function that operates on a row

```
get_contain_date <- function(row) {          # defines a row as its parameter
  if (is.na(row[["ContainDate"]])) {          # double bracket syntax for columns
    return(row[["DiscoveryDate"]])            # uses return() to return values
  } else {
    return(row[["ContainDate"]])
  }
}
```

How to test the function on a single row

```
fires %>% head(1) %>% get_contain_date()     # uses head() to get first row
[1] "2004-10-21"
```

How to apply the function to all rows

```
fires <- fires %>% mutate(                              # uses apply() to apply to all rows
  ContainDate = apply(X = fires, MARGIN = 1, FUN = get_contain_date))
```

How to convert the column back to the Date type

```
fires <- fires %>% mutate(ContainDate = as.Date(ContainDate))
```

How to view the date columns

```
fires %>% select(DiscoveryDate, ContainDate)
# A tibble: 238,955 x 2
   DiscoveryDate ContainDate
   <date>        <date>
 1 2004-10-06    2004-10-21
 2 2004-10-13    2004-10-17
 3 2005-01-27    2005-01-28
 4 2005-02-12    2005-02-13
 5 2005-04-16    2005-04-16
...
```

Description

- To define a function that operates on a row, define a parameter for the row. You can give this parameter any name you want.
- To access columns within a row by name, use the double bracket syntax.
- If necessary, you can fix the data type for a column after applying a function that changes the data type for a column.

Figure 7-11   How to define functions that operate on rows

column. For those fires, it doesn't make sense to replace the NA value in the ContainDate column with the value in the DiscoveryDate column. As a result, you would probably want to keep the NA value or calculate a more realistic ContainDate value based on the size of the fire.

How to define functions that operate on columns

Figure 7-12 starts by defining a function designed to operate on a column. Then, it shows how to apply that function to all columns in the fires data set. These skills work much like the skills presented in the previous figure.

The first example defines a function named count_na() that accepts a column as a parameter. To make this clear, the parameter is named col, but you could use any name for the parameter. When you pass a column, it is passed as a vector containing all the values in that column.

Within the braces for this function, the first statement uses the is.na() and sum() functions get a count of the NA values for the column. Then, it assigns the result to a variable named count. The second statement uses the return() function to return the count value.

The second example shows how to test this count_na() function on a single column. To do that, you can use the $ operator to access and pass a single column from a data set. Running the function shows that the FireName column of the fires data set contains over 122,000 NA values.

The third example shows how to apply this count_na() function to all columns in the fires data set. To do that, this code sets the X parameter to the fires data set, the MARGIN parameter to 2 so it applies the function to columns (not rows), and the FUN parameter to the count_na() function.

The table that's returned by the third example shows the count of NA values for each column. It turns out most columns in the fires data set don't have any NA values, but a few columns have a large number of NA values. Note that the ContainDate column doesn't have any NA values. That's because the previous figure applied the custom get_contain_date() function to all rows to remove these NA values from the data set.

How to define a function that operates on a column

```
count_na <- function(col) {
  count <- sum(is.na(col))
  return(count)
}
```

How to test the function on a single column

```
count_na(fires$FireName)
[1] 122870
```

How to apply the function to all columns in a data set

```
apply(X = fires, MARGIN = 2, FUN = count_na)
    FireName         Year        State      Size DiscoveryDate  ContainDate
      122870            0            0         0             0            0
       Month  DaysBurning     FullName AcresPerDay       Quarter    SizeGroup
           0       114856            0      114856             0            0
     Outlier
           0
```

Description

- To define a function that operates on a column, define a parameter for the column.
- Columns are passed to functions as vectors.

Figure 7-12   How to define functions that operate on columns

How to use lambda expressions instead of functions

If your custom function is short and only used once, it may make sense to use a *lambda expression,* or *lambda*, instead. That's because a lambda provides a more concise way to code a function.

The syntax for a lambda is similar to the syntax of a regular function as shown by the top of figure 7-13. However, lambdas typically contain only a single statement and don't use braces.

The first example shows a lambda expression that provides the same functionality as the count_na() function presented in the previous figure. Here, the lambda defines a parameter named col. Then, it uses the sum() and is.na() functions to count the number of NA values in the column.

The second example shows how to apply this lamba expression to all columns. To do that, you code this lambda expression for the FUN parameter of the apply() function.

The third example shows a lambda expression that provides the same functionality as the get_contain_date() function presented in figure 7-11. Here, the lambda defines a parameter named row. Then, it uses the ifelse() function to check if the ContainDate column contains an NA value. If so, the ifelse() function returns the value for the DiscoveryDate column. Otherwise, it returns the value for the ContainDate column.

The fourth example shows how to apply this lamba expression to all rows. To do that, you code this lambda expression for the FUN parameter of the apply() function.

While lambda expressions are shorter to code, they do have several disadvantages. First, they are harder to read and debug. Second, the only way to reuse them is to copy them into multiple places in your scripts. This violates the *DRY (Don't Repeat Yourself) principle* of coding that states that you should avoid using the same chunk of code in multiple places. That's because copying the same code in multiple places makes your code harder to maintain since any changes to the code need to be applied to multiple places.

As a result, if you think you may need to call a function multiple times, you should code it as a normal custom function. However, if a custom function is short and you only need to call it once, it may be a good candidate for a lambda expression.

The syntax of a lambda expression

```
function(parameters) statement
```

A lambda expression that operates on a column

```
function(col) sum(is.na(col))
```

How to apply the lambda expression to all columns

```
apply(X = fires, MARGIN = 2, FUN = function(col) sum(is.na(col)))
```

A lambda expression that operates on a row

```
function(row)
    ifelse(is.na(row[["ContainDate"]]),
           row[["DiscoveryDate"]],
           row[["ContainDate"]])))
```

How to apply a lambda expression to rows

```
fires <- fires %>% mutate(
  ContainDate = apply(X = fires, MARGIN = 1, FUN = function(row)
    ifelse(is.na(row[["ContainDate"]]),
           row[["DiscoveryDate"]],
           row[["ContainDate"]])))
```

Description

- A *lambda expression* provides a concise way to code a function.
- If a function is very short and will only be used once, it often makes sense to use a lambda expression instead.
- Lambda expressions often use the ifelse() function instead of if/else statements.

Figure 7-13   How to use lambda expressions instead of functions

How to reshape tibbles

When you prepare your data, you may want to combine two or more tibbles into a single tibble.

How to add columns by joining tibbles

Figure 7-14 begins by presenting the functions that you can use to *join* two tibbles based on one or more columns. These functions provide a way to add columns to a tibble even when one tibble has more rows than another tibble. If you're familiar with using SQL statements to code joins, you'll be glad to know that R provides functions for joining data that are based on the joins provided by SQL. As a result, they use the same terminology and concepts.

To start, an *inner join* only returns rows where the specified column or columns match in both tables. A *left join* returns all rows in the tibble that's coded on the left but only the rows that match for the tibble on the right. A *right join* is similar, but it returns all rows from the tibble that's coded on the right and only the rows that match for the tibble on the left. And an *outer join*, also known as a *full join*, returns all rows from both tibbles, matching them when possible.

Since an inner join is the most common type of join, part 1 of this figure shows how to code one. To start, the first two examples present the two tibbles to join. Here, the fires tibble has more than 238K rows. Meanwhile, the state_names tibble has only 53 rows.

Both of these tibbles have a column named State that contains a two-letter abbreviation that uniquely identifies each state. Since both of these columns have the same name, this column can be used as the *index* for this join. This index can be a single column as shown in this figure. Or, if you need to use multiple columns to uniquely identify each row, you can use a vector to specify the columns to use for the index.

The third example shows how to use an inner join to join the tibbles in the first two examples. To do that, the code calls the inner_join() function and passes the two tibbles to it as the first two arguments. Then, the third argument specifies that the data should be joined by the State column.

For rows with matching indexes, all non-index columns from the second tibble are added to the first tibble. In this example, there's only one non-index column (StateName) in the second tibble. As a result, this example only adds that column to the first tibble.

The tibble that's returned by the inner_join() function shows that the StateName column has been joined to the columns of the fires tibble. That's why all of the rows for CA have a state name of California, and all of the rows for NC have a state name of North Carolina, and so on. Since the state_names tibble contains a name for every state abbreviation, this adds a state name to every row in the fires tibble and doesn't add any extra rows, which is usually what you want when joining tibbles.

Functions for joining tibbles

Function	Description
inner_join()	Uses an *inner join* to only returns rows where the specified column or columns match in both tibbles.
left_join()	Uses a *left join* to return all rows in the tibble on the left but only the rows that match for the tibble on the right.
right_join()	Uses a *right join* to returns all rows in the tibble on the right but only the rows that match for the tibble on the left.
full_join()	Uses an *outer join*, also known as a *full join*, to return all rows from both tibbles, matching them when possible.

Parameter	Description
x	The tibble that's on the left.
y	The tibble that's on the right.
by	One or more columns to join the tibbles by.

The fires tibble

```
fires
# A tibble: 238,955 x 6
   FireName        Year State    Size DiscoveryDate ContainDate
   <chr>          <int> <chr>   <dbl> <date>        <date>
 1 POWER           2004 CA      16823 2004-10-06    2004-10-21
 2 FREDS           2004 CA       7700 2004-10-13    2004-10-17
 3 HOWARD GAP      2005 NC       50.3 2005-01-27    2005-01-28
 4 AUSTIN CREEK    2005 NC        125 2005-02-12    2005-02-13
...
```

The state_names tibble

```
state_names
# A tibble: 53 x 2
   StateName      State
   <chr>          <chr>
 1 Alabama        AL
 2 Alaska         AK
 3 Arizona        AZ
...
```

How to code an inner join

```
inner_join(fires, state_names, by = "State")
# A tibble: 238,955 x 7
   FireName        Year State    Size DiscoveryDate ContainDate   StateName
   <chr>          <int> <chr>   <dbl> <date>        <date>        <chr>
 1 POWER           2004 CA      16823 2004-10-06    2004-10-21    California
 2 FREDS           2004 CA       7700 2004-10-13    2004-10-17    California
 3 HOWARD GAP      2005 NC       50.3 2005-01-27    2005-01-28    North Carolina
 4 AUSTIN CREEK    2005 NC        125 2005-02-12    2005-02-13    North Carolina
...
```

Figure 7-14 How to add columns by joining tibbles (part 1)

Although you typically use an inner join to join tibbles, you may sometimes need to use one of the other join types. That's why part 2 of figure 7-14 shows how all four types of joins work. To start, the first two examples define two tibbles that only have two rows each.

The third example shows how to code an inner join that joins the two tibbles by their State column. Since these two tibbles only contain one row that matches, this returns a single row.

The fourth example shows how to code a left join. Here, the fires_min tibble is coded to the left of the names_min tibble. As a result, this statement returns all rows from the fires_min tibble and only returns rows from the names_min tibble that match. This returns the same matching row returned by the inner join. However, since the states_min table doesn't have a matching row for NC, it sets the StateName column for the second row in the fires_min tibble to NA.

The fifth example shows how to code a right join. Here, the names_min tibble is coded on the right. As a result, this returns all rows from the names_min tibble. Since the fires_min tibble doesn't have any matching rows for AK, this sets the first three columns for AK to NA values.

The sixth example shows how to code a full join. This returns three rows where the first row contains data from both tables, the second row contains data from the left table, and the third row contains data from the right table. In other words, it keeps all rows from both tables, even if the index doesn't match.

The state_names_min tibble

```
state_names_min
# A tibble: 2 x 2
  StateName  State
  <chr>      <chr>
1 Alaska     AK
2 California CA
```

The fires_min tibble

```
fires_min
# A tibble: 2 x 4
  FireName    Year     Size State
  <chr>       <int>   <dbl> <chr>
1 POWER        2004   16823 CA
2 HOWARD GAP   2005    50.3 NC
```

An inner join with the new tibbles

```
inner_join(fires_min, state_names_min, by = "State")
# A tibble: 1 x 5
  FireName  Year  Size State StateName
  <chr>    <int> <dbl> <chr> <chr>
1 POWER     2004 16823 CA    California
```

How to code a left join

```
left_join(fires_min, state_names_min, by = "State")
# A tibble: 2 x 5
  FireName    Year     Size State StateName
  <chr>       <int>   <dbl> <chr> <chr>
1 POWER        2004   16823 CA    California
2 HOWARD GAP   2005    50.3 NC    NA
```

How to code a right join

```
right_join(fires_min, state_names_min, by = "State")
# A tibble: 2 x 5
  FireName  Year  Size State StateName
  <chr>    <int> <dbl> <chr> <chr>
1 POWER     2004 16823 CA    California
2 NA          NA    NA AK    Alaska
```

How to code a full join

```
full_join(fires_min, state_names_min, by = "State")
# A tibble: 3 x 5
  FireName    Year     Size State StateName
  <chr>       <int>   <dbl> <chr> <chr>
1 POWER        2004   16823 CA    California
2 HOWARD GAP   2005    50.3 NC    NA
3 NA             NA      NA AK    Alaska
```

Figure 7-14   How to add columns by joining tibbles (part 2)

How to add rows

Figure 7-15 presents the rbind() function, which you can use to append rows from one tibble to another tibble. The first two examples show the data for the fires data set and a new tibble named new_fires. This fires tibble contains six columns and over 238K rows, and the new_fires tibble only contains a single row.

Both tibbles have the exact same number of columns and these columns are named identically. That's important because the rbind() function only works if both tibbles have identical columns. If they don't, the rbind() function displays an error message and doesn't let you add the rows.

You can use the identical() function to check whether both tibbles have the same columns before calling the rbind() function. This is shown in the third example, with the result that both tibbles have the same column names.

The fourth example begins by calling the rbind() function to add the row in the new_fires tibble to the end of the fires tibble. Then, it shows that the new row has been added by using the tail() function to display the last six rows of the fires tibble.

Although both tibbles must have the same column names, the columns don't have to be of the same type. If the columns are of different types, the rbind() function coerces the data types to be the same. As a result, after using rbind() to add rows, you may need to convert some columns back to their correct types.

Functions for binding rows to the end of a tibble

Function	Description
`rbind(data, ...)`	Adds the rows in one or more data objects to the end of a data object. The objects must have the same number of columns and the same names for those columns.
`identical(x, y)`	Checks if two vectors store the same values.

The fires data

```
fires
# A tibble: 238,955 x 6
    FireName        Year State   Size DiscoveryDate ContainDate
    <chr>          <int> <chr>  <dbl> <date>        <date>
  1 POWER           2004 CA     16823 2004-10-06    2004-10-21
  2 FREDS           2004 CA      7700 2004-10-13    2004-10-17
  3 HOWARD GAP      2005 NC      50.3 2005-01-27    2005-01-28
...
```

The new_fires data

```
new_fires
# A tibble: 1 x 6
  FireName  Year State  Size DiscoveryDate ContainDate
  <chr>    <dbl> <chr> <dbl> <chr>         <chr>
1 MURACH    2022 CA       1 2022-01-01    2022-01-01
```

How to compare the tibble column names

```
identical(names(fires), names(new_fires))
[1] TRUE
```

How to add the new row to the fires data

```
fires <- rbind(fires, new_fires)
tail(fires)
# A tibble: 6 x 6
    FireName         Year State   Size DiscoveryDate ContainDate
    <chr>           <dbl> <chr>  <dbl> <date>        <date>
  1 SODA             2015 CA        24 2015-07-04    2015-07-04
  2 GILMAN           2009 CA        73 2009-05-03    2009-05-03
  3 TERRA            2009 CA        23 2009-06-19    2009-06-19
  4 SLAUGHTERHOUSE   2010 CA        50 2010-05-29    NA
  5 NA               2015 CA        11 2015-12-07    NA
  6 MURACH           2022 CA         1 2022-01-01    2022-01-01
```

Description

- While the column names must be identical to use rbind(), the columns don't need to contain the same data types. R will coerce the types in the columns to be the same.
- You may need to convert the column data types back to their original types after adding rows.

Figure 7-15   How to add rows

Perspective

This chapter presented some essential skills for preparing data for analysis. However, it's common to continue to prepare the data as you analyze it. That's because, in many cases, you don't know exactly how to prepare the data until you start trying to analyze it. Similarly, as you prepare your data, you may find that some of it is dirty and needs to be cleaned. This just shows that there's almost always overlap in the phases of a data analysis.

Summary

- The *standard deviation* measures the amount of variation for the values in a column.

- A *cumulative sum*, also known as a *running total*, displays the total sum for a column as it grows with each row.

- You can create categories by putting continuous data into groups known as *bins*, or *buckets*. This is known as *binning* data.

- The values that specify the edges of the bins are sometimes referred to as *bin edges*.

- A *lambda expression,* or *lambda*, provides a concise way to code a short function.

- The *DRY (Don't Repeat Yourself) principle* of coding states that you should avoid using the same chunk of code in multiple places because it makes your code harder to maintain.

- To add columns to a tibble, you can *join* two tibbles by one or more columns that are known as the *index* for the join.

- An *inner join* returns rows where the specified column or columns match in both tibbles.

- A *left join* returns all rows in the tibble on the left but only the rows that match for the tibble on the right.

- A *right join* returns all rows in the tibble on the right but only the rows that match for the tibble on the left.

- An *outer join*, also known as a *full join*, returns all rows from both tibbles, matching them when possible.

Exercise 7-1   Adding and mutating columns

In this exercise, you'll create two tibbles from the Lake Huron water levels and Nile annual flow data sets, join them, and mutate the columns.

Create the tibbles

1. Create a new R script.

2. Load the tidyverse and datasets packages.

3. View the data for the data sets named LakeHuron and Nile. These are both time series and viewing them shows the years they cover.

    The Lake Huron series contains the recorded water level for Lake Huron in feet.

    The Nile series contains the annual flow of the river Nile in 10^8 cubic meters.

4. Convert both data sets to tibbles.

5. Add a column named Year to both tibbles that stores the year. To do that, you can use the seq() function to generate the values for the column.

6. Rearrange the columns in each tibble so that the Year column is first. To do that, you can use the select() function.

7. Rename the column named x in the Lake Huron tibble to LHWaterLevel.

8. Rename the column named x in the Nile tibble to NAnnualFlow.

9. Add a column to the Lake Huron tibble that calculates the water level for each year in meters. To do that, you need to convert feet to meters.

10. Add a column to the Nile tibble that calculates the annual flow in 10^8 gallons. To do that, you need to convert cubic meters to gallons.

11. View the median, minimum, maximum, and standard deviation for the Lake Huron water levels. To do that, use the median(), min(), max(), and sd() functions.

Join the tibbles

12. Use an inner join to join the tibbles with by the Year column. Then, view the first and last six rows. Were all of the rows included in the resulting tibble?

13. Use a full join to join the tibbles. Then, view the first and last six rows of the resulting tibble. Does it include all of the rows? Are any in an unexpected position?

14. Add a column to the new tibble named Decade that stores the decade for the row. To calculate the decade, you can use integer division like this:

    ```
    (Year %/% 10) * 10
    ```

15. Group the tibble by decade.

16. Use summarize() to view the average water level and annual flow for each decade plus a count for how many values are in each group. Are any results missing?

17. Use summarize() to view the same summary statistics as the previous step, but make sure to use the na.rm parameter to remove NA values from the mean() functions.

18. Ungroup the data.

19. Use tapply() to get the same calculated averages as summarize() in the previous steps.

Create and use functions

20. Create a function with the following signature.

```
lh_high_or_low <- function(row, high, low)
```

21. Within this function, use an if-else statement to check the value for the LHWaterLevel column in the row. If it is NA, return NA. If it is higher than the high parameter, return "High". If it is lower than the low parameter, return "Low". Otherwise, return "Medium".

22. Calculate the average water level for Lake Huron plus the standard deviation for the water level and store the result in a variable named lh_high.

23. Calculate the average water level minus the standard deviation and store the result in a variable named lh_low.

24. Use the variables from the previous two steps to test the function on the first row of the Lake Huron tibble. It should return "High."

25. Apply the function to all rows in the Lake Huron tibble to add a column named Description that indicates whether the water level is high, low, or medium. To pass arguments to the function, you can add them after the FUN argument like this:

```
FUN = lh_high_or_low, high = lh_high, low = lh_low
```

Create a bin column and a rank column

26. Add a column named DescriptionBin to the Lake Huron tibble. This column should use the cut() function to store the same water level descriptions as the Description column. To do that, you can use the lh_low and lh_high variables as two of the bin edges, and you can specify literal values for the other bin edges.

27. Display the Lake Huron tibble and note that the Description column uses the character type but the DescriptionBin column uses the factor type.

28. Add a column named DenseRank to the Niles tibble. This column should use the dense_rank() function to rank the NAnnualFlow column in descending order.

29. Sort the Niles tibble by the DenseRank column. This should display the years with the largest annual flow first. Note how the DenseRank column handles ties.

Chapter 8

More skills
for data visualization

In chapter 4, you learned how to create most basic plot types with ggplot2. However, data visualization is a critical part of data analysis, and there's still much more to learn about it. That's why this chapter presents more skills for working with data visualization. These skills include working with shapes, plotting geographic data on maps, fine-tuning a plot, and working with a grid of plots.

More skills for working with plots

This chapter starts by presenting some more skills for working with basic plot types such as scatter plots, line plots, and bar plots. But first, it shows how to get the data that's used by the plots presented in this chapter.

Get the data

Figure 8-1 starts by showing how to load three packages that this chapter uses. By now, you should be familiar with the first two packages. However, this chapter also uses the ggforce package, which contains functions such as geom_circle(), geom_arc(), and facet_matrix().

This figure continues by showing how to get the data for this chapter. This data includes the tibbles for the irises, chicks, mortality, and California cities data. Here, the code that gets the chicks data fixes the capitalization of the Weight column and also fixes the order of the numbers in the Chick column. This makes the capitalization for the columns consistent, and it fixes the order of the levels in the Chick column, which is of the ordered factor type.

By now, you should be familiar with all of these tibbles except the California cities data. This data set includes the city name, rank by population, population, latitude, and longitude for 1,252 cities in California. While the first three data sets use title case for the column names, the California cities data set uses lowercase.

The packages for this chapter

```
library("tidyverse")    # the core tidyverse packages (tibble, ggplot2, etc.)
library("datasets")     # the sample datasets (iris, ChickWeight, etc.)
library("ggforce")      # needed for geom_circle(), geom_arc(), and facet_matrix()
```

The irises data

```
irises <- as_tibble(iris)
irises
A tibble: 150 x 5
# Groups:    Species [3]
   Sepal.Length Sepal.Width Petal.Length Petal.Width Species
          <dbl>       <dbl>        <dbl>       <dbl> <fct>
1           5.1         3.5          1.4         0.2 setosa
2           4.9         3            1.4         0.2 setosa
...
```

The chicks data

```
chicks = as_tibble(ChickWeight)
chicks <- chicks %>%
  rename(Weight = weight) %>%          # fix capitalization
  mutate(Chick = as.numeric(Chick),    # fix order of Chick column
         Chick = factor(Chick, ordered = TRUE))
# A tibble: 578 x 4
   Weight  Time Chick Diet
    <dbl> <dbl> <ord> <fct>
1      42     0 15    1
2      51     2 15    1
3      59     4 15    1
...
```

The mortality data

```
mortality_long <- readRDS("mortality_long.rds")
mortality_long
# A tibble: 476 x 4
    Year AgeGroup    DeathsPer100K Decade
   <dbl> <chr>               <dbl>  <dbl>
1   1900 01-04 Years         1984.   1900
2   1900 05-09 Years          466.   1900
...
```

The California cities data

```
ca_cities <- readRDS("california_cities.rds")
ca_cities
# A tibble: 1,252 x 5
   city             rank population   lat   lng
   <chr>           <int>      <int> <dbl> <dbl>
1 Acalanes Ridge   1072        875  37.9 -122.
2 Acampo           1266        261  38.2 -121.
...
```

Figure 8-1   Get the data

More skills for working with scatter plots

Figure 8-2 presents two additional parameters for working with scatter plots. The first example shows how to work with the shape parameter. This parameter sets the shape of the points in the plot based on the value of a specified column. For this plot, the code sets the shape parameter to the Species column from the irises data set. As a result, the plot uses a different shape for each species.

In the first example, the code sets both the shape and color parameters to the Species column. As a result, the plot uses a different color and a different shape for each species. However, the shape parameter can be especially useful when the color parameter is set to another categorical column.

The shape parameter doesn't work for continuous columns because it only provides for a limited number of shapes. As a result, it's typically used with categorical columns like the Species column.

The second example shows how to work with the alpha parameter. This changes the opacity of a point. This can be useful because some scatter plots have points that display on top of each other. This is known as *overplotting*, and it makes it difficult to interpret the plot because you can't see all of the points for the plot. Since the alpha parameter makes the points transparent, the more points in an area, the darker the point. In the second plot, for example, the points have 50% transparency. As a result, when two or more dots display on top of each other, the dot is displayed in a darker color.

When you set a parameter to a literal or static value, you set it directly in the plotting function, not in the aes() function. In the second plot, for example, the code sets the alpha parameter to a value of .5 in the geom_point() function. Conversely, when you set a parameter to a column, you set it in the aes() function, not the plotting function. In the first plot, for example, the code sets the shape parameter to the Species column in the aes() function. It's also possible to set the alpha parameter to a column. But first, you would need to move it into an aes() function.

Two more parameters for working with a scatter plot

Parameter	Description
shape	Sets the shape of the datapoints based on a categorical column or literal value.
alpha	Sets the transparency of the datapoints based on a column or literal value.

How to change the shape of the datapoints

```
ggplot(irises, aes(x = Sepal.Length, y = Sepal.Width, color = Species,
        shape = Species)) +
    geom_point(size = 3)
```

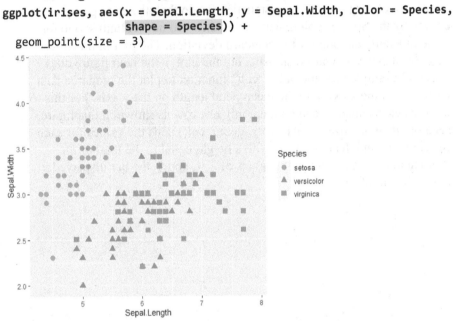

How to change the transparency of the datapoints

```
ggplot(irises, aes(x = Sepal.Length, y = Sepal.Width, color = Species)) +
    geom_point(size = 5, alpha = .5)
```

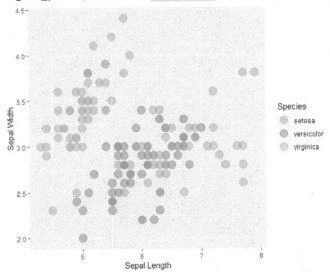

Figure 8-2   More skills for working with scatter plots

More skills for working with bar plots

In chapter 4, you learned how to use a bar plot to display the count of the rows in a category. Now, you'll learn how to use bar plots to show other summary statistics for a category.

To start, you need to prepare the data for the plot by calculating the summary statistics you want to display with one row for each category. In figure 8-3, the first example calculates two summary statistics. To do that, the code groups the irises data by the Species column and calculates two new columns: one for the mean petal length and one for the standard deviation. The plot in this figure doesn't use the standard deviation statistic, but the plot in the next figure does.

The second example uses the geom_col() function to plot the prepared data with the species on the x axis and the mean petal length on the y axis. For this to work correctly, each category can contain only one row as shown in this figure. If there is more than one row per category, geom_col() adds the values for each row in the category and displays a bar with a height equal to the total of the values. For example, given rows with values of 2, 4, and 5, the bar displayed would have a height of 11.

A function for creating bar plots

Function	Description
geom_col()	Works like geom_bar(), but lets you specify the value of a summary statistic on the y axis instead of the count.

How to calculate two summary statistics

```
iris_stats <- irises %>% group_by(Species) %>%
  summarize(Mean.Petal.Length = mean(Petal.Length),
            SD.Petal.Length = sd(Petal.Length))
```

```
iris_stats
# A tibble: 3 x 3
  Species    Mean.Petal.Length SD.Petal.Length
  <fct>                  <dbl>           <dbl>
1 setosa                  1.46           0.174
2 versicolor              4.26           0.470
3 virginica               5.55           0.552
```

How to plot one summary statistic on the y axis

```
ggplot(iris_stats,
       aes(x = Species, y = Mean.Petal.Length, fill = Species)) +
  geom_col()
```

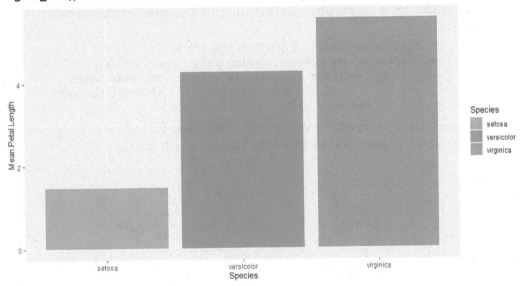

Description

- The geom_col() function can be used to display summary statistics, but there must be only one row per category for this to work correctly.

Figure 8-3   More skills for working with bar plots

How to add an error bar to a bar plot

If you're unsure about the results that are displayed in a bar plot, you may want to add an *error bar*. Error bars typically indicate a margin of error, uncertainty, or the standard deviation of the data.

To add an error bar to your plots, you can use the geom_errorbar() function as shown in figure 8-4. Here, the example adds the geom_error() function to the bar plot shown in this previous figure. Then, it sets the x parameter to the Species column and uses the ymin and ymax parameters to set the top and bottom of the error bar to the petal length plus or minus the standard deviation.

Adding the error bars to the visualization makes it clear that there is more standard deviation in the petal lengths of versicolor and virginica irises than setosa irises. In addition, this code uses the width parameter to set the width of the error bars to 25% of the width of the bar. This makes the plot more visually appealing.

In this figure, the example codes the data for the plot in the ggplot() function. That way, it's available to both the geom_col() and geom_errorbar() functions. However, it codes the two aes() functions separately, one in the geom_col() function and one in the geom_errorbar() function. That way, each of these plotting functions can specify its own parameters.

Since both plotting functions set the x parameter to the Species column, you could reduce some code duplication by coding the x parameter in the ggplot() function like this:

```
ggplot(iris_petal_length, aes(x = Species)) +
  geom_col(aes(y = Mean.Petal.Length, fill = Species)) +
  geom_errorbar(aes(ymin = Mean.Petal.Length - SD.Petal.Length,
                    ymax = Mean.Petal.Length + SD.Petal.Length),
                width = .25)
```

However, this involves adding another aes() function to the code. As a result, it's questionable whether this is an improvement that's worth making.

A function for creating error bars

Function	Description
geom_errorbar()	Create error bars for the specified column on the x axis.

Parameter	Description
ymin	Sets the minimum y value for the error bar.
ymax	Set the maximum y value for the error bar.
width	Set the width of the error bar to the specified percentage of the bar.

How to create a bar plot with an error bar

```
ggplot(iris_stats) +
  geom_col(aes(x = Species, y = Mean.Petal.Length, fill = Species)) +
  geom_errorbar(aes(x = Species,
                ymin = Mean.Petal.Length - SD.Petal.Length,
                ymax = Mean.Petal.Length + SD.Petal.Length),
            width = 0.25)
```

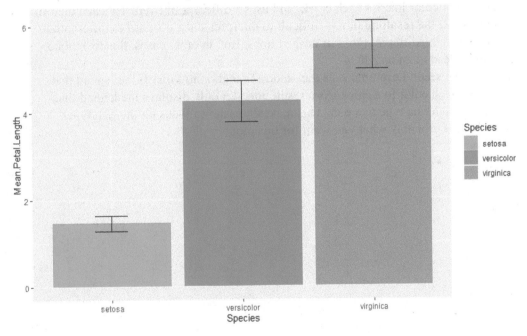

Description

- An *error bar* can be used to indicate a margin of error, uncertainty, or standard deviation.

Figure 8-4   How to add an error bar to a bar plot

More skills for working with line plots

When working with line plots, you may want to use another parameter besides color to differentiate between the lines for different categories. One way to do that is to set the linetype parameter in the aes() function to a categorical column. In figure 8-5, for instance, the first example sets the linetype parameter to the AgeGroup column. As a result, each age group uses a different type of line.

Since it's typically easier to distinguish different colors than different line types, you usually want to use color to differentiate between categories. Also, there are only thirteen types of lines, so you can only use the linetype function for columns that contain thirteen or less values.

However, if a plot is already using color to differentiate between one group, you may want to use line types to differentiate between another subgroup. In this figure, for instance, the second example sets the color parameter to the Chick column, and it sets the linetype parameter to the Diet column. This displays a different color line for each chick, and uses a different line type for each diet. In this case, the resulting plot is difficult to interpret since it's hard to differentiate between the different types of dashed lines. But, in other cases, this technique might lead to an insight.

The second example uses the guides() function to turn off the legend that maps each color to a chick. As a result, the plot only displays the legend that maps each line type to a diet. This makes it easier to focus on the line types, which is probably what you want for this plot.

A parameter for the geom_line() function

Parameter	Description
linetype	Sets the column that determines the line type. There are only thirteen types of lines, so this column should have thirteen or less values.

How to use line types instead of color

```
ggplot(mortality_long, aes(x = Year, y = DeathsPer100K, linetype = AgeGroup)) +
    geom_line(size = 1)
```

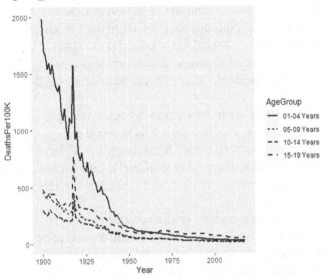

How to use line types in addition to color

```
ggplot(chicks, aes(x = Time, y = Weight, color = Chick, linetype = Diet)) +
    geom_line(size = 1) +
    guides(color = "none")   # turns off color legend
```

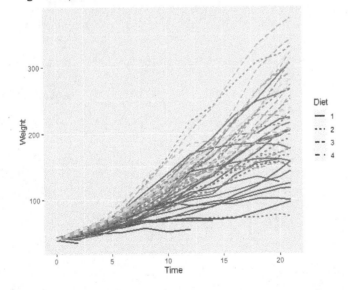

Figure 8-5   More skills for working with line plots

How to create a smooth line plot

Sometimes, you're more interested in the overall trends in the data than the details. For example, if you want to examine the overall trend of the mortality data, you might want to draw a smooth line through the data points instead of the jagged line that's shown in the previous figure. To do that, you can use the geom_smooth() function that's presented in figure 8-6. Since the lines drawn by the geom_smooth() function show the general trend of the data, they are sometimes called *trendlines*.

The first example shows how use the geom_smooth() function to draw a smooth line through the data points. By default, this function displays a shaded area around the line known as a *confidence interval*, or *standard error interval*. This means that values have a 95% chance of falling within the shaded area. However, this example doesn't display the confidence interval because it sets the se parameter to FALSE.

When the geom_smooth() function in the first example executes, it displays the message shown in the second example. This message tells you that the function is using the LOESS smoothing method with a formula of 'y ~ x' for this data. The *LOESS* (*Locally Estimated Scatterplot Smoothing*) method works well for creating a smooth line through scatter plot data, and it handles outliers particularly well.

For now, you can ignore this warning message. That's because the geom_smooth() function does its best to set the method and formula parameters based on the data that it's plotting. More on how to use geom_smooth(), including how to set these parameters, is presented in chapter 12.

The geom_smooth() function

Function	Description
geom_smooth()	Generates a smooth line for the specified data.
Parameter	Description
se	Whether to display the standard error interval, also known as the confidence interval, around the line. Default is TRUE.

A smooth line plot without a confidence interval

```
ggplot(mortality_long, aes(x = Year, y = DeathsPer100K, color = AgeGroup)) +
    geom_smooth(se = FALSE)
```

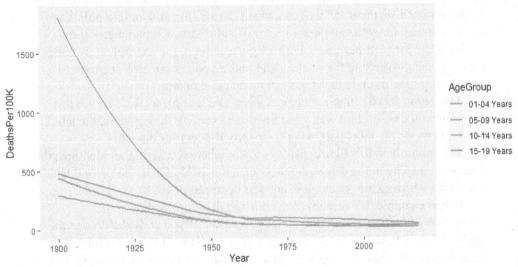

The message that's displayed when you run the code

```
`geom_smooth()` using method = 'loess' and formula 'y ~ x'
```

Description

- *LOESS* (*Locally Estimated Scatterplot Smoothing*) is a method for drawing a *trendline* through a scatter plot.
- The *confidence interval*, also known as the *standard error interval*, indicates the area in which 95% of the values fall.

Figure 8-6   How to create a smooth line plot

How to add labels to plots

When you present your data to others, it sometimes makes sense to add *labels* to your plot to make the visualization more self-explanatory. For example, you may want to label spikes in a line plot or outliers in a scatter plot.

To create a label, you can use the two functions summarized in figure 8-7. The geom_segment() function draws line segments that you can use to point to locations on your plot. To use this function, you use the x and y parameters to code the starting point for the segment, and the xend and yend parameters to code the ending point for the segment.

If you are drawing a lot of segments, you may want to pass a data set to the geom_segment() function that has columns that store the starting and ending values for each segment. Then, you can set the starting and ending points to those columns. Or, you can pass a vector of values to each parameter if that works better for your purposes. In this figure, though, the example only codes two segments, so setting the starting and ending points for each segment to literal values yields code that's easy to read and maintain.

The geom_label() function creates a box and text for a label. To use this function, you can set the x and y parameters to specify the center of the label, and you can set the label parameter to specify the text for the label.

The example in this figure displays a plot where the first label identifies the reason for a spike in the death rate that occurs in 1918. Then, the second label highlights a historically relevant event that occurred in 1928.

In this example, it's important to code the data for the plot in the geom_line() function, not in the ggplot() function. If you code the data in the ggplot() function, the geom_segment() and geom_label() functions inherit the data. As a result, they are executed once for each row in the data set. Obviously, that's not efficient. Even with a relatively small data set like this one, it takes the plot noticeably longer to display if you code the data in the ggplot() function.

However, if you code the data in the geom_line() function as shown in this example, the geom_segment() and geom_label() functions don't inherit the data. As a result, they're only drawn once, and the plot displays quickly.

Functions for creating plot labels

Function	Description
geom_segment()	Draws a straight line from the x and y point to the xend and yend point.
geom_label()	Creates a label centered at the specified x and y location.

Some parameters for these functions

Parameter	Description
xend	Sets the ending x value for the segment.
yend	Sets the ending y value for the segment.
label	Sets the text for the label.

A plot with two labels and two lines

```
ggplot() +
  geom_line(data = mortality_long,
            aes(x = Year, y = DeathsPer100K, color = AgeGroup)) +
  geom_segment(aes(x = 1918, y = 1550, xend = 1918, yend = 1700)) +
  geom_label(aes(x = 1918, y = 1700, label = "1918 Spanish Flu")) +
  geom_segment(aes(x = 1928, y = 650, xend = 1928, yend = 750)) +
  geom_label(aes(x = 1940, y = 800, label = "Penicillin Invented"))
```

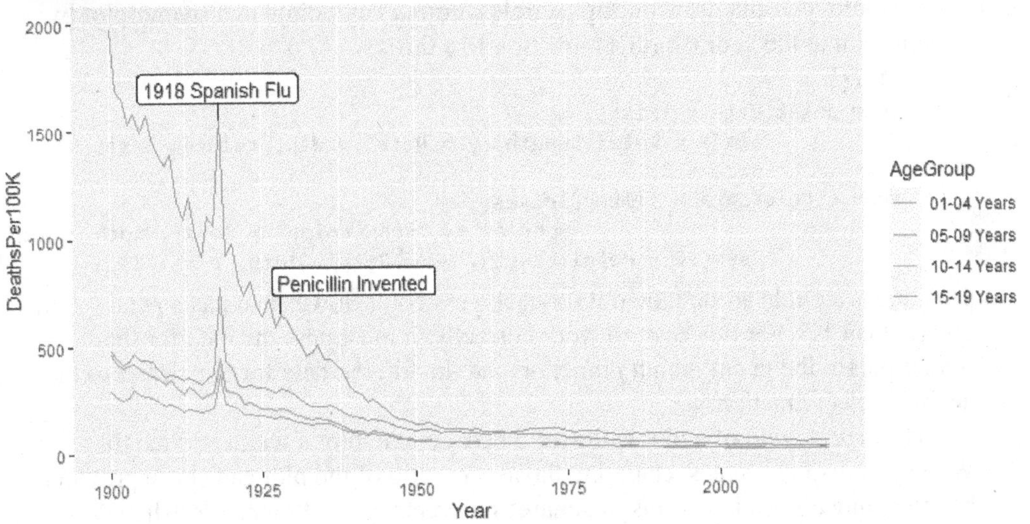

Description

- *Labels* allow you to highlight points of interest or explain data discrepancies.
- When using labels and segments, you don't typically want to code the data for your plot in the ggplot() function.

Figure 8-7   How to add labels to plots

How to work with shapes

The ggplot2 and ggforce packages provide several functions that you can use to plot shapes. These functions can be used to highlight interesting parts of your plot or to create custom shapes such as the baseball field that's presented later in this chapter.

How to plot shapes

The top of figure 8-8 summarizes several functions that you can use to plot shapes. In addition to using geom_segment() to draw lines, you can use geom_rect() and geom_tile() to draw rectangles. These functions are available from the ggplot2 package.

Furthermore, you can use geom_circle() to draw circles, geom_arc() to draw partial circles, and geom_ellipse() to draw ovals. These functions are available from the ggforce package that isn't part of the tidyverse. As a result, if you want to use these functions you must load the ggforce package.

When you use these functions, you can draw a single shape by setting the function's parameters to literal values as shown by the example in this figure. Alternately, you can draw multiple shapes by providing a vector for each value. For example, you can draw multiple circles around key points in a scatter plot by passing data to the geom_circle() function like this:

```
ggplot() +
  geom_point(data = irises,
             aes(x = Petal.Length, y = Petal.Width, color = Species),
             size = 3) +
  geom_circle(data = filter(irises,
                            Species == "versicolor" & Petal.Width > 1.5),
              aes(x0 = Petal.Length, y0 = Petal.Width, r = .1))
```

This draws a circle around the dot for each versicolor flower that has a petal width greater than 1.5. For this code to work correctly, it must pass the data for the scatter plot to the geom_point() function and the filtered data for the circles to the geom_circle() function.

The *aspect ratio* of a plot is the ratio between the plot's width and height. When you display shapes, changing the aspect ratio of the plot can cause problems by stretching circles into ovals or squares into rectangles. As a result, when you display shapes, you typically want to call the coord_fixed() function to make sure the aspect ratio of the plot isn't changed when the plot is displayed.

If you need to plot a polygon, you can use the geom_polygon() function summarized in this figure. However, this figure doesn't show an example of this function because it's typically used to draw maps as shown later in this chapter.

Functions for plotting shapes

Function	Description
geom_rect()	Draws a rectangle for two points.
geom_tile()	Draws a rectangle centered on a point.
geom_circle()	Draws a circle.
geom_ellipse()	Draws an oval.
geom_arc()	Draws an arc.
geom_polygon()	Draws a polygon, such as a map.
coord_fixed()	Fixes the ratio between the plot's height and width so shapes aren't distorted.

Some parameters for these functions

Parameter	Description
x, y	Sets the x and y values.
xend, yend	Sets the ending x and y values.
xmin, ymin	Sets the minimum x and y values.
xmax, ymax	Sets the maximum x and y values.
x0, y0	Sets the value for the center of a circle, arc, or ellipse.
start, end	Sets the start and end degrees for an arc.
width, height	Sets the width and height for a tile.
angle	Sets the angle for an ellipse.

How to plot shapes

```
ggplot() +
  geom_segment(aes(x = 5, y = 5, xend = 10, yend = 10)) +
  geom_rect(aes(xmin = 15, xmax = 20, ymin = 5, ymax = 10),
            color = "black", fill = NA) +
  geom_tile(aes(x = 27.5, y = 7.5, width = 5, height = 10)) +
  geom_circle(aes(x0 = 40, y0 = 7.5, r = 5)) +
  geom_arc(aes(x0 = 50, y0 = 7.5, r = 5, start = 0, end = pi)) +
  geom_ellipse(aes(x0 = 60, y0 = 7.5, a = 3, b = 5, angle = 0)) +
  coord_fixed()
```

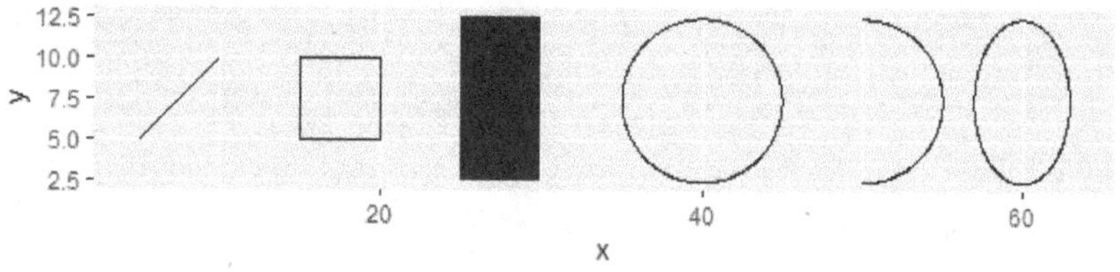

Description

- The geom_circle(), geom_ellipse(), and geom_arc() functions are in the ggforce package.

- When you plot shapes, you shouldn't pass data to the ggplot() function. If you do, the functions for plotting shapes inherit the data passed to the ggplot() function.

Figure 8-8   How to plot shapes

How to plot a baseball field

If necessary, you can combine shape components together to create a custom shape. For example, let's say you need to display a plot of where the hits in a baseball game landed. However, you have searched the internet and can't find any code for plotting a baseball field. In that case, you can plot your own baseball field as shown in figure 8-9.

The first step to creating a custom shape is to get the measurements and decide on a scale to use. The measurements shown at the top of this figure were taken from the MLB (Major League Baseball) website. To keep things simple, this plot uses 1 foot per unit for the scale.

When you create a plot like this, it's usually easiest to start with the straight lines. In this case, these are the foul lines and the base lines. To calculate the correct length for these lines, you need to use a little geometry. You can view the foul lines as 45-45-90 right triangles where the hypotenuse is the foul line. Then, to get the coordinates you can use the formula *hypotenuse = leg * √2*. This gives you roughly 230, so that is the endpoint for each foul line. You can use a similar technique to get the endpoints for the base lines.

After you plot the straight lines, you can move on to the circles. In this case, you need to use a circle to plot the pitching mound. The pitching mound is 59.5 feet from home plate which is at 0,0 on this plot, so you can use this information to specify the x0 and y0 parameters for the center of the circle. The diameter of the pitching mound is 18 feet, so you can divide this by 2 to get the radius for the circle.

The trickiest part of most custom plots is configuring the arcs, so you should do this last if possible. In this plot, you need to use an arc for the back fence. Since 0 is the center point for the arc, you can set its x0 parameter to 0. To get the y0 value, you need to use the endpoints of the of the foul lines and then add the radius to reach 240. From here, you can use the formula *degrees/180*pi* to get the start and end degrees for the arc. When you do this, the arc won't line up with the foul line, but you can fix it by changing the y0 and radius values until the arc lines up correctly. With arcs, it's common to need some adjustment, which is why it typically makes sense to code them last.

The specifications for a baseball field provided by the MLB association

- No baseball field is exactly the same, but certain aspects of the field are.
- The infield is always a square that is 90 feet on each side.
- The center of the pitching mound is 59.5 feet away from the back point of home plate.
- The pitching mound is 18 feet in diameter.
- There must be at least 325 feet between home base and the back fence at the foul lines.
- There must be at least 400 feet between home base and the back fence at center field.

The code for the baseball field

```
ggplot() +
  # infield + foul lines
  geom_segment(aes(x = 0, y = 0, xend = -230, yend = 230)),
  geom_segment(aes(x = 0, y = 0, xend = 230, yend = 230)),
  geom_segment(aes(x = -63.6, y = 63.6, xend = 0, yend = 127.2)),
  geom_segment(aes(x = 63.6, y = 63.6, xend = 0, yend = 127.2)),
  # pitchers mound
  geom_circle(aes(x0 = 0, y0 = 59.5, r = 9)) +
  # outfield arc
  geom_arc(aes(x0 = 0, y0 = 160, r = 240,
            start = -73/180*pi, end = 73/180*pi)) +
  # coordinate fixing
  coord_fixed()
```

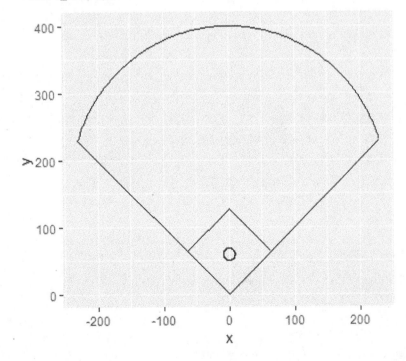

Figure 8-9   How to plot a baseball field

How to return plot components from a function

When you create a custom shape, there's often a lot of code that you don't want to rewrite every time you use that custom shape. To make this code easy to reuse, you can return the components for the custom shape from a function. To do that, you can define a function. Then, within the function, you can store each component of the plot in a vector and return the vector. This allows you to add the function call to an existing plot just like you would any other plotting function.

The first example in figure 8-10 shows how to return the plot components for the baseball field from a function. Then, the second example calls the function to create the plot. This mimics the syntax that's used by other functions in the ggplot2 package, and it makes your code easy to reuse. Furthermore, if you need to edit the custom shape you only need to edit it in one place, even if you call this function from multiple places.

How to return plot components from a function

```
plot_baseball_field <- function() {
  bball_field <- c(coord_fixed(),
    geom_segment(aes(x = 0, y = 0, xend = -230, yend = 230)),
    geom_segment(aes(x = 0, y = 0, xend = 230, yend = 230)),
    geom_segment(aes(x = -63.6, y = 63.6, xend = 0, yend = 127.2)),
    geom_segment(aes(x = 63.6, y = 63.6, xend = 0, yend = 127.2)),
    # pitchers mound
    geom_circle(aes(x0 = 0, y0 = 59.5, r = 9)),
    # outfield arc, 73/180pi == 74 degrees
    geom_arc(aes(x0 = 0, y0 = 160, r = 240, start = -73/180*pi,
              end = 73/180*pi)))
  return(bball_field)
}
```

How to add the components to an existing plot

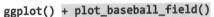

```
ggplot() + plot_baseball_field()
```

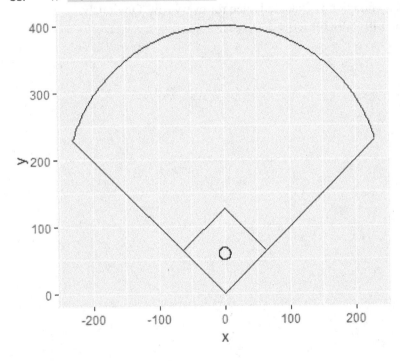

Description

- To return plot components from a function, you can store them in a vector and return the vector.

Figure 8-10   How to return plot components from a function

How to plot hits on a baseball field

If you want to plot the locations of hits on the baseball field from the previous figure, you need to create some data to plot. The first example in figure 8-11 shows how to do that. Here, the code creates a data frame that contains four rows where each row specifies the x and y location for a hit.

The second example shows how to plot this data on the field shape. To do that, this example adds a geom_point() function to the custom plot_baseball_field() function. Then, it passes the data for the hits to the geom_point() function. As a result, it displays a point for each hit on the shape for the baseball field.

When working with the plot_baseball_field() function, it's important to pass the data to the geom_point() function rather than the ggplot() function. That's because the plot_baseball_field() function inherits the data from the ggplot() function. As a result, if you pass data to the ggplot() function, the plot_baseball_field() function is executed once for every row. Obviously, that's not efficient, especially for large data sets.

How to create some hit locations to plot

```
baseball_hits <- data.frame(hitx = c(-100,-25, 100, 150),
                            hity = c(200, 50, 350, 275))

baseball_hits
   hitx hity
1 -100  200
2  -25   50
3  100  350
4  150  275
```

How to add the hit locations to the field

```
ggplot() +
  plot_baseball_field() +
  geom_point(data = baseball_hits,
             aes(x = hitx, y = hity))
```

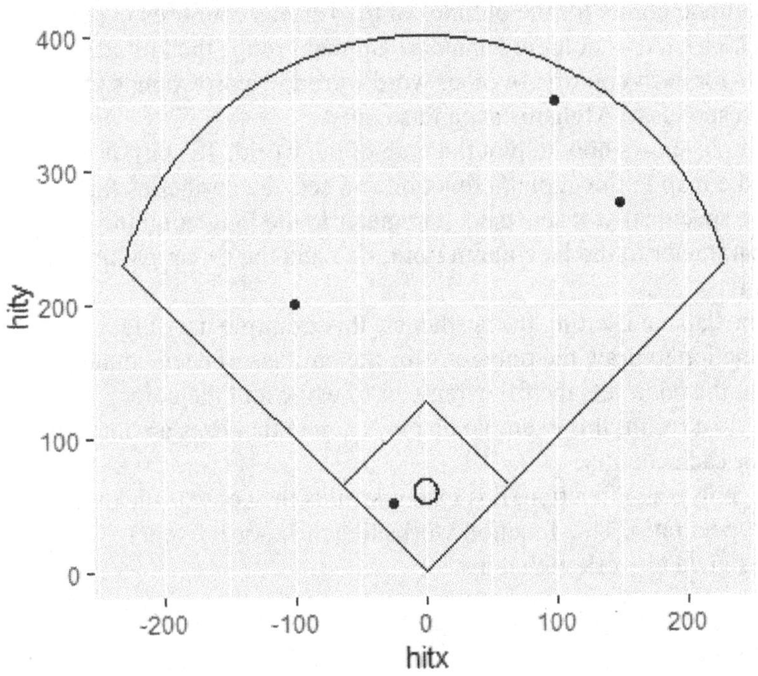

Figure 8-11   How to plot hits on a baseball field

Description

- Custom shapes can be combined with scatter plots to create visualizations.
- When you add geom functions to a plot that contains shapes, you should pass the data to the geom functions, not the ggplot() function.

How to work with maps

Geospatial data can be plotted on maps. Typically, geospatial data includes *longitude* and *latitude* measurements that correspond with the x and y axes of a plot. To plot maps and geospatial data with R, you can use the maps package.

How to plot maps

Figure 8-12 begins by showing how to use the map_data() function to get the data for different types of maps. This shows that the map_data() function provides map data for a variety of different countries, states, and even counties. You can get a list of available maps by entering this code into the Console pane:

```
help(package='maps')
```

The third example shows the data for the world map. This data frame stores the longitude and latitude points for the outlines of the various countries in the world. In addition, it provides a column named group that groups the latitude and longitude points for each country. In other words, group 1 corresponds to Aruba, group 2 corresponds to Afghanistan, and so on.

The fourth example shows how to plot the map of the world. To start, it passes the data for the map to the ggplot() function and sets the aesthetics for the map there. More specifically, it sets the x parameter to the long column (longitude), the y parameter to the lat column (latitude), and the group parameter to the group column.

After passing the data and setting the aesthetics, this example uses the geom_polygon() function to draw the polygons for the outlines of each country. Within this function, the code sets the fill parameter to white and the color parameter to black. As a result, this example displays a map that uses a white fill and black outline for each country.

After the geom_polygon() function, this example adds the coord_quickmap() function to fix the aspect ratio. This function works like the coord_fixed() function, but it's designed to work with maps.

Two functions for plotting maps

Function	Description
map_data(map, region)	Gets the data needed to create a map based on the passed map and region.
coord_quickmap()	Fixes the aspect ratio for a map so its shapes aren't distorted.

How to use the map_data() function

```
map_data("world")                     # all countries in world
map_data("usa")                       # only USA
map_data("france")                    # only France
map_data("state")                     # all states in USA
map_data("state", "california")       # only California
map_data("county")                    # all counties in USA
map_data("county", "california")      # only California counties
```

The map data for the world

```
map_data("world")
          long     lat group order    region subregion
1    -69.89912 12.45200     1     1     Aruba      <NA>
2    -69.89571 12.42300     1     2     Aruba      <NA>
3    -69.94219 12.43853     1     3     Aruba      <NA>
...
```

How to plot a map of the world

```
ggplot(data = map_data("world"),
       aes(x = long, y = lat, group = group)) +
  geom_polygon(fill = "white", color = "black") +
  coord_quickmap()
```

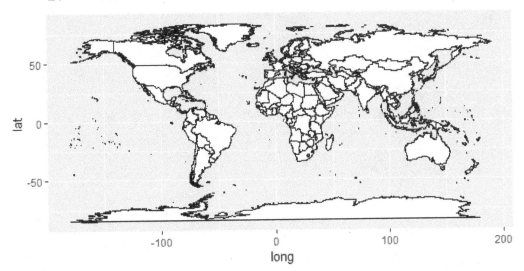

Figure 8-12   How to plot maps

How to add data to a map

After you display a map, you typically want to add data to it as shown in figure 8-13. To start, the first example gets the map data for California by passing "state" and "california" as arguments to the map_data() function. Then, the second example displays the California cities data that was loaded in figure 8-1. This tibble contains data for 1,252 cities in California including the city's name, rank by population, population, longitude, and latitude.

The third example uses the geom_polygon() function to display a map of California. Then, it uses the geom_point() function to add a point for each city to the map using its longitude and latitude. In addition, it uses the alpha parameter to make the points more or less transparent depending on the population of the city. This makes the areas with higher populations darker than the areas with lower populations.

How to get the data for the map of California

```
ca_map <- map_data("state", "california")
ca_map
          long      lat group order      region subregion
1    -120.0060 42.00927     1     1 california      <NA>
2    -120.0060 41.20139     1     2 california      <NA>
3    -120.0060 39.70024     1     3 california      <NA>
...
```

The California cities data

```
ca_cities
# A tibble: 1,252 x 5
   city            rank population   lat   lng
   <chr>          <int>      <int> <dbl> <dbl>
 1 Acalanes Ridge  1072        875  37.9 -122.
 2 Acampo          1266        261  38.2 -121.
 3 Acton            575       7054  34.5 -118.
...
```

How to plot the map of California with the cities on it

```
ggplot(data = ca_map, aes(x = long, y = lat)) +
  geom_polygon(fill = "white", color = "black") +
  geom_point(data = ca_cities,
             aes(x = lng, y = lat, alpha = population)) +
  coord_quickmap()
```

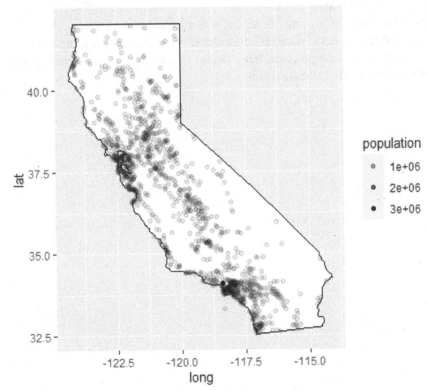

Figure 8-13   How to add data to a map

How to tune plots

At this point, you should have most of the skills you need for creating plots you for your analyses. Now, this chapter shows some techniques for further fine-tuning your plots.

How to zoom in on part of a plot

One way to fine-tune your plots is to zoom in on one part, or *facet*, of the plot while still presenting the plot as a whole. This is often helpful with geographic plots like the one shown in figure 8-14 and with relational plots like line plots and scatter plots.

The top of this figure summarizes some parameters of the facet_zoom() function that you can use to zoom in on part of a plot. Then, the example shows how to use this function.

The example begins by displaying the map of California with population data plotted on it. Then, it uses the facet_zoom() function to zoom in on part of the map. Here, the code uses vectors to pass the x and y ranges for the zoomed window. Next, it uses the zoom.size parameter to specify the size of the zoomed window relative to the original window's width. Since this code sets the size parameter to 1, the zoomed window and the original plot have the same width. However, if you set the size parameter to 2, the zoomed window would be twice as wide as the original window.

When you use the facet_zoom() function, you should know that the coord_fixed() and coord_quickmap() functions don't work with it. Adding the these functions doesn't cause an error, but it doesn't maintain the aspect ratio for the zoomed window or the original window.

The facet_zoom() function

Function	Description
facet_zoom()	Zooms in on one facet of a plot while leaving the original plot visible.
Parameter	**Description**
xlim	Specifies a vector for the starting and ending x values for the zoom window.
ylim	Specifies a vector for the starting and ending y values for the zoom window.
zoom.size	Specifies the size of the zoom window. By default, the size is specified as a percentage of the width of the original plot.

How to zoom in on part of a plot

```
ggplot(ca_map, aes(long, lat)) +
  geom_polygon(fill = "white", color = "black") +
  geom_point(data = ca_cities,
             aes(x = lng, y = lat, alpha = population)) +
  facet_zoom(xlim = c(-123,-121.5), ylim = c(37,38.5), zoom.size = 1)
```

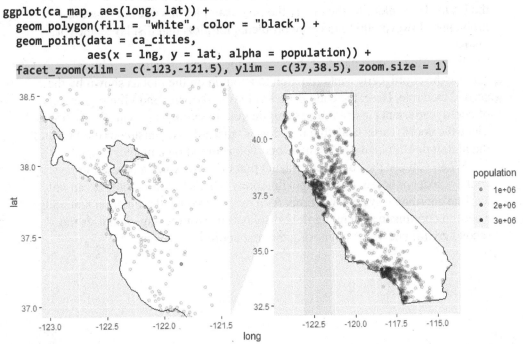

Description

- The coord_quickmap() and coord_fixed() functions don't work with the facet_zoom() function.

Figure 8-14   How to zoom in on part of a plot

How to adjust the limits of a plot

Another way to focus on a certain region of a plot is to change the *plot limits*. To do this, you can use the xlim() and ylim() functions as shown in figure 8-15.

The first example shows how to use the xlim() and ylim() functions to set the limits for a plot of the chicks data. This provides a way to focus on the data for just the first 10 days, which you might want to do in some situations. You can use these functions together to limit both the x and y axis, or either one individually to limit just one axis.

When you use these functions to adjust the plot limits, the plot still extends slightly beyond the limits. Also, the plot discards the data that lies outside the limits. For a plot like the one in the first example, this doesn't cause any problems. However, this can cause problems for plots that display shapes or maps.

If zooming in on a plot that displays shapes or maps causes problems, you can use the coord_cartesian() function to zoom in on the plot as shown by the second example. Here, the code passes values to the xlim and ylim parameters of coord_cartesian() instead of passing them to the xlim() and ylim() functions. Since the coord_cartesian() function doesn't discard data outside of the limits of the plot, it doesn't cause problems when you zoom in on a map or shape.

When you use the coord_cartesian() function, you can't also use the coord_fixed() or coord_quickmap() functions. That's because ggplot2 can only use one coordinate system at a time. As a result, if you call a function that adds a second coordinate system, ggplot2 displays an error message and uses the second coordinate system, ignoring the first coordinate system.

Functions for setting the x and y limits of a plot

Function	Description
xlim(min, max)	Sets the x limits of the plot.
ylim(min, max)	Sets the y limits of the plot.
coord_cartesian(xlim, ylim)	Zoom in on part of a plot without removing data.

How to create a chart with x and y limits

```
ggplot(filter(chicks, Diet == 1), aes(x = Time, y = Weight, color = Chick)) +
  geom_line() +
  xlim(0,10) +
  ylim(0, 150)
```

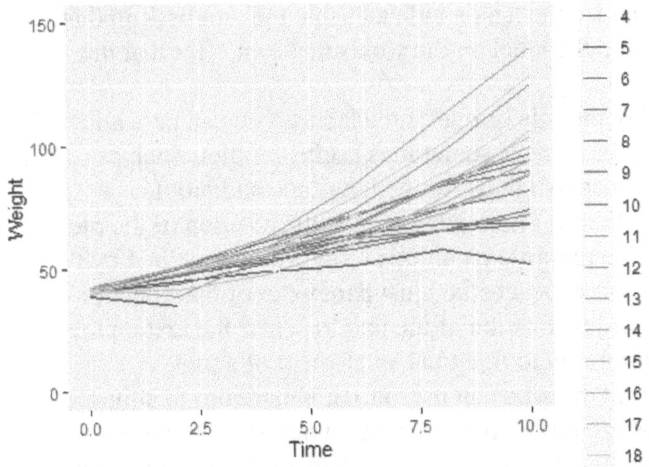

How to zoom in on a map

```
ggplot(ca_map, aes(long, lat)) +
  geom_polygon(fill = "white", color = "black") +
  geom_point(data = ca_cities, aes(x = lng, y = lat, alpha = population)) +
  coord_cartesian(xlim = c(-123,-121.5), ylim = c(37,38.5))
```

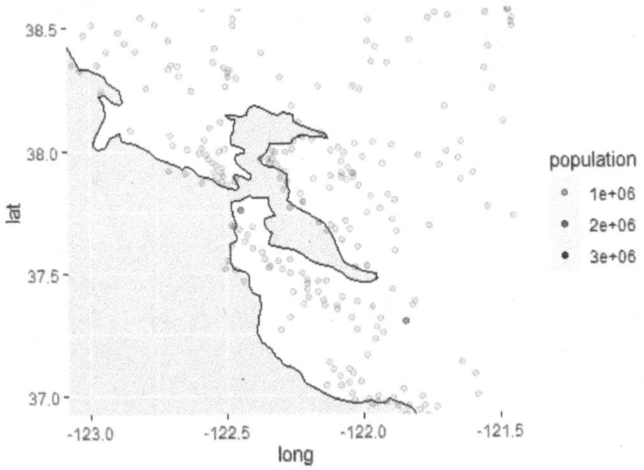

Figure 8-15   How to adjust the limits of a plot

How to work with the plot title and axes labels

When working with plots, you often want to set the *plot title* and *axes labels* for a plot to make them easier for others (and yourself!) to interpret. By default, a plot doesn't include a title, and the text for the axes labels is typically set to the column names for the x and y axes, which isn't always what you want.

Figure 8-16 begins by summarizing three functions that you can use to work with plot titles and axes labels. Then, the example shows how to use these functions to display a plot title that's centered at the top of the plot and to provide descriptive labels for the x and y axes.

To start, the labs() function sets the text for the plot title as well as the labels for the x and y axes. Here, the title indicates that the lines in the plot are for the chicks on the first diet, the label for the x axis indicates that the numbers on that axis are for the days elapsed, and the label on the y axis makes it clear that the weight is measured in grams.

If you don't want to display the title or one of the labels, you can pass an empty string ("") to the parameter. This is sometimes useful for plots such as maps where you don't need to display labels for the longitude and latitude.

The example also uses the theme() function to adjust the position of the plot title. To do that, the code sets the plot.title parameter to the value specified by the element_text() function. In this case, it sets the hjust parameter of that function to .5. This adjusts the horizontal justification of the text to center the text on the top of the plot. This is a common way to fine tune an element of a plot.

If you dig deeper, you'll find that you can use similar techniques to adjust all aspects of a plot. To do that, you often use element_ functions. For instance, element_blank() draws nothing and assigns no space, element_text() works with text, element_rect() handles borders and backgrounds, and element_line() works with lines. You can learn more about these options by typing ? and the name of the function in the Console pane.

Functions for working with the plot title and axes labels

Function	Description
`labs(title, x, y)`	Sets the title as well as the labels for the x and y axes.
`theme()`	Set a theme or individual non-data elements of the plot.
`element_text()`	In conjunction with the theme system, specifies the display of text. Used with parameters like hjust (horizontal justification), vjust (vertical justification), family (font family), face (font face, like bold or italic), and color.

A parameter of the theme() function

Parameter	Description
`plot.title`	Adjusts the title element. Typically used with element_text().

A chart that sets the plot title and the axes labels

```
ggplot(filter(chicks, Diet == 1),
        aes(x = Time, y = Weight, color = Chick)) +
    geom_line() +
    labs(title = "Diet 1", x = "Days Elapsed", y = "Weight (Grams)") +
    theme(plot.title = element_text(hjust = 0.5))
```

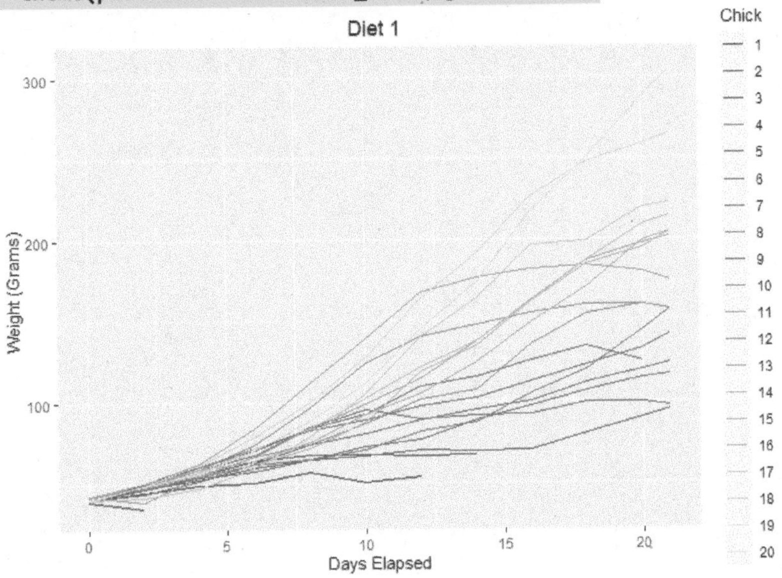

Description

- The labs() function allows you to set the labels and title of a plot, while theme() can be used to customize nearly every non-data element in a plot.

Figure 8-16   How to work with the plot title and axes labels

How to change the position of the legend

When you display a plot, the plot automatically generates one or more *legends* that describe some aesthetics such as color and size. However, these legends aren't always formatted the way you want. In addition, by default the legends are displayed to the right of the plot. Again, that might not be what you want. Fortunately, ggplot2 gives you control to customize your legends.

In some cases, the legend for a plot doesn't display useful information. In that case, hiding the legend typically improves the appearance of the plot and makes it easier for your target audience to interpret. To hide a legend, you can call the theme() function and set its legend.position parameter to "none" as shown in the first example of figure 8-17. Here, the legend doesn't display useful information, so hiding the legend makes the plot cleaner by focusing on the main point of the plot, which is how the various chicks gained weight over time.

In other cases, you may find that you can improve the appearance of a plot by changing the position of the legend. To do that, you can call the theme() function and set its legend.position parameter to "top", "bottom", or "left". For instance, the second example displays the legend at the bottom of the plot instead of to the right. This gives the plot more horizontal space.

Another parameter for the theme() function

Parameter	Description
legend.position	Sets the position of the legend element. Possible values are "top", "bottom", "left", "right", and "none".

How to hide the legend

```
ggplot(chicks, aes(x = Time, y = Weight, color = Chick)) +
  geom_line() +
  theme(legend.position = "none")
```

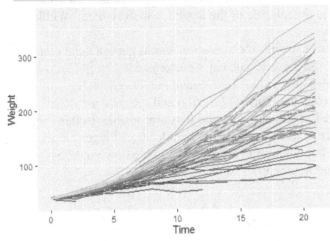

How to change the position of the legend

```
ggplot(irises, aes(x = Petal.Length, y = Petal.Width, color = Species)) +
  geom_point(size = 3) +
  theme(legend.position = "bottom")
```

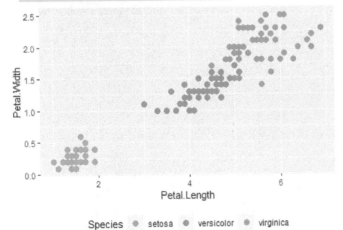

Description

- The theme() function has over 90 parameters available to customize plot elements.

Figure 8-17   How to change the position of the legend

How to edit the legend

In some cases, you may want to edit the data-related elements of a plot, like the components of a legend. For example, you may need to change the title of a legend. Or, when a plot displays multiple legends, you may want to hide one of the legends. To do that, you can use the guides() and guide_legend() functions presented in figure 8-18.

The first example shows how to change the title of a legend. To do that, it calls the guides() function. Then, it uses the guide_legend() function to change the title of the legend for the size to "Petal Size". That's more visually pleasing and easier to interpret than the default title of the legend, which is "Petal.Width * Petal.Length".

Note that if the code doesn't specify the size argument in ggplot(), the plot doesn't display a size legend. As a result, you can't customize it.

The second example shows how to hide one legend but leave another legend visible. To do that, it calls the guides() function and sets size to "none". This hides the legend for size but leaves the legend for color visible. In this case, hiding the legend for size makes sense if you want to show bigger points for bigger petals on the plot, but you don't care about showing how the sizes correspond with the calculated numbers.

Two functions for working with the legend

Parameter	Description
guides()	Modifies data-related elements of a plot. Used with guide_ functions like guide_legend(), guide_axis(), and guide_colorbar().
guide_legend()	Used with guides() to specify aspects of the plot legend. Available parameters include title, title.position, label, label.hjust, direction, and so on.

How to change the title of a legend

```
ggplot(irises, aes(x = Petal.Length, y = Petal.Width, color = Species,
                   size = Petal.Width * Petal.Length)) +

geom_point() +
guides(size = guide_legend(title = "Petal Size"))
```

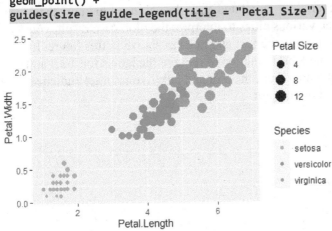

How to hide a legend

```
ggplot(irises, aes(x = Petal.Length, y = Petal.Width, color = Species,
                   size = Petal.Width)) +

geom_point() +
guides(size = "none")
```

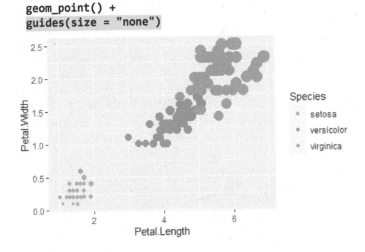

Figure 8-18   How to edit the legend

How to hide the text and ticks for each axis

By default, a plot displays the values for each axis using *ticks* (the small lines indicating values at intervals on the axes) and *text* (the text for the value at each interval). In most cases, that's what you want. However, for some plots, you might not want to display the text for an axis. And if you don't want to show the text, you probably don't want to show the ticks either. In that case, you can use the theme() function to hide the text and ticks for one or both axes. To do that, you can set the parameters shown in figure 8-19 to the element that's returned by the element_blank() function.

In the plot of California cities shown earlier in this chapter, the x axis displays the text and ticks for the longitude, and the y axis displays the text and ticks for the latitude. However, your target audience doesn't need to know the latitude and longitude for the various cities to interpret the plot. As a result, you can clean up your plot by removing the text and ticks as shown in this figure. In addition, this example uses the labs() function to remove the labels for the x and y axes. This removes visual clutter from the plot and helps your target audience focus on the insight provided by the plot.

More parameters for the theme() function

Parameter	Description
axis.ticks	Sets the ticks on the x and y axes.
axis.ticks.x	Sets the ticks on the x axis.
axis.ticks.y	Sets the ticks on the y axis.
axis.text	Sets the text on the x and y axes.
axis.text.x	Sets the text on the x axis.
axis.text.y	Sets the text on the y axis.

A plot that hides the text and ticks for the x and y axes

```
ggplot(data = ca_map, aes(x = long, y = lat)) +
  geom_polygon(fill = "white", color = "black") +
  geom_point(data = ca_cities,
             aes(x = lng, y = lat, alpha = population)) +
  coord_quickmap() +
  labs(x = "", y = "") +
  theme(axis.ticks = element_blank(),
        axis.text = element_blank())
```

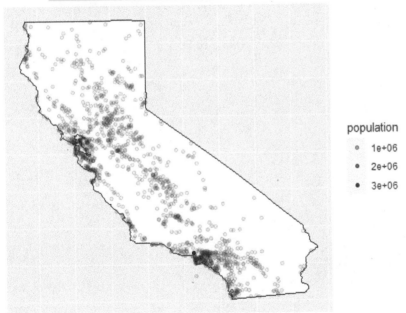

Description

- Sometimes you may want to hide the ticks and text on the x axis, y axis, or both to create a better visualization.
- In addition to hiding the ticks and text for an axis with element_blank(), you can customize the ticks and text for an axis using element_line() and element_text().

Figure 8-19   How to hide the text and ticks for each axis

How to set the colors for the plot

R allows you to set the colors for your plots both manually and to a pre-defined set. To use a pre-defined set of colors, you can choose from groups of colors called *palettes*.

To display a list of the available palettes, you first have to load the RColorBrewer package as shown in the first example in part 1 of figure 8-20. Then, you call the display.brewer.all() function to display all of the available palettes along with their names as shown by the second example.

To use these palettes, you can call the scale_color_brewer() or scale_fill_brewer() functions. Alternatively, you can set the colors for your plots manually with the scale_color_manual() or scale_fill_manual() functions. When you do that, you may want to use the colors() function to view a complete list of the names of the colors available in R as shown in the third example.

How to load the package for the color brewer

```
library("RColorBrewer")
```

Some functions for working with colors

Function	Description
display.brewer.all()	Displays all of the available color palettes.
scale_color_brewer(palette)	Sets the colors for the plot to the specified palette.
scale_fill_brewer(palette)	Sets the fill color for the plot to the specified palette.
scale_color_manual(values)	Sets the colors for the plot to the values in the specified vector.
scale_fill_manual(values)	Sets the fill colors for the plot to the values in the specified vector.

How to view the available color palettes

```
display.brewer.all()
```

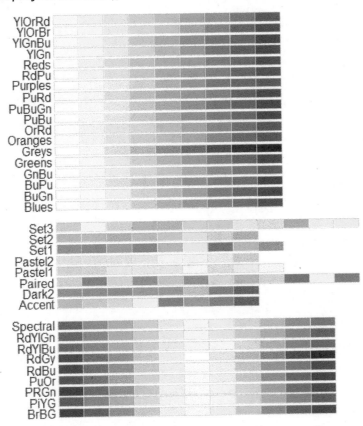

How to view the available names for colors

```
colors()
[1] "white"          "aliceblue"       "antiquewhite"
[4] "antiquewhite1"  "antiquewhite2"   "antiquewhite3"
...
```

Figure 8-20   How to set the colors for the plot (part 1)

In part 2 of figure 8-20, the first example shows how to use the scale_color_brewer() function to make a scatterplot use the Set2 palette. This automatically sets the colors for the scatterplot using the first three colors from that palette.

The second example uses the scale_color_manual() function to make the scatterplot use the specified vector of colors. When you pass the vector of colors, you can pass the colors as strings such as "blue", or you can pass hex color codes such as "#0000FF" if you want more control.

When you set the colors for a plot, you should realize that they work best with columns that store categorical data. This is because each palette has a finite number of colors. For example, the Set2 palette only provides eight colors. As a result, if a plot tries to use more than the number of colors in the palette, the colors won't display correctly. It's also worth remembering that colorblind members of your target audience may have trouble distinguishing between certain colors.

How to use a different color palette

```
ggplot(irises, aes(x = Petal.Length, y = Petal.Width, color = Species)) +
  geom_point(size = 3) +
  scale_color_brewer(palette = "Set2")
```

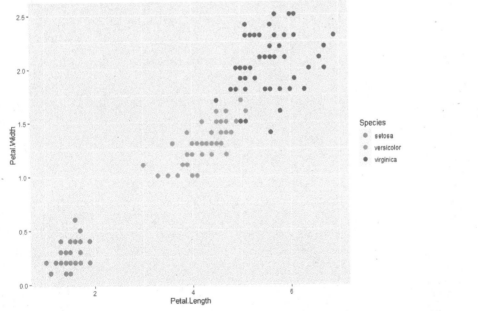

How to set the plot colors manually

```
ggplot(irises, aes(x = Petal.Length, y = Petal.Width, color = Species)) +
  geom_point(size = 3) +
  scale_color_manual(values = c("red","green","blue"))
```

Figure 8-20   How to set the colors for the plot (part 2)

How to change the theme of the plot

Sometimes you may want to quickly change the overall look of the plot without affecting how the data is displayed. To do this, you can use *themes*. The ggplot2 package includes several pre-set themes that change the background color, the lines in the background of the plot, and how the exterior lines are drawn.

The table at the top of figure 8-21 summarizes the functions for most of the pre-set themes. The default is theme_gray(). As a result, if you add a call to any of the other theme functions, it changes the theme for your plot.

To demonstrate, the example in this figure adds a call to the theme_bw() function. This theme sets the background of the plot to white, the grid lines in the plot to grey, and adds black outer lines. You can use themes when you need your plots to match the themes used by your organization, or if you don't like the way the default theme looks.

Some themes for ggplot2

Theme	Description
theme_gray()	The default ggplot2 theme.
theme_bw()	Sets the background to white, grid lines to grey, and outer lines to black.
theme_linedraw()	Sets the background to white, grid lines to black, and outer lines to black.
theme_light()	Sets the background to white and all lines to grey.
theme_dark()	Sets the background and all lines to dark grey.
theme_minimal()	Sets the background and outer lines to white and the grid lines to grey.
theme_classic()	Removes the grid lines and sets the outer line to black.

How to add a theme to a plot

```
ggplot(irises, aes(x = Petal.Length, y = Petal.Width, color = Species)) +
  geom_point() +
  theme_bw()
```

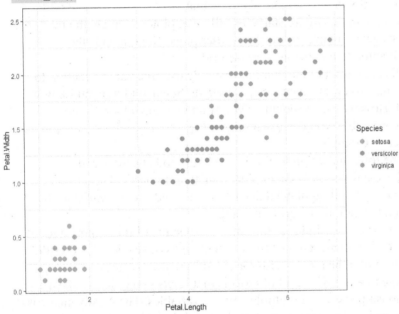

Description

- Pre-set themes are included with R to make it easy to customize several features of your plots at once.

Figure 8-21   How to change the theme of the plot

How to work with a grid of plots

When you're first starting an analysis, you may sometimes want to quickly examine the relationships between multiple variables in different ways and see if there are any relationships worth looking at more closely. For example, you may want to compare sepal and petal lengths and widths and see if there are any correlations between various measurements. To quickly examine these relationships, you can create a grid of plots.

How to create a pairwise grid of scatter plots

A quick way to compare multiple variables is by using the facet_matrix() function to create a *pairwise grid* of scatter plots as shown in figure 8-22. A pairwise grid compares each variable to each other in x-y pairs. Note that it uses a single statement to create 16 subplots. However, you need to install and load the ggforce package if you haven't already done that.

The first example in this figure starts by calling ggplot() to create the base plot. Then, it calls geom_point() to display scatter plots. Next, it adds the facet_matrix() function to create the pairwise grid.

Within the facet_matrix() function, this example uses the vars() function to specify four columns to be compared: Sepal.Length, Sepal.Width, Petal.Length, and Petal.Width. However, if the columns you want to compare are located sequentially in your tibble, you can use a range to specify the columns instead as shown in the second example.

To specify the rows and columns for the grid, this code uses special arguments for the x and y axes in the ggplot() function. In particular, it uses .panel_x and .panel_y to tell ggplot() that each plot should use the variables that correspond to that location in the grid for x and y.

If you study the generated plots, you can see that some of them show a lot of separation between the measurements for different types of species. As a result, those relationships might be most useful for classifying the species.

The advantage of a pairwise grid of plots is that it lets you quickly take a look at the relationships between multiple pairs of variables. The disadvantage is that the plots can end up being small and hard to read. In addition, generating the plots can be slow. Both of these disadvantages are especially true if you want to examine a large number of variables. But, because they are so easy to generate and let you compare a lot of data at once, they can be a great starting point for some analyses.

Functions for creating a grid of plots

Function	Description
facet_matrix(vars)	Allows you to put columns into a grid of plots. Typically used with vars().
vars()	When used with facet_matrix(), specifies the variables to use in the plot. If you use this parameter to specify the variables, you can use .panel_x and .panel_y in aes() to access the values for the x and y axes.

Code that creates a pairwise grid of scatter plots

```
ggplot(irises, aes(x = .panel_x, y = .panel_y, color = Species)) +
  geom_point() +
  facet_matrix(vars(Sepal.Length, Sepal.Width, Petal.Length, Petal.Width))
```

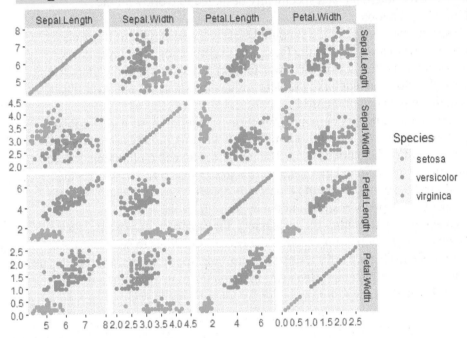

Another way to create this grid of plots

```
ggplot(irises, aes(x = .panel_x, y = .panel_y, color = Species)) +
  geom_point() +
  facet_matrix(vars(Sepal.Length:Petal.Width))
```

Description

- You can use the facet_matrix() function to create a pairwise grid of scatter plots that quickly plots several pairs of variables.
- The facet_matrix() function is available from the ggforce package, not ggplot2.

Figure 8-22   How to create a pairwise grid of scatter plots

How to use other plot types in the grid

If you use the same type of plot for all of the plots in a pairwise grid of plots like the previous figure, the plots in the upper right and lower left parts are redundant, and the plots on the diagonal might not be useful for your analysis. Alternately, you can specify one type of plot for the upper right part of the grid, another type for the diagonal, and a third type for the lower left part of the grid.

The example in figure 8-23 is similar to the previous figure, but instead of adding one plot function, it adds three: a scatter plot, a stacked KDE plot, and a box plot. Note that the stacked KDE plot in this example uses geom_autodensity(), not geom_density(). This function is a version of the geom_density() function that's specifically designed to work with the facet_matrix() function.

To specify which type of plot should be used on which part of the grid, you use the layer parameters in the facet_matrix() function. The layer.lower parameter refers to the plots in the lower left, layer.diag refers to the plots on the diagonal (the plots that compare the same values for x and y), and layer.upper refers to the upper right of the grid.

In the example, the geom_point() function is coded first, so it corresponds to the literal value 1. As a result, setting layer.lower to 1 displays scatter plots in the lower left of the pairwise grid. The geom_autodensity() function is coded second. As a result, setting layer.diag to 2 displays stacked KDE plots on the diagonal. And the geom_boxplot() function is coded third. As a result, setting layer.upper to 3 displays box plots in the upper right of the grid.

To make it easy to control the appearance for each plot, this example uses a separate aes() function within each of the geom_ functions to set the color or fill, as necessary. This provides a way to control the appearance of each plot type individually. By contrast, using the aes() function that's in ggplot() to set the color or fill would set the same aesthetics for all plots in the grid.

More facet_matrix() parameters

Parameter	Description
`layer.lower`	The plot type to use for the lower left half of the grid.
`layer.diag`	The plot type to use for the diagonal of the grid.
`layer.upper`	The plot type to use for the upper right half of the grid.

How to create pairwise grid of plots

```
ggplot(irises, aes(x = .panel_x, y = .panel_y)) +
  geom_point(aes(color = Species)) +
  geom_autodensity(aes(fill = Species)) +
  geom_boxplot(aes(fill = Species)) +
  facet_matrix(vars(Sepal.Length:Petal.Width),
               layer.lower = 1,
               layer.diag = 2,
               layer.upper = 3)
```

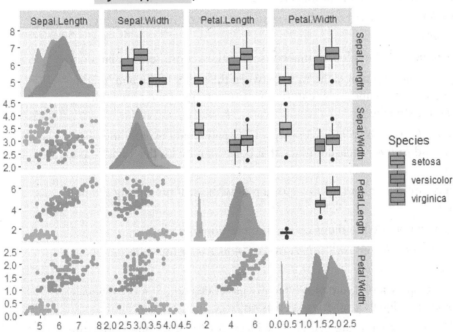

Description

- You can use the facet_matrix() function to create a grid of three different types of plots.
- The geom_autodensity() function is a version of the geom_density() function that's designed to work with the facet_matrix() function.

Figure 8-23   How to use other plot types in the grid

Perspective

Now that you've completed this chapter, you should be able to create and fine-tune the plots that you need for most of your analyses. That's important because data visualization is a critical part of data analysis. As you work with data visualization, you'll find that ggplot2 provides a huge number of functions for creating and tuning plots. As a result, there's always more to learn about it. But if you understand the skills presented in this chapter, you have a solid foundation for learning more about data visualization.

Summary

- When a scatter plot displays points on top of each other, it's called *overplotting*. Changing the alpha parameter for the points can make it easier to see how dense the points are.

- An *error bar* can be used to indicate a margin of error, uncertainty, or standard deviation.

- *Labels* can be added to plots to make visualizations more self-explanatory.

- The *aspect ratio* of a plot is the ratio between the plot's width and height. When you display shapes, you can fix the aspect ratio of the plot so the shapes aren't stretched.

- *Geospatial data* can be plotted on maps. Geospatial data usually includes *longitude* and *latitude* measurements that correspond with the x and y axes of a plot.

- You can zoom in on a part, or *facet*, of a plot while still presenting the plot as a whole.

- *Plot limits* can be used to focus on a certain region of a plot.

- Setting the *plot title* and *axes labels* makes plots easier for your audience to interpret.

- When you display a plot, the plot typically generates one or more *legends* that provide a guide to aesthetics such as color or size.

- The values for each axis are displayed with *ticks* (the small lines indicating values at intervals on the axes) and *text* (the text for each value interval).

- R allows you to set the colors for your plots both manually and to pre-defined sets of colors called *palettes*.

- To quickly change the overall look of a plot, you can use a pre-defined *theme*.

- A *pairwise grid* compares multiple variables to each other in x-y pairs.

Exercise 8-1   Plot world cities on a map

In this exercise, you'll create some plots for world cities and fine-tune them.

Plot some data about world cities

1. Create a new R script.

2. Load the tidyverse, ggforce, and maps packages.

3. View the documentation for the world.cities data set. In the capital column, 0 is a regular city, 1 is a capital city, and 2 and 3 are special types of cities in China.

4. Get the data for this exercise by converting the world.cities data set into a tibble named cities. This data set is available via the maps package.
```
cities <- as_tibble(world.cities)
```

5. Convert the capital column to the factor type.

6. Generate a small tibble named city_stats using the summarize() function to calculate the average population and standard deviation when grouped by capital status.

7. Use geom_col() to generate a bar plot that displays the calculated average population per city type. Set the fill palette to Dark2.

8. Add error bars to the plot that center the standard deviation on the average population. In the case of the capital cities, the error bar is bigger than the column!

9. Create a scatter plot comparing population and latitude of the world cities. To make the points easier to see, set their size to 2 and set the alpha parameter to .5. Set the color and shape of the points based on the capital.

10. Create a facet matrix comparing population, longitude, and latitude of the world cities with scatter plots. Set the size of the points to .5 to make them a little easier to see. As this is a very large data set, it may take a while to render.

11. Create another facet matrix, but this time put box plots in the lower left corner, density plots on the diagonal, and scatter plots in the upper right corner. Are you surprised to see some correlations in the data? Which graphs are the most useful for understanding the data set?

Plot the cities on a map of the world

12. View the data for the map of the world.
```
map_data("world")
```

13. Use the geom_point() function to plot the world as a scatter plot. To do that, set x to latitude and y to longitude, but don't use aes() to customize the map in any other way. It will take a while to render. Don't forget to add coord_fixed() to your ggplot() function.

14. Plot the world again. This time, use geom_polygon() to plot the world map. Again, use aes() to set x and y but don't use aes() for any other customizations. The map should render faster, but it won't look right. The shapes of the countries are there, but there are also many extra lines. Why?

15. Plot the world again. This time, use aes() to set the group parameter to the group column. The map should now look correct. This is because the map data contains polygons for individual countries. Using group keeps the country polygons individual instead of turning them into one connected polygon.

16. Plot the cities that have a population greater than 2 million on the map of the world. To do that, you can use the filter() function to filter the cities data set. For the map, set fill to white and color to black. This should outline the individual countries. For the points, set shape and color to capital and the size to 2.

17. Plot all cities on the map of the world. For the points, set shape and color to the capital category and the size and alpha to the population. This shows that there are a number of cities in the middle of the Pacific Ocean. Use facet_zoom() to zoom in on this section of the map.

18. Use coord_cartesian() to view the same area. Which do you find more useful?

19. Label the x and y axes "Latitude" and "Longitude", and title the plot "Pacific Islands".

20. Use guides() to change the titles of the two legends to "Population" and "City Type". Start by using guide_legend() to specify the title for size and color. Note that this causes the alpha and shape legends to be separated out. When you specify the appropriate title for alpha and shape, the plot combines the legends again.

21. View the documentation for the theme() function. This shows that there are many options for customizing a plot's theme. Scroll until you find the entry for panel.background.

22. Since panel.background uses element_rect(), view the documentation for the element_rect() function to learn about options for customizing the background of your plot.

23. Use the theme() and element_rect() functions to fill in the background of the Pacific Islands plot with light blue to indicate the ocean. Remember that fill, not color, is the parameter for filling in a shape with a color.

Exercise 8-2   Work with shapes

In this exercise, you'll use shapes to draw a human face.

Draw a face

1. Create a new script.

2. Load the tidyverse and ggplot packages.

3. Create a plot that uses geom_ellipse() to draw an oval for your face. You can make it as long or wide as you want, but keep in mind that you need to add additional features in the next steps.

4. Use the fill parameter to fill in your oval with a color. You can see a list of colors by typing colors() into the Console pane.

5. Add two eyes to your face using geom_circle(). They should be evenly placed to have horizontal symmetry. It may take some experimenting to get them in about the right place. Use the fill parameter to color them in.

6. To create a triangle nose, create two vectors, one named xcoord and one named ycoord. Use these to store the x and y coordinates for the three points of your nose. Put the statements creating these vectors prior to your code for the face.

7. Add a call to geom_polygon() and set the x and y arguments to xcoord and ycoord, like this:

    ```
    geom_polygon(aes(x = xcoord, y = ycoord))
    ```

8. Color the outline for the nose in black or another color that contrasts with your face color, but set fill to NA to make the nose the same color as the face.

9. Add a mouth using geom_arc(). It can smile or frown as you like. Set the color to whatever you like, linetype to a number between 2 and 6 (the default of 1 is a solid line), and the size to something larger than 1 so the mouth is more visible.

10. Add a segment and label to your plot indicating where the left eye is.

11. Use xlim() and ylim() to adjust the size of your plot.

Create a custom function for drawing a face

12. Store the components for your face in a function named my_face().

13. Code a statement that uses ggplot() and my_face() to draw your face components on a plot.

14. Add the theme_void() function to the plot to remove the axes, grid lines, and other plot elements, leaving just your face and the label.

Bonus

15. Use additional geom_ functions to add more features to your face, like hair, ears, or freckles.

Section 3

Three case studies

This section presents three real-world analyses that illustrate how the skills presented in the first two sections of this book can be applied in the context of a complete analysis. In addition, they show examples of thought processes that you can use when you analyze data.

Since the scripts for these analyses are included in the download for this book, we recommend that you open them and run the code as you read the descriptions of the code. That way, you can experiment with the code as you work your way through the chapters. You can also copy snippets of the code into your own scripts and modify them for your own analyses.

Chapter 9

The Polling analysis

This chapter presents an analysis of the polling data for the 2016 United States presidential election. This particular analysis focuses on the two leading candidates, Donald Trump and Hillary Clinton, and gets its data from the FiveThirtyEight website, which is a statistical site that focuses on opinion polls, politics, economics, and sports.

Get and examine the data

In all analyses, the first step is to get the data and then examine that data, so that's what figure 9-1 shows how to do. For this analysis, the data is available as a CSV (comma-separated values) file that you can download from the FiveThirtyEight website. It's also available from the download for this book.

Load the packages

The first example in figure 9-1 loads all of the packages needed by this analysis. In this case, that's just the tidyverse package. That's because the core tidyverse contains all the packages needed by this analysis.

Get the data

The second example shows how to get the data for this analysis. To do that, it uses the read_csv() function to read the polls.csv file that's stored in the data directory of the download for this book. This assumes that the working directory is set to the same directory that contains the script for this analysis. Then, the code assigns the tibble that's returned by read_csv() to a variable named polls.

When you run the read_csv() function, it displays this warning message:

```
One or more parsing issues, see `problems()` for details
```

If you run the problems() function, it displays a tibble that shows that the read_csv() function encountered 36 problems that all involved parsing the 22nd column (multiversions) of the CSV file. Since this analysis doesn't use the multiversions column, you can ignore this message.

Examine the data

After getting the data, you need to examine it to determine what parts to use. Displaying the tibble shows that the data set has 12,624 rows and 27 columns. However, displaying it in the console only displays some of the rows and columns. As a result, it doesn't give a complete idea of what the data set contains.

To learn more about the data, you can display it in RStudio by clicking the the tibble's name in the Environment pane. This lets you use the scroll bars to view all columns and rows. If you study the data, you should begin to get some ideas about what parts of it might be useful for this analysis.

Load the packages

```
library("tidyverse")
```

Get the data

```
polls <- read_csv("../../data/polls.csv")
```

View the data on the console

```
polls
# A tibble: 12,624 x 27
   cycle branch    type     matchup forecastdate state startdate enddate
   <int> <chr>     <chr>    <chr>   <chr>        <chr> <chr>     <chr>
 1  2016 President polls-p~ Clinto~ 11/8/16      U.S.  11/3/2016 11/6/2~
 2  2016 President polls-p~ Clinto~ 11/8/16      U.S.  11/1/2016 11/7/2~
 3  2016 President polls-p~ Clinto~ 11/8/16      U.S.  11/2/2016 11/6/2~
 4  2016 President polls-p~ Clinto~ 11/8/16      U.S.  11/4/2016 11/7/2~
 5  2016 President polls-p~ Clinto~ 11/8/16      U.S.  11/3/2016 11/6/2~
 6  2016 President polls-p~ Clinto~ 11/8/16      U.S.  11/3/2016 11/6/2~
 7  2016 President polls-p~ Clinto~ 11/8/16      U.S.  11/2/2016 11/6/2~
 8  2016 President polls-p~ Clinto~ 11/8/16      U.S.  11/3/2016 11/5/2~
 9  2016 President polls-p~ Clinto~ 11/8/16      New ~ 11/6/2016 11/6/2~
10  2016 President polls-p~ Clinto~ 11/8/16      U.S.  11/4/2016 11/7/2~
# ... with 12,614 more rows, and 19 more variables: pollster <chr>,
#   grade <chr>, samplesize <int>, population <chr>, poll_wt <dbl>,
#   rawpoll_clinton <dbl>, rawpoll_trump <dbl>, rawpoll_johnson <dbl>,
#   rawpoll_mcmullin <dbl>, adjpoll_clinton <dbl>, adjpoll_trump <dbl>,
#   adjpoll_johnson <dbl>, adjpoll_mcmullin <dbl>, multiversions <chr>,
#   url <chr>, poll_id <int>, question_id <int>, createddate <chr>,
#   timestamp <chr>
```

View the data in RStudio

	cycle	branch	type	matchup	forecastdate	state	startdate	enddate	pollster
1	2016	President	polls-plus	Clinton vs. Trump vs. Johnson	11/8/16	U.S.	11/3/2016	11/6/2016	ABC
2	2016	President	polls-plus	Clinton vs. Trump vs. Johnson	11/8/16	U.S.	11/1/2016	11/7/2016	Goo(
3	2016	President	polls-plus	Clinton vs. Trump vs. Johnson	11/8/16	U.S.	11/2/2016	11/6/2016	Ipso:
4	2016	President	polls-plus	Clinton vs. Trump vs. Johnson	11/8/16	U.S.	11/4/2016	11/7/2016	YouC
5	2016	President	polls-plus	Clinton vs. Trump vs. Johnson	11/8/16	U.S.	11/3/2016	11/6/2016	Grav
6	2016	President	polls-plus	Clinton vs. Trump vs. Johnson	11/8/16	U.S.	11/3/2016	11/6/2016	Fox I
7	2016	President	polls-plus	Clinton vs. Trump vs. Johnson	11/8/16	U.S.	11/2/2016	11/6/2016	CBS
8	2016	President	polls-plus	Clinton vs. Trump vs. Johnson	11/8/16	U.S.	11/3/2016	11/5/2016	NBC
9	2016	President	polls-plus	Clinton vs. Trump vs. Johnson	11/8/16	New Mexico	11/6/2016	11/6/2016	Zia P
10	2016	President	polls-plus	Clinton vs. Trump vs. Johnson	11/8/16	U.S.	11/4/2016	11/7/2016	IBD/
11	2016	President	polls-plus	Clinton vs. Trump vs. Johnson	11/8/16	U.S.	11/4/2016	11/6/2016	Selze
12	2016	President	polls-plus	Clinton vs. Trump vs. Johnson	11/8/16	U.S.	11/1/2016	11/4/2016	Angt
13	2016	President	polls-plus	Clinton vs. Trump vs. Johnson	11/8/16	U.S.	11/3/2016	11/6/2016	Mon
14	2016	President	polls-plus	Clinton vs. Trump vs. Johnson	11/8/16	Virginia	11/3/2016	11/4/2016	Publ
15	2016	President	polls-plus	Clinton vs. Trump vs. Johnson	11/8/16	U.S.	11/1/2016	11/3/2016	Marl

Showing 1 to 16 of 12,624 entries, 27 total columns

Figure 9-1   Get and examine the data (part 1)

The first example in part 2 of figure 9-1 shows how to check the unique number of values stored in each column. Here, five columns have only one unique value. For example, the cycle column has a value of 2016 for every row. Plainly, these columns aren't useful to the analysis and can be removed from the data set. Meanwhile, the state column has 57 unique values, but the United States has only 50 states plus the District of Columbia. That's something to look into.

Several other columns have a small number of unique values. For example, the type column has just three unique values, the population column has four, and the timestamp column has three as well. Curiously, the poll_id and question_id columns have 4208 values, which is one-third of the number of rows in the data set. All of these results raise questions that need to be answered.

At this point, you have identified some columns that aren't needed, but you should also be able to identify a few columns that are definitely needed for this analysis as well, such as the state and date columns. For this analysis, we decided to focus on the end dates and not start dates for each poll.

The columns that store the poll percentages have two clear types, one set prefixed with "rawpoll_" and one with "adjpoll_". The difference is important to know, and even if you're a hardcore politics watcher who knows the difference between raw polling data and adjusted data, you won't know *how* the data has been adjusted unless you do some research.

The best way to find out what the data represents is to check with the source of the data set. If you read about the methodology posted on FiveThirtyEight's website, you'll learn that these adjusted numbers are meant to take into account factors like the "house effect" of different polling businesses or the grade FiveThirtyEight has assigned the pollster.

Interestingly, the "raw" data given by pollsters is usually already adjusted. For example, if a poll seems to have under-sampled likely Hispanic voters, the pollster might multiply the Hispanic responses to better match their model of the electorate. The exact methodology for how each pollster does this is generally a carefully guarded secret, as having an accurate model is key to making accurate polls.

This is all important information to keep in mind when performing an analysis. In addition, it's a great example for real-world uses for data analysis! For this analysis, we decided to use the rawpoll data instead of the adjpoll data.

View the unique item count for each column

```
apply(X = polls, MARGIN = 2, FUN = unique) %>%
  lapply(FUN = length) %>% str()
List of 27
 $ cycle           : int 1
 $ branch          : int 1
 $ type            : int 3
 $ matchup         : int 1
 $ forecastdate    : int 1
 $ state           : int 57
 $ startdate       : int 352
 $ enddate         : int 345
 $ pollster        : int 196
 $ grade           : int 11
 $ samplesize      : int 1767
 $ population      : int 4
 $ poll_wt         : int 4399
 $ rawpoll_clinton : int 1312
 $ rawpoll_trump   : int 1385
 $ rawpoll_johnson : int 585
 $ rawpoll_mcmullin: int 17
 $ adjpoll_clinton : int 12569
 $ adjpoll_trump   : int 12582
 $ adjpoll_johnson : int 6630
 $ adjpoll_mcmullin: int 58
 $ multiversions   : int 1
 $ url             : int 1305
 $ poll_id         : int 4208
 $ question_id     : int 4208
 $ createddate     : int 222
 $ timestamp       : int 3
```

A source for the data set methodology

https://fivethirtyeight.com/features/a-users-guide-to-fivethirtyeights-2016-general-election-forecast/

Question

- Should the analysis use the start date, end date, or both for each poll?
- What's the difference between the rawpoll and adjpoll columns?

Investigation

- The amount of time it took to conduct a poll is unlikely to affect the data very much.
- The adjpoll data has been adjusted by FiveThirtyEight according to various factors.

Decision

- Use only the end date for each poll to simplify the analysis.
- Use the rawpoll data for the analysis and discard the adjpoll columns.
- Drop any columns that only have one unique value because they aren't useful.
- Drop columns with specific pollster or question data, like poll_id or question_id, because they aren't relevant to our analysis.

Figure 9-1   Get and examine the data (part 2)

The first example in figure 9-1 part 3 displays the unique value counts for the type column by using the table() function. The three different values ("now-cast", "polls-only", and "polls-plus") each have 4208 rows. Which, if any, are useful for our analysis?

Again, the answer can be found from the data source. The FiveThirtyEight website tells us that each type refers to a different model for weighting and adjusting the polling data. To show how this works, the second example selects the type column along with one column with raw data and another with adjusted data. Since we decided to use the raw data, not the adjusted data, we can pick any one of the type values and use it to select the rows for our analysis. Then, we can drop the rest of the rows.

The third example shows how to view the unique values in the state column. This shows that there's a value of "U.S." for national polls and a value of "District of Columbia" for Washington D.C. It also shows that in addition to polls for Maine and Nebraska as a whole, some polls are for those states' individual districts. This is because those two states award their electoral votes by district, so there is value in knowing if, say, Maine's electoral votes will be split between candidates. As neither state has very many electoral votes, however, and to keep things simple, this analysis drops the polls for individual congressional districts.

View value counts for the type column

```
table(polls$type)
  now-cast polls-only polls-plus
      4208       4208       4208
```

Compare the type column with the poll data columns

```
polls %>%
  select(type, state, enddate, rawpoll_clinton, adjpoll_clinton) %>%
  arrange(state, enddate, type)
# A tibble: 12,624 x 5
    type        state    enddate    rawpoll_clinton adjpoll_clinton
    <chr>       <chr>    <chr>                 <dbl>           <dbl>
  1 now-cast    Alabama  1/12/2016                32            31.0
  2 polls-only  Alabama  1/12/2016                32            30.9
  3 polls-plus  Alabama  1/12/2016                32            30.9
  4 now-cast    Alabama  10/13/2016             38.3            37.3
  5 polls-only  Alabama  10/13/2016             38.3            37.2
  6 polls-plus  Alabama  10/13/2016             38.3            37.2
...
```

View the unique values for the state column

```
unique(polls$state)
 [1] "U.S."                  "New Mexico"        "Virginia"
 [4] "Iowa"                  "Wisconsin"         "North Carolina"
 [7] "Georgia"               "Florida"           "Oregon"
[10] "Ohio"                  "South Carolina"    "New York"
[13] "Michigan"              "Pennsylvania"      "Missouri"
[16] "New Hampshire"         "Arizona"           "Nevada"
[19] "Colorado"              "California"        "Washington"
[22] "Texas"                 "Utah"              "Illinois"
[25] "Indiana"               "Tennessee"         "Connecticut"
[28] "Massachusetts"         "New Jersey"        "Kansas"
[31] "Kentucky"              "Minnesota"         "Oklahoma"
[34] "Maryland"              "Alabama"           "Nebraska"
[37] "Louisiana"             "Maine"             "Arkansas"
[40] "Alaska"                "Vermont"           "Idaho"
[43] "Mississippi"           "West Virginia"     "South Dakota"
[46] "Montana"               "Hawaii"            "Maine CD-1"
[49] "Maine CD-2"            "Rhode Island"      "Nebraska CD-3"
[52] "Nebraska CD-1"         "Delaware"          "North Dakota"
[55] "District of Columbia"  "Nebraska CD-2"     "Wyoming"
```

Questions

- Why does the type column contain three unique values, each a third of the rows?
- Why does the state column contain 57 unique values when there are 50 states?

Investigation

- Each version of now-cast, polls-plus, and polls-only weights the data differently.
- The state column includes the U.S., congressional districts, and Washington D.C.

Decision

- Keep the rows with "now-cast" in the type column and drop the rest.
- Drop the polls for individual congressional districts to simplify the analysis.

Figure 9-1   Get and examine the data (part 3)

The population column contains four unique values: "lv", "rv", "a", and "v". Consulting the methodology, the first two definitely stand for Likely Voters and Registered Voters. The value "a" is presumably [All] Adults, which is mentioned in the methodology along with likely and registered voters, but what is "v"? Perhaps [All] Voters? Before we spend more time looking into it, however, let's see how many rows use these values.

The first example in part 4 of figure 9-1 uses a bar plot to generate a visual representation of the value counts for the population column. The "a" and "v" values account for very few rows compared to "lv" and "rv", and are unlikely to be statistically significant. Because of that, this analysis drops them.

It's possible to get the same data by using the table() function. However, using a bar plot makes it easier to compare the number of "a" and "v" values to the "lv" and "rv" values.

In addition to bar plots, it's common to examine data by using distribution plots to visualize how the values for columns with continuous data are distributed. For example, you might want to use the geom_density_2d() function to view the distribution of the values in the samplesize or poll_wt columns.

It's also common to use scatter plots to visualize the relationship between columns. For example, the second example here shows how to visualize the relationship between the values in the samplesize, poll_wt, and grade columns. This shows that most of the values in the samplesize and poll_wt columns are clustered on the low end of the scale but that there are some outliers on the high end. Also, it doesn't show an obvious correlation between the grade, sample size, and poll_wt columns.

Furthermore, FiveThirtyEight uses the poll_wt column (or poll weight) to calculate its adjusted data, which we have already decided not to use. As a result, this analysis doesn't use any of these columns.

Use a plot to examine the population column

```
ggplot(polls, aes(x = population, fill = population)) +
    geom_bar()
```

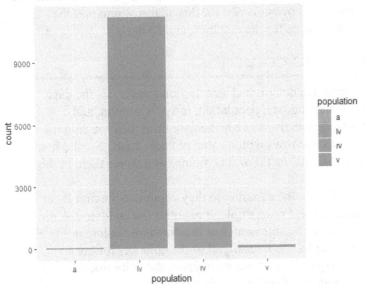

Use a scatter plot to examine the relationships between three columns

```
ggplot(polls, aes(x = poll_wt, y = samplesize, color = grade)) +
    geom_point(size = 3)
```

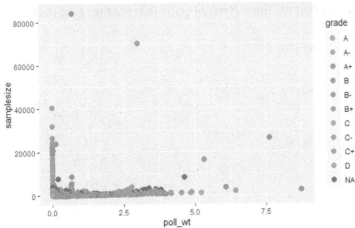

Question

- Are the population, grade, samplesize, and poll_wt columns useful for this analysis?

Investigation

- These columns are used by FiveThirtyEight to determine data adjustments.

Decision

- Keep the rows with "rv" and "lv" in the population column and drop the rest.

Figure 9-1   Get and examine the data (part 4)

Clean the data

Figure 9-2 shows how to clean the data for the Polling analysis. To do that, this code selects the columns and rows needed for this analysis, renames the columns, sets the correct data types for the columns, and so on.

Select and rename the columns

After examining the data, we determined that this analysis needs the data from these five columns: state, enddate, population, rawpoll_clinton, and rawpoll_trump. In addition, this analysis needs the data from the type column so we can use it to select only the rows with a value of "now-cast". So, the first example in part 1 of figure 9-2 selects these six columns and stores them in the polls tibble.

The second example renames the columns so they capitalize the first letter in each word of the column name. For example, it renames the enddate column to EndDate. In addition, it simplifies the names of the "rawpoll" columns by dropping the "rawpoll_" prefix. As a result, all of the columns follow a consistent naming scheme. By contrast, before renaming the columns, they used a mix of all lowercase (enddate) and snake case (rawpoll_clinton).

Sort the rows

The third example sorts the rows by the State column first and then the EndDate column. This shows that the three types of polls provide the same data for our selected columns.

Select the columns to use

```
polls <- polls %>%
  select(type, state, enddate, population,
         rawpoll_clinton, rawpoll_trump)
```

Rename the columns

```
polls <- polls %>% rename(
  Type = type, State = state, EndDate = enddate,
  Population = population,
  Clinton = rawpoll_clinton, Trump = rawpoll_trump)
```

Sort the rows

```
polls <- polls %>% arrange(State, EndDate)
```

The data so far

```
polls
# A tibble: 12,624 x 6
   Type       State    EndDate    Population Clinton Trump
   <chr>      <chr>    <chr>      <chr>        <dbl> <dbl>
 1 polls-plus Alabama  1/12/2016  rv              32    68
 2 now-cast   Alabama  1/12/2016  rv              32    68
 3 polls-only Alabama  1/12/2016  rv              32    68
 4 polls-plus Alabama  10/13/2016 lv            38.3  53.8
 5 now-cast   Alabama  10/13/2016 lv            38.3  53.8
 6 polls-only Alabama  10/13/2016 lv            38.3  53.8
 7 polls-plus Alabama  10/14/2016 lv            19.6  54.8
 8 now-cast   Alabama  10/14/2016 lv            19.6  54.8
 9 polls-only Alabama  10/14/2016 lv            19.6  54.8
10 polls-plus Alabama  10/19/2016 lv            23.7  53.4
# ... with 12,614 more rows
```

Figure 9-2   Clean the data (part 1)

Select the rows

During the data examination phase, we determined that several sets of rows wouldn't be needed for our analysis. The first example in part 2 of figure 9-2 selects only rows with "now-cast" in the Type column, dropping the "polls-plus" and "polls-only" rows from the tibble. Then, it drops the Type column since it's no longer needed.

The second example selects the rows in the Population column that have values or "lv" or "rv", dropping the rows that have values of "a" or "v". The third example removes rows where the State column specifies a congressional district, not a state. To do that, this code uses the filter() and str_detect() functions to select all rows where the State column does not contain a string of "CD-". Then, the fourth example checks the number of unique values now in the State column. This returns a count of 52, which is what we want: 50 states, plus the District of Columbia and national polls.

Improve some columns

The fifth example improves the Population column for readability. At this point, this column contains two values: "rv" for registered voter and "lv" for likely voter. To make what they stand for clearer, this code uses the mutate() and str_replace() functions to replace "lv" with "Likely" and "rv" with "Registered". Then, it pipes the resulting tibble into the rename() function to rename the Population column to VoterType. This makes it clear that the column contains data about the type of voter included in each poll.

The sixth example converts the EndDate column to the Date type. This allows us to use this column in range-based filtering operations and to easily extract components of the date with the format() function.

The seventh example displays the data set so far. This tibble now has only 4,116 rows and 5 columns. Despite the small number of columns, there is still plenty of data to analyze.

The eighth example shows how to save the data in an RDS file named polls_clean.rds. This is a good practice because it provides an easy way for you to return to the clean data if you need it again. For example, if you accidentally mess up the data that's stored in the polls variable during the preparation phase, you don't have to rerun the parts of the script that get and clean the data. Instead, you can start again from this point by reading the clean data.

Select only "now-cast" rows and drop the Type column

```
polls <- polls %>%
  filter(Type == "now-cast") %>%
  select(-Type)
```

Select only rows with "lv" or "rv" in the Population column

```
polls <- polls %>% filter(Population %in% c("lv", "rv"))
```

Remove the rows for congressional districts

```
polls <- polls %>% filter(!str_detect(State, "CD-"))
```

Check the number of states

```
length(unique(polls$State))
[1] 52
```

Rename the Population column and improve its values

```
polls <- polls %>%
  mutate(Population = str_replace(Population,"lv", "Likely"),
         Population = str_replace(Population, "rv", "Registered")) %>%
  rename(VoterType = Population)
```

Convert the EndDate column to the Date type

```
polls <- polls %>% mutate(EndDate = as.Date(EndDate, format="%m/%d/%Y"))
```

The data so far

```
polls
# A tibble: 4,116 x 5
    State   EndDate    VoterType  Clinton Trump
    <chr>   <date>     <chr>        <dbl> <dbl>
 1 Alabama 2016-01-12 Registered    32    68
 2 Alabama 2016-10-13 Likely        38.3  53.8
 3 Alabama 2016-10-14 Likely        19.6  54.8
 4 Alabama 2016-10-19 Likely        23.7  53.4
 5 Alabama 2016-10-20 Likely        38.4  52.4
 6 Alabama 2016-10-24 Likely        22.4  56.8
 7 Alabama 2016-10-25 Likely        36    52
 8 Alabama 2016-10-26 Likely        36    52
 9 Alabama 2016-10-27 Likely        38.9  51.3
10 Alabama 2016-10-28 Likely        36    53
# ... with 4,106 more rows
```

Save the data

```
saveRDS(polls, file = "../../data/polls_clean.rds")
```

Figure 9-2   Clean the data (part 2)

Prepare the data

Figure 9-3 shows how to prepare the polling data now that it's been cleaned. To start, the first example reads the clean polling data created by the previous figure.

Add columns

The second example shows how to add two calculated columns. Here, the first statement uses the mutate() function to create the Gap column. This column calculates the gap (difference) between Trump and Clinton in the polls by subtracting the Trump column from the Clinton column. If the result is positive, Clinton is polling higher than Trump. If it's negative, Trump is polling higher than Clinton.

Then, the second statement calculates the mean (average) gap for each state. To calculate this value, the code groups the data by the State column and uses the mutate() and mean() functions to add a new column named StateGap that stores the mean gap for each state. Finally, this statement uses the ungroup() function to remove the group from the polls data set.

The third example adds a column to determine if a state is a swing state. In this case, the code considers a state to be a swing state if the State column isn't "U.S." and the absolute value of the mean state gap is less than 7. It's possible to find a better way to determine if a state is a swing state, but this technique works well enough for this analysis.

The fourth example drops the StateGap column. That's because this column is only needed to determine if a state is a swing state.

Pivot the data

The fifth example pivots the wide data to long data. To do this, the code uses the pivot_longer() function. This combines the Clinton and Trump columns into a Candidate column and a Percent column. This makes the data easier to plot later in the analysis.

The data so far shows that the Swing column contains a logical (Boolean) value of TRUE or FALSE. In addition, the data set now has twice as many rows due to the pivot operation. In other words, the data is longer.

The seventh example saves the data in a new RDS file. That way, you can easily get the prepared polling data if you need it again.

Read the clean data

```
polls <- readRDS("../../data/polls_clean.rds")
```

Create two gap columns

```
polls <- polls %>% mutate(Gap = Clinton - Trump)

polls <- polls %>% group_by(State) %>%
  mutate(StateGap = mean(Gap)) %>%
  ungroup()
```

Create a column for swing states

```
polls <- polls %>% mutate(
  Swing = ifelse(State != "U.S." & (abs(StateGap) < 7),
  TRUE, FALSE))
```

Drop the unneeded StateGap column

```
polls <- select(polls, -StateGap)
```

Create a long version of the data

```
polls <- pivot_longer(polls, cols = c("Clinton", "Trump"),
                      names_to = "Candidate", values_to = "Percent")
```

The data so far

```
polls
# A tibble: 8,232 x 7
    State    EndDate    VoterType    Gap Swing Candidate Percent
    <chr>    <date>     <chr>      <dbl> <lgl> <chr>       <dbl>
 1 Alabama 2016-01-12 Registered   -36   FALSE Clinton      32
 2 Alabama 2016-01-12 Registered   -36   FALSE Trump        68
 3 Alabama 2016-10-13 Likely      -15.5  FALSE Clinton      38.3
 4 Alabama 2016-10-13 Likely      -15.5  FALSE Trump        53.8
 5 Alabama 2016-10-14 Likely      -35.2  FALSE Clinton      19.6
 6 Alabama 2016-10-14 Likely      -35.2  FALSE Trump        54.8
 7 Alabama 2016-10-19 Likely      -29.7  FALSE Clinton      23.7
 8 Alabama 2016-10-19 Likely      -29.7  FALSE Trump        53.4
 9 Alabama 2016-10-20 Likely      -14.0  FALSE Clinton      38.4
10 Alabama 2016-10-20 Likely      -14.0  FALSE Trump        52.4
# ... with 8,222 more rows
```

Save the data

```
saveRDS(polls, file = "../../data/polls_prepared.rds")
```

Figure 9-3   Prepare the data

Analyze the data

Figure 9-4 shows how to analyze the data. Most of this analysis involves plotting the data so you can visualize the relationships between the variables. To start, the first example in part 1 reads the prepared polling data saved by the previous figure.

Plot the national polls

The second example shows how to create a line plot for polls where the State column is "U.S.". In other words, it plots the national polls. In this plot, the line for Clinton is displayed in blue, which is the color of her political party, the Democratic party. By contrast, the line for Trump is displayed in red, which is the color of his party, the Republican party.

The resulting plot shows the lines get compressed as the x axis approaches the election date. This indicates that there were increasingly more polls released closer to election day, November 9th. The lines become especially compressed and hard to read around September.

This plot shows that Clinton leads consistently in the national polls towards the start of the polling period but the race becomes closer as election day approaches. The pattern becomes more difficult to see at the end, however, because of the vertical lines in the plot. These vertical lines are due to several polls ending on the same day.

One way to make it easier to visualize the trends for the two candidates is to use a smooth line plot instead of a regular line plot. That way, the plot attempts to draw a smooth line through the data points as shown by the second example. This plot provides a different way to visualize the same data as the first plot.

In the second example, two plots have been layered over each other. The scatter plot shows the individual data points for the polls, and the smooth line provides a trend line.

Read the prepared data

```
polls <- readRDS("../../data/polls_prepared.rds")
```

The national polls with a line plot

```
ggplot(filter(polls, State == "U.S."),
       aes(x = EndDate, y = Percent, color = Candidate)) +
geom_line() +
scale_color_manual(values = c("blue", "red")) +
labs(title = "Polls for the U.S.", x = "Date", y = "") +
theme(plot.title = element_text(hjust = 0.5),
      legend.position = "bottom")
```

The national polls with a scatter plot and a smooth line

```
ggplot(filter(polls, State == "U.S."),
       aes(x = EndDate, y = Percent, color = Candidate)) +
geom_point() +
geom_smooth(se = FALSE, size = 2) +
# other functions are same as previous plot
```

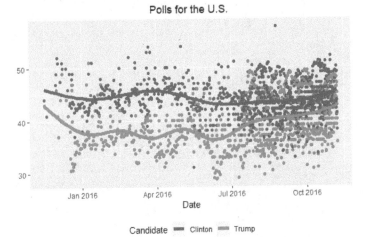

Figure 9-4   Analyze the data (part 1)

The first example in part 2 of figure 9-4 shows another plot of the national polls. However, this plot only shows a smooth line, not a scatter plot. As a result, it's easier to visualize how Clinton had a significant polling lead in the spring and summer that got smaller closer to election day.

One interesting thing to note is that there are times when both Clinton and Trump fall or rise in the polls. These correlate with events like the party conventions, positive or negative news stories, the rise or fall of other candidates such as Johnson and McMullin in the polls, and so on.

Plot the polls for swing states

The national polls give some insight into how the nation felt about the candidates over the course of the election. However, in the U.S., elections aren't decided by the national popular vote. Instead, they're decided by an electoral system in which each state has a certain number of electoral votes.

Many states are considered very likely for one candidate or another. Because of that, they don't get polled as often as the swing states. It doesn't matter if Clinton wins California by 10% or 25% because she gets the same number of electoral votes either way. And if Clinton is losing in California, she's probably losing everywhere else, too. This is another example of why it's important to understand the data you're analyzing.

For this reason, gaining insights into who might win the so-called "swing states" are most important for predicting who might win the election. These are the states that are most likely to have close results.

The second example in part 2 of figure 9-4 plots the polls for only the swing states using the geom_smooth() function. Like the national plot, this plot shows that Clinton had a significant polling lead in the spring and summer that got smaller as the election got closer to election day. However, this plot also shows that the polls got extremely close for swing states in the week before the election.

The national polls with a smooth line (no scatter plot)

```
ggplot(filter(polls, State == "U.S."),
       aes(x = EndDate, y = Percent, color = Candidate)) +
  geom_smooth(se = FALSE) +
  scale_color_manual(values = c("blue", "red")) +
  labs(title = "Polls for the U.S.", x = "Date", y = "") +
  theme(plot.title = element_text(hjust = 0.5),
        legend.position = "bottom")
```

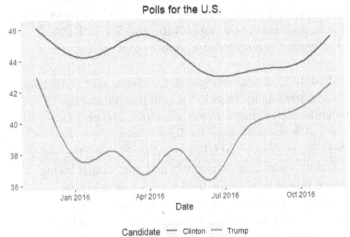

The swing state polls with a smooth line

```
ggplot(polls %>% filter(Swing == TRUE),
       aes(x = EndDate, y = Percent, color = Candidate)) +
  geom_smooth(se = FALSE) +
  scale_color_manual(values = c("blue", "red")) +
  labs(title = "Polls for the Swing States", x = "Date", y = "") +
  theme(plot.title = element_text(hjust = 0.5),
        legend.position = "bottom")
```

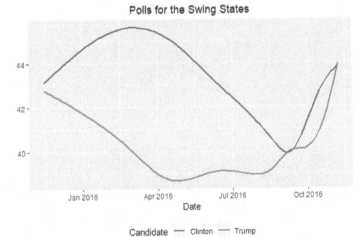

Figure 9-4   Analyze the data (part 2)

Since the polls got so close for swing states in the weeks before the election, you might want to only plot the last three months before the election. One way to do this is shown by the first example in part 3 of figure 9-4. In addition, this plot only displays polls of likely voters, not registered voters. That's because likely voters are by definition more likely to vote than registered voters. Of course, how the pollster determines a voter is "likely" varies from pollster to pollster and is not exact science.

Although the first plot shows that the polling is close for swing states as a group, U.S. elections are ultimately decided state by state. As a result, to analyze which candidate might win in each state, you can plot the data for each state individually. That's why the second example shows how to plot the data for selected swing states.

To save space, this code only plots two swing states, Arizona and Wisconsin. However, you can easily add other swing states to the plot just by adding their names to the vector named states shown in this example. The plot for Arizona shows that Clinton led in the polls until the final weeks of the election. Meanwhile, the plot for Wisconsin shows that Clinton led in the polls the entire time. However, Trump ended up winning in Wisconsin and many other swing states where the polls showed him behind in the final week of the election.

One thing to consider when making plots like this is the scale. Small differences can be exaggerated when the scale is focused on just a small part of the plot. Notice the y axis in these two plots only shows values from 30% to 45%.

Also note that using a smooth line obscures how many data points are being used to generate the plot. One common complaint after the election was that certain states were under polled. That is, there weren't enough polls conducted to get a truly accurate picture of the electorate.

Furthermore, remember that we're using only the end dates for polls in our analysis. If a poll took place over a week or more, it's possible that a shift taking place in the last two days of the poll wouldn't be accurately reflected.

Likely voters in swing states three months prior to the election

```
ggplot(polls %>% filter(Swing == TRUE &
                        VoterType == "Likely" &
                        EndDate > "2016-08-01"),
       aes(x = EndDate, y = Percent, color = Candidate)) +
  geom_smooth(se = FALSE) +
  scale_color_manual(values = c("blue", "red")) +
  labs(title = "Likely voters in Swing States", x = "Date", y = "") +
  theme(plot.title = element_text(hjust = 0.5),
        legend.position = "bottom")
```

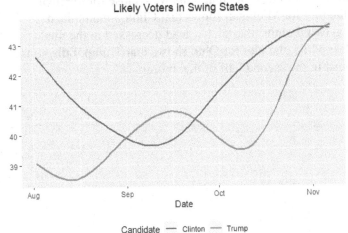

Two swing states

```
states <- c("Arizona","Wisconsin") # add more if you want

ggplot(polls %>% filter(State %in% states & EndDate > "2016-08-01"),
       aes(x = EndDate, y = Percent, color = Candidate)) +
  geom_smooth(se = FALSE) +
  scale_color_manual(values = c("blue", "red")) +
  facet_grid(vars(State)) +
  theme(legend.position = "bottom")
```

Figure 9-4   Analyze the data (part 3)

Part 4 of figure 9-4 shows how to plot the gap data for selected states. In this example, the code begins by setting the starting date for the plot to September 1, which was approximately two months before election day. Then, the code creates a vector for the states of Florida, Michigan, and Ohio. Next, the code plots the gap data for these three states beginning at the start date.

To make these plots easier to read, the geom_hline() function adds a horizontal black line to each plot at 0 on the y axis. This line makes it easy to determine which candidate is leading in the polls. If the blue line for the gap data goes above the black line, Clinton is leading. Otherwise, Trump is leading.

For example, the plot for Florida shows that polls were close for the final two months. However, the plot for Michigan shows that Clinton maintained a polling lead for the final two months, though her lead decreased in the final week of the election. Meanwhile, the plot for Ohio shows that Trump trailed on October 1 but took the lead in the second half of that month.

Plot the gap data for selected states

```
start_date <- as.Date("2016-09-01")

states <- c("Florida","Michigan","Ohio")

ggplot(polls %>% filter(State %in% states & EndDate > start_date),
        aes(x = EndDate, y = Gap)) +
  geom_point() +
  geom_smooth(se = FALSE) +
  geom_hline(yintercept = 0, size = 1) +
  facet_grid(vars(State))
```

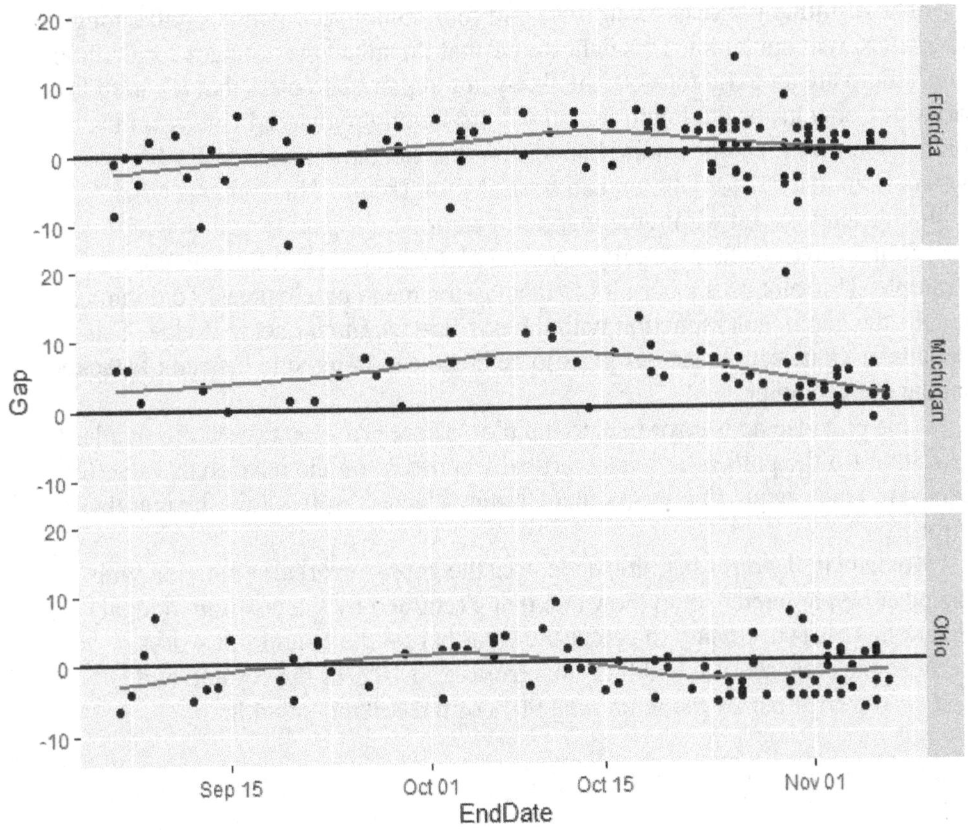

Description

- The three swing states in this figure were selected from the fifteen swing states identified by this analysis.

- In the plot, the black line represents a gap of 0. Dots above the line indicate that Clinton is polling higher. Dots below the line indicate that Trump is polling higher.

Figure 9-4   Analyze the data (part 4)

Analyze the polls by voter type

So far, the plots have focused on analyzing the data over time with smooth line plots. Of course, there are many other ways to analyze the polling data. For example, you can summarize statistics to analyze some of the categorical data such as the VoterType column.

The first example in part 5 of figure 9-4 shows how to do this. First, the code selects only polls from swing states. Then, it groups the polls by the VoterType and Candidate columns. Next, it calculates the mean percent and the standard deviation for each group.

The resulting tibble has four rows and four columns, which is small enough to analyze without a plot. This data shows that the mean percentages for Clinton and Trump are so close for both the likely and registered voters that it's hard to draw any conclusions. Clinton leads by 2.4% among registered voters but by only 1.1% among likely voters. In addition, the standard deviation for likely voters is almost 5% for Clinton and over 5% for Trump. This indicates that Clinton's lead is well within the margin of error.

To interpret this data visually, you can plot it as shown in the second example. This plot uses a bar plot to display the mean percentages. To do this, it calls the geom_col() function with the position parameter set to dodge. This parameter value causes the bar plots to be created side-by-side instead of stacked on top of each other.

This plot also adds error bars to the plot. These error bars show the standard deviation for the polls after being vertically centered on the maximum value for each bar. Once again, this shows that Clinton's lead is well within the margin of error.

To display the error bar, this code uses the geom_errorbar() function with the position parameter set to the value that's returned by the position_dodge() function. This is necessary to center the error bar on the column. It works because the code sets the width of the error bar to 10% of the column and dodges the error bar by the remaining 90% of the column's width.

Summarize the data by voter type and candidate

```
voter_types <- polls %>%
  filter(Swing == TRUE) %>%
  group_by(VoterType, Candidate) %>%
  summarize(MeanPercent = mean(Percent), SD = sd(Percent))
voter_types
# A tibble: 4 x 4
# Groups:   VoterType [2]
  VoterType  Candidate MeanPercent    SD
  <chr>      <chr>           <dbl> <dbl>
1 Likely     Clinton          42.3  4.67
2 Likely     Trump            41.2  5.50
3 Registered Clinton          42.8  4.43
4 Registered Trump            40.4  4.56
```

Plot the data

```
ggplot(voter_types,
       aes(x = Candidate, y = MeanPercent, fill = VoterType)) +
  geom_col(position = "dodge") +
  geom_errorbar(aes(ymin = MeanPercent - SD, ymax = MeanPercent + SD),
                width = 0.1, position = position_dodge(0.9)) +
  labs(y = "", x = "")
```

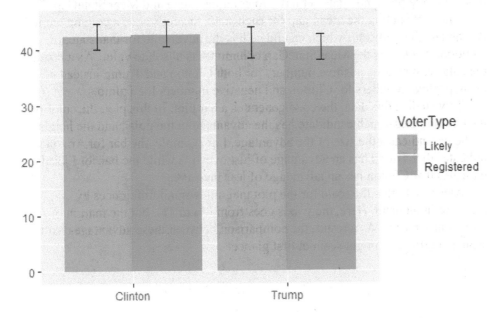

Description

- The error bars show that the polls for swing states were within the margin of error for both likely and registered voters.

Figure 9-4　Analyze the data (part 5)

More preparation and analysis

At this point, you've done enough exploratory analysis to get some ideas for more in-depth analysis. For example, you might want to look at the polling gap in the last week of the election in key swing states. Or, you might want to look at how the polling gap changed each week in the swing states over the last two months leading up to the election.

Figure 9-5 shows that ideas like these require more preparation and further analysis. That's to be expected because data analysis often requires going back and forth between the preparation and analysis phases.

Plot the gap for the last week of the election

Part 1 of figure 9-5 begins by showing how to clean and prepare the data for the last week of the election. To start, this example creates a vector of the swing states to be plotted. Then, it selects the polls for those states. In addition, it selects the polls with an end date after November 1. Since election day is November 9, this selects polls from approximately the last week of the election. Next, it groups the data by state and calculates the mean gap for each state. At this point, the polls_nov data set has two columns, State and MeanStateGap.

After calculating the mean gap for each state, the code adds a column named Advantage that indicates which candidate has the advantage for that state. In addition, it converts the MeanStateGap column to its absolute value. As a result, the data set now uses positive numbers for both Clinton and Trump instead of using positive numbers for Clinton and negative numbers for Trump.

To visualize this data, the code generates a bar plot. In this plot, the color of the bar indicates which candidate has the advantage in the polls, and the height of the bar indicates the size of the advantage. For example, the bar for Arizona indicates that Trump has an advantage of just over 1% while the bar for Florida indicates that Clinton has an advantage of less than 1%.

Again, note that the scale for the plot magnifies small differences by zooming in on them. Here, the y axis goes from 0% to 4%, but the margin of error is around 5%. As a result, the comparison between these advantages isn't as meaningful as it might seem at first glance.

Prepare the data for the last week of the election

Get the mean gap for selected swing states

```
states <- c("Arizona","Florida","Iowa","Nevada","North Carolina",
            "Ohio","Pennsylvania","Wisconsin")
polls_nov <- polls %>%
  filter(State %in% states & EndDate > as.Date("2016-11-01")) %>%
  group_by(State) %>%
  summarize(MeanStateGap = mean(Gap))
```

Add an Advantage column and get the absolute value for the mean gap

```
polls_nov <- polls_nov %>% mutate(
  Advantage = ifelse(MeanStateGap >= 0, "Clinton","Trump"),
  MeanStateGap = abs(round(MeanStateGap, 3)))
```

The prepared data

```
polls_nov
# A tibble: 8 x 3
  State          MeanStateGap Advantage
  <chr>                 <dbl> <chr>
1 Arizona               1.38  Trump
2 Florida               0.573 Clinton
3 Iowa                  4.24  Trump
4 Nevada                0.414 Clinton
5 North Carolina        2.47  Clinton
6 Ohio                  1.65  Trump
7 Pennsylvania          2.61  Clinton
8 Wisconsin             4.71  Clinton
```

Plot the data

```
ggplot(polls_nov) +
  geom_col(aes(x = State, y = MeanStateGap, fill = Advantage)) +
  scale_fill_manual(labels = c("Clinton","Trump"),
                    values = c("blue","red")) +
  labs(title = "Results for Final Week of Election",
       x = "", y = "Mean Percent") +
  theme(plot.title = element_text(hjust = 0.5))
```

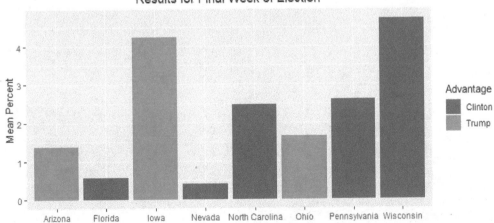

Figure 9-5   More preparation and analysis (part 1)

Plot the weekly gap over time

This analysis finishes by plotting the gap for each swing state for each week over the last two months of the election. But first, this analysis needs to prepare the data for this plot.

To start, the first example in part 2 of figure 9-5 gets the polls for the swing states and selects three columns (State, EndDate, and Gap). Then, it selects only the unique rows. This drops the duplicate rows that resulted from creating one row for Clinton and another for Trump when we previously pivoted the data to create long data.

After selecting the initial data set, the code needs to put each poll in a weekly bin for the weeks leading up to the election. To do that, the second example defines a custom function named get_next_sunday() that gets the last day of the week (Sunday) for the EndDate value for each poll. This function defines a parameter for a row, gets the Date object that's stored in the second column, extracts the numerical day of the week from that object, and returns the Sunday for that week. Or, if the EndDate value is a Sunday, it returns that date.

The third example uses the get_next_sunday() function to add a new column named Week to the polls_weekly data set. Adding the new column converts the date to an integer value that represents the number of days since Jan 1, 1970. To convert this integer value back to the Date type, this code calls the as.Date() function and sets its origin parameter to "1970-01-01". The resulting tibble displayed by the fourth example shows that the Week bin groups each poll by week.

Prepare the data for the weekly gap of the swing states

Select only the rows and columns that are needed for this plot

```
polls_weekly <- polls %>% filter(Swing == TRUE) %>%
  select(State, EndDate, Gap) %>%
  unique()
```

Create a function for putting each end date into a weekly bin

```
get_next_sunday <- function(row) {
  date <- as.Date(row[2])
  day_of_week <- as.integer(format(date, "%w"))
  next_sunday <- ""
  if(day_of_week == 0) {
    next_sunday <- date
  } else {
    next_sunday <- date + (7 - day_of_week)
  }
  return(next_sunday)
}
```

Use the function to add a column to hold the date bins

```
polls_weekly <- polls_weekly %>%
  mutate(Week = apply(MARGIN = 1, FUN = get_next_sunday),
         Week = as.Date(Week, origin = "1970-01-01"))
```

The data so far

```
polls_weekly
# A tibble: 1,283 x 4
   State   EndDate       Gap Week
   <chr>   <date>      <dbl> <date>
 1 Arizona 2016-10-12   1    2016-10-16
 2 Arizona 2016-10-13  -5.64 2016-10-16
 3 Arizona 2016-10-14   2    2016-10-16
 4 Arizona 2016-10-14   9.33 2016-10-16
 5 Arizona 2016-10-15   5.6  2016-10-16
 6 Arizona 2016-10-16  -3    2016-10-16
 7 Arizona 2016-10-18   0    2016-10-23
 8 Arizona 2016-10-19  10.4  2016-10-23
 9 Arizona 2016-10-20  -2    2016-10-23
10 Arizona 2016-10-24  -1    2016-10-30
# ... with 1,273 more rows
```

Figure 9-5   More preparation and analysis (part 2)

In part 3 of figure 9-5, the first example begins by selecting rows where the Week column is later than September 1. Then, it groups the polls_weekly data set by state and week, calculates the mean gap for each group, and ungroups the tibble.

After calculating the mean gap, this analysis uses the ifelse() function to determine which candidate has the advantage and assigns that candidate to a column named Advantage. Then, it modifies the MeanGap column so it stores the absolute value for the mean gap. This is similar to the previous Advantage column except now it's for each week, not each poll.

At this point, the polls_weekly tibble has all of the data that's needed for the plot. Each row contains the name of the state, the bin for the week, the absolute mean gap for the state, and the leading candidate.

Prepare the data for the weekly gap of the swing states (continued)

Get the mean poll for each date bin

```
polls_weekly <- polls_weekly %>%
  filter(Week > as.Date("2016-09-01")) %>%
  group_by(State, Week) %>%
  summarize(MeanGap = mean(Gap)) %>%
  ungroup()
```

Add an Advantage column and get the absolute value for the mean gap

```
polls_weekly <- polls_weekly %>%
  mutate(Advantage = ifelse(MeanGap >= 0, "Clinton", "Trump"),
         MeanGap = abs(round(MeanGap, 1)))
```

The prepared data

```
polls_weekly
# A tibble: 165 x 4
   State   Week        MeanGap Advantage
   <chr>   <date>        <dbl> <chr>
 1 Arizona 2016-09-04      0.7 Clinton
 2 Arizona 2016-09-11      2.7 Trump
 3 Arizona 2016-09-18      3.7 Trump
 4 Arizona 2016-09-25      1   Trump
 5 Arizona 2016-10-02      0.8 Clinton
 6 Arizona 2016-10-09      0.8 Clinton
 7 Arizona 2016-10-16      1.5 Clinton
 8 Arizona 2016-10-23      2.8 Clinton
 9 Arizona 2016-10-30      0.4 Trump
10 Arizona 2016-11-06      1.8 Trump
# ... with 155 more rows
```

Figure 9-5   More preparation and analysis (part 3)

In part 4 of figure 9-5, the code plots the prepared data for the weekly gap of the swing states. This plot uses the fill color to show which candidate has the advantage for each week, and it uses a label to display the percentage for the weekly gap. The fill color is added with the geom_tile() function, which is used to generate each rectangle on the plot. The boundaries of the rectangles are automatically set which makes it easy to draw a grid of adjacent rectangles.

The geom_tile() function also sets the alpha parameter to the MeanGap column. As a result, the tile appears darker when the mean gap is high and lighter when it's low. Then, the geom_text() function adds labels for each tile that display the value for the mean gap.

This plot makes it easy to visualize how the polling numbers changed in each state over time. In addition, since each state only takes a small portion of the plot, it makes it possible to compare many states over time, which isn't possible with other types of plots such as bar plots or line plots.

This plot uses two colors, blue and red, to indicate which candidate has the advantage each week. In addition, the alpha parameter makes an advantage of .1 look much lighter than an advantage of 11.9. This makes it easier to interpret the data.

However, like the previous plots in this analysis, there is no indication of how many electoral votes each state has. New Hampshire's 4 electoral votes take up the same amount of visual space as Florida's 29 votes. As a result, the plot doesn't provide any visual representation for the weight of each state in determining the winner of the election.

Plot the weekly gap data of the swing states

```
ggplot(polls_weekly, aes(x = Week, y = State)) +
  geom_tile(aes(fill = Advantage, color = "", alpha = MeanGap)) +
  geom_text(aes(label = MeanGap, color = "")) +
  scale_fill_manual(values = c("blue","red")) +
  scale_color_manual(values = "white", guide = "none") +
  labs(title = "Weekly Gap for the Swing States") +
  theme(plot.title = element_text(hjust = 0.5),
        legend.position = "bottom") +
  scale_alpha(range = c(0.25,0.95)) +
  guides(alpha = "none")
```

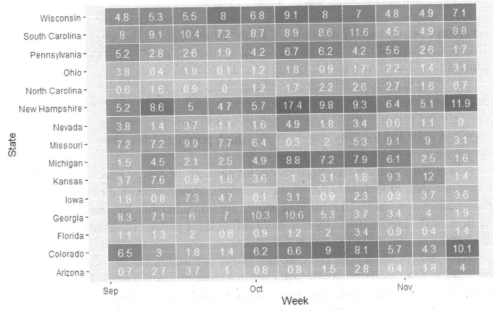

Figure 9-5   More preparation and analysis (part 4)

Perspective

This chapter shows how you can use R and the tidyverse to perform a real-world data analysis. Along the way, it shows that cleaning the data can be a time-consuming phase of data analysis and that the preparing and analyzing phases are closely related.

In addition, it shows that data visualization is critical for the analysis phase because it makes it easy to see trends and relationships in the data that would be extremely difficult to see otherwise. However, it's possible to rely too much on simplified visualizations or focus on small differences that appear large due to scale.

As it turns out, Clinton won the popular vote in 2016 by a couple percentage points, exactly as predicted by the national polls. However, Trump won most of the swing states, even ones that the polls predicted Clinton would win. As a result, Trump won the 2016 election.

So, what happened? Pollsters have been debating this ever since 2016. There are several possible explanations, including a late-breaking story about Clinton's email server that there wasn't time to properly survey in polls and the "shy voter" effect, which postulates that some people lie about their preferences when polled.

Since 2016, pollsters have put enormous amounts of energy into correcting their models to better predict elections. However, even the best polls still have a margin of error, and people are unpredictable. In a race as close as the one in 2016, it's extremely difficult for pollsters and data analysts to predict the winner with confidence.

This doesn't mean polls and data analysis aren't useful. On the contrary, they're more important than ever in a close election. The trick is to figure out which factors are important and which aren't while getting as clean and accurate a data set as possible. The fact that this is so difficult to do well is why polling companies continue to be in business and good data analysists are in high demand.

Exercise 9-1   Another way to determine swing states

In this exercise, you'll look at another way to determine which states are "swing states" and use that information to compare how many polls were conducted for each state.

Create the tibble

1. Create a new R script and load the tidyverse library.

2. Read the polls data from the CSV file and create a tibble with only the state, enddate, rawpoll_clinton, and rawpoll_trump columns. Rename the columns to use title case as in this chapter.

    Note: This exercise doesn't use the actual EndDate data, but combined with the rawpoll data it serves to uniquely identify each row.

3. Remove duplicate rows to account for FiveThirtyEight's three poll types.

4. Add a column for the gap between Clinton and Trump in each poll.

5. Add a column for the average gap for each state. To do that, you can use the group_by(), mutate(), and ungroup() functions.

6. Add a column named Candidate that contains the value "Clinton" if the gap for that row is positive and "Trump" otherwise.

Add some columns to calculate how often the polls changed

7. Create a new tibble named polls_by_state that groups the polls tibble by state and code the beginning of a summarize() function.

8. Within the summarize() function, add a column named MeanGap that stores the mean value of the Gap column for each state.

9. Still within the call to summarize(), add a column named PollCount that stores the number of polls conducted for each state. To do that, you can use the n() function.

10. Add two more columns named ClintonCount and TrumpCount. Each column should contain the total number of polls that went for Clinton and Trump respectively in each state. To do that, you can use the sum() function to count the number of TRUE values returned by an expression like this:

    ```
    ClintonCount = sum(Candidate == "Clinton")
    ```

11. Add a column named MinCount that stores how many times the candidate who won the fewest number of polls was in the lead. To do that, get the lower value between the Clinton count and the Trump count. For example, if a state had 50 polls with 2 for Clinton and 48 for Trump, the MinCount value should be 2.

    ```
    MinCount = min(ClintonCount, TrumpCount)
    ```

12. Add a column named MinFreq that stores how frequently the candidate with the minimum count led each state. To do that, you can divide by the minimum count by the total poll count and round the result to two decimal places.

13. Close the call to the summarize() function, pipe it to ungroup(), and view the resulting tibble. It should have 57 rows, one for each state/region. Arrange them in descending order by the MinFreq column. Note that there are over 1,000 national polls (U.S.).

Plot the state polling data

14. Generate a histogram that displays the distribution of the values in the MinFreq column. You can use a range of numbers for bins, but 15 works fine for this exercise.

15. Note that nearly half of the states/regions polled never had a different winning candidate! Furthermore, the trailing candidate led in a poll more than 20% of the time in only nine states/regions.

16. Use filter() to view a list of all states where the minimum frequency was higher than 15%. The resulting tibble contains 10 states plus the national (U.S.) polls and one of Maine's congressional districts (CD).

17. Create a scatter plot with x set to MinFreq and y to PollCount. Note that the PollCount for the national polls is an outlier, with well over a 1,000 polls.

18. Create another scatter plot that excludes the national polls. Do you notice any trends?

19. Create another scatter plot that excludes the national polls and only includes states with a MinFreq greater than 15%. Do you notice any trends?

20. Create yet another scatter plot. This time, set the size for each point to be absolute value of the MeanGap. To do that, you may need to modify the MeanGap column to make all of its values positive. Do you see any new correlations in the data?

Exercise 9-2   Run the Polling analysis with third-party votes

In this exercise, you'll run the analysis from this chapter, after adjusting the data to split the third-party votes between Clinton and Trump. The 2016 election had an abnormally high number of votes cast for a third party, and there is evidence that many of those votes were made as "protest votes" under the assumption that Clinton would win. In a strictly two-way race, many of these votes may have gone to Clinton instead of Trump. How would that have affected the results?

Create a tibble that includes Johnson's poll data

1. Open the script named exercise_9-2.R in the exercises/ch09 folder.
2. Find the "Clean the data" section.
3. Modify the code that selects the columns to include the rawpoll_johnson column.
4. Rename the rawpoll_johnson column to Johnson.
5. Replace NA values in the Johnson column with 0.
6. Find the "Prepare the data" section.
7. Modify the Clinton and Trump columns by giving Clinton two thirds of Johnson's data and Trump one third. This is an easy way to split Johnson's votes and isn't based on any data in particular.
8. Run the rest of the analysis. With the third-party votes distributed this way between Trump and Clinton, do you notice any difference in the results?

Chapter 10

The Wildfires analysis

This chapter presents an analysis of wildfires in the United States from 1992 through 2015. The data for this analysis comes from the U.S. Forest Service, which is part of the U.S. Department of Agriculture. This data is available in several forms, but this analysis reads the data from a SQLite database.

Get the data

The data for this analysis is stored in a SQLite database that's in a zip file in the download for this book. Figure 10-1 shows how to get the data out of this database and into a tibble.

Load the packages for this analysis

The script for this analysis begins by loading the packages it needs as shown in the first example in part 1. To start, it loads the tidyverse package. Then, it loads two packages needed to read the SQLite database (DBI and RSQLite).

Unzip the database file

After loading the packages, the code unzips the download file. To start, the second example lists the files in the zip file named fires.zip that's in the data directory. The resulting list of files shows that the zip file contains a SQLite file whose name begins with FPA_FOD. This file is stored in a subdirectory named Data.

The third example unzips this SQLite file. However, it doesn't unzip any of the other files in the zip file.

The fourth example renames the SQLite file to give it the more user-friendly name of fires.sqlite. This statement also moves the fires.sqlite file from the data/Data directory to the data directory. As a result, after you run this statement, the fires.sqlite file is stored in the same directory that stores the fires.zip file.

Load the packages for this analysis

```
library("tidyverse")
library("DBI")
library("RSQLite")
```

List the files in the zip file

```
filename  <- "../../data/fires.zip"
unzip(filename, exdir = "../../data", list = TRUE)
                        Name     Length              Date
1        Data/FPA_FOD_20170508.sqlite 795785216 2017-05-17 16:12:00
2                            Data/         0 2017-05-24 12:15:00
3      _metadata_RDS-2013-0009.4.xml     51400 2017-08-09 12:22:00
4 Supplements/FPA_FOD_Source_List.pdf    109336 2017-05-09 14:20:00
5                    Supplements/         0 2017-05-09 14:20:00
6     _fileindex_RDS-2013-0009.4.html      4398 2017-05-22 12:44:00
7     _metadata_RDS-2013-0009.4.html     89005 2017-08-09 12:22:00
```

Unzip the SQLite file from the zip file

```
unzip(filename, exdir = "../../data",
      files = c("Data/FPA_FOD_20170508.sqlite"))
```

Rename the SQLite file and move it to the data directory

```
file.rename(c("../../data/Data/FPA_FOD_20170508.sqlite"),
            c("../../data/fires.sqlite"))
```

The directory and filename for the resulting SQLite file

```
~/murach_r/data/fires.sqlite
```

Description

- The data for this analysis is stored in a SQLite file in a zip file.

Figure 10-1   Get the data (part 1)

Read the data from the database

Part 2 of figure 10-1 shows how to read the data from the unzipped database file. To start, the first example shows how to connect to the database. Here, the first statement creates a variable that stores the path to the database file. Then, the second statement creates a connection object by passing a SQLite driver and the path to the database file to the dbConnect() function.

The second example uses the database connection to list the tables in the database. This makes it possible to find the table that contains the data for the analysis. To save space, this figure doesn't show all of the table names. However, just by reading the names of the tables you can determine that most of them aren't needed for this analysis. In fact, of the 33 tables listed, this analysis only uses the Fires table.

The third example lists the columns in the Fires table. Again, you can determine which columns might be needed by this analysis largely just by reading their names. To save space, this figure doesn't list all of the 39 columns. However, reading through the complete list of column names reveals many columns that might be useful including STATE, FIRE_NAME, FIRE_SIZE, LATITUDE, LONGITUDE, FIRE_YEAR, DISCOVERY_DATE, and CONT_DATE.

The fourth example begins by showing a SQL statement that selects 8 of the 39 columns from the Fires table. In this statement, the SQL keywords and SQL functions are capitalized but the column names aren't. That's possible because SQL isn't case-sensitive. As a result, you can use whatever case you want when coding the names of columns with a SQLite database.

After the code stores the SQL statement in the fires_sql variable, it passes this variable and the connection object to the dbSendQuery() function to send the SQL query to the database and to get a response object. Then, it passes the response object to the dbFetch() function to get the data. In addition, it uses the as_tibble() function to convert the data to a tibble. Next, it closes the response and connection objects to free system resources.

The resulting tibble stores more than 1.8 million rows and 8 columns. This is a large data set. As a result, some operations may take quite a while to run on it. Because of that, this analysis begins by looking for ways to reduce the number of rows that it uses.

Connect to the database

```
sql_file <- "../../data/fires.sqlite"
con <- dbConnect(SQLite(), sql_file)
```

List the tables in the database

```
dbListTables(con)
[1] "ElementaryGeometries"
[2] "Fires"
[3] "KNN"
...
[31] "virts_geometry_columns_auth"
[32] "virts_geometry_columns_field_infos"
[33] "virts_geometry_columns_statistics"
```

List the columns in the Fires table

```
dbListFields(con, "Fires")
[1] "OBJECTID"              "FOD_ID"
[3] "FPA_ID"               "SOURCE_SYSTEM_TYPE"
[5] "SOURCE_SYSTEM"        "NWCG_REPORTING_AGENCY"
...
[35] "STATE"               "COUNTY"
[37] "FIPS_CODE"           "FIPS_NAME"
[39] "Shape"
```

Select columns from the Fires table

```
fires_sql <- "
  SELECT fire_name, fire_size, state, latitude,
         longitude, fire_year,
         DATETIME(discovery_date) AS discovery_date,
         DATETIME(cont_date) AS contain_date
  FROM Fires
"
response <- dbSendQuery(con, fires_sql)
fires <- as_tibble(dbFetch(response))
dbClearResult(response)
dbDisconnect(con)
```

View the tibble

```
fires
# A tibble: 1,880,465 x 8
   FIRE_NAME  FIRE_SIZE STATE LATITUDE LONGITUDE FIRE_YEAR
   <chr>          <dbl> <chr>    <dbl>     <dbl>     <int>
 1 FOUNTAIN        0.1  CA        40.0     -121.      2005
 2 PIGEON          0.25 CA        38.9     -120.      2004
 3 SLACK           0.1  CA        39.0     -121.      2004
 4 DEER            0.1  CA        38.6     -120.      2004
 5 STEVENOT        0.1  CA        38.6     -120.      2004
 6 HIDDEN          0.1  CA        38.6     -120.      2004
 7 FORK            0.1  CA        38.7     -120.      2004
 8 SLATE           0.8  CA        41.0     -122.      2005
 9 SHASTA          1    CA        41.2     -122.      2005
10 TANGLEFOOT      0.1  CA        38.5     -120.      2004
# ... with 1,880,455 more rows, and 2 more variables:
#   discovery_date <chr>, contain_date <chr>
```

Figure 10-1   Get the data (part 2)

Clean the data

The data for this analysis comes from the Forest Service Research Data Archive, which has already gone to great lengths to clean and merge a variety of sources to create a comprehensive data set. However, figure 10-2 shows how to further clean this data for the purpose of our analysis. In addition, it shows how to examine this data to make sure that it's been cleaned.

Improve column names and data types

The first example in part 1 of this figure shows the structure of the fires tibble. The first six column names use all caps and underscores, but the last two use lowercase and underscores because they were renamed by the query shown in the previous figure.

To make these columns consistent and easy to work with, the second example shows how to rename them all to use title case. To do that, it uses the names() function to get the vector of column names for the data set, and it assigns a new vector of column names to the data set.

The first example also shows that the two date columns use the character type, not the Date type. However, it's generally a good practice to use the Date type for dates. That's why the third example converts the DiscoveryDate and ContainDate columns to the Date type.

The fourth example shows the structure of the data set again. This shows the new column names and data types.

At this point, you could also convert the State and Year columns to the factor type. However, the character and integer types work fine for those columns for this analysis. As a result, this code doesn't bother converting these data types. Furthermore, if you later discovered a need for them to be the factor type it would be easy to convert them at that time.

Drop duplicate rows

After improving the column names and data types, this analysis drops duplicate rows. Since this data set was compiled from multiple government agencies, it's likely that duplicate data was introduced. So, the fifth example displays the count of duplicate rows. This shows that the data set contains 1,867 duplicate rows. This is a small percentage of the total number of rows, but it still makes sense to drop these rows as shown by the sixth example.

View the structure of the data set

```
fires %>% str(strict.width = "cut")
tibble [1,880,465 x 8] (S3: tbl_df/tbl/data.frame)
 $ FIRE_NAME    : chr [1:1880465] "FOUNTAIN" "PIGEON" "SLACK" "DEER"..
 $ FIRE_SIZE    : num [1:1880465] 0.1 0.25 0.1 0.1 0.1 0.1 0.1 0.8 1..
 $ STATE        : chr [1:1880465] "CA" "CA" "CA" "CA" ...
 $ LATITUDE     : num [1:1880465] 40 38.9 39 38.6 38.6 ...
 $ LONGITUDE    : num [1:1880465] -121 -120 -121 -120 -120 ...
 $ FIRE_YEAR    : int [1:1880465] 2005 2004 2004 2004 2004 2004 2004..
 $ discovery_date: chr [1:1880465] "2005-02-02 00:00:00" "2004-05-12"..
 $ contain_date : chr [1:1880465] "2005-02-02 00:00:00" "2004-05-12"..
```

Rename the columns

```
names(fires) <- c("FireName","Size","State","Latitude","Longitude",
                  "Year","DiscoveryDate","ContainDate")
```

Convert the date columns to the Date type

```
fires <- fires %>% mutate(
  DiscoveryDate = as.Date(DiscoveryDate),
  ContainDate = as.Date(ContainDate))
```

View the new column names and data types

```
fires %>% str(strict.width = "cut")
tibble [1,878,598 x 8] (S3: tbl_df/tbl/data.frame)
 $ FireName     : chr [1:1880465] "FOUNTAIN" "PIGEON" "SLACK" "DEER" ..
 $ Size         : num [1:1880465] 0.1 0.25 0.1 0.1 0.1 0.1 0.1 0.8 1 ..
 $ State        : chr [1:1880465] "CA" "CA" "CA" "CA" ...
 $ Latitude     : num [1:1880465] 40 38.9 39 38.6 38.6 ...
 $ Longitude    : num [1:1880465] -121 -120 -121 -120 -120 ...
 $ Year         : int [1:1880465] 2005 2004 2004 2004 2004 2004 2004 ..
 $ DiscoveryDate: Date[1:1880465], format: "2005-02-02" ...
 $ ContainDate  : Date[1:1880465], format: "2005-02-02" ...
```

View the count of duplicate rows

```
fires %>% duplicated() %>% sum()
[1] 1867
```

Drop duplicate rows

```
fires <- fires %>% unique()
```

Description

- The column names for this tibble use title case to make them easy to work with.
- The duplicate rows from the data set make up less than .1% of the total number of rows, but this analysis drops them anyway.

Figure 10-2   Clean the data (part 1)

Select rows for large fires

At this point, the data set no longer has duplicate data, but it still contains over 1.8 million rows. To narrow down the data set, we decided to focus this analysis on large fires.

To examine the sizes of the fires in this data set, the first example in part 2 of figure 10-2 generates the summary statistics for the Size column. Since the median fire size is 1.0, half of all fires are smaller than 1 acre.

The second example shows that 54,093 fires in the data set are 100 acres or larger. Since 100 is a nice round number that still leaves plenty of large fires to analyze, that's what this analysis uses for its cutoff. The third example selects these large fires and drops the smaller ones.

This approach to reducing the data set works well for experimentation and learning. Later, if you want your analysis to include all fires, you can run this analysis again without dropping the smaller fires. Or, you might be interested in comparing an analysis of only the smaller fires to the results of this one.

Examine NA values

When cleaning the data, it's a good idea to examine the NA values in the data set and see what information might be missing. That's why the fourth example displays a count of NA values for each column. This shows that the FireName and ContainDate columns both contain a large number of NA values. So, does it make sense to make the data set even smaller by dropping some or all of the rows that contain NA values?

To examine this issue, the fifth example views the FireName, State, Size, and ContainDate columns for all fires that have an NA value in the FireName column. In addition, it sorts the tibble by fire size with the largest fire first. This shows that even some very large fires have NA values for their names and contain dates, but the other pertinent information is all there. In other words, these rows still contain useful information. As a result, it doesn't make sense to drop the rows for these fires.

If you look at the metadata included with the database, it says that each fire had to have a discovery date, final fire size, and location for inclusion in the set. This explains why none of those values are missing and why so many rows with NA values for the other columns are still included. It further suggests that you can't draw any conclusions about which fires have NA values without looking into the individual contributing sources, which is beyond the scope of this analysis. However, those sources are listed in the metadata if you are interested in further research.

The sixth example shows how to save the data in an RDS file named fires_clean.rds. This is a good practice because it provides an easy way for you to read the clean data if you need it again. For example, if you mess up the data that's stored in the fires variable during the preparation phase, you don't have to rerun the parts of the script that get and clean the data. Instead, you can start again from this point by reading the clean data.

View the summary data for the Size column

```
fires %>% select(Size) %>% summary()
       Size
 Min.   :      0.0
 1st Qu.:      0.1
 Median :      1.0
 Mean   :     74.6
 3rd Qu.:      3.3
 Max.   :606945.0
```

View a count of fires 100 acres or larger

```
nrow(fires %>% filter(Size >= 100))
[1] 54093
```

Select rows for large fires

```
fires <- fires %>% filter(Size >= 100)
```

View NA values for all columns

```
apply(fires, MARGIN = 2, function(col) sum(is.na(col)))
     FireName         Size        State     Latitude    Longitude
        14047            0            0            0            0
         Year DiscoveryDate  ContainDate
            0            0        20334
```

View the largest fires with missing names

```
fires %>% filter(is.na(FireName)) %>%
  select(FireName, State, Size, ContainDate) %>%
  arrange(desc(Size))
# A tibble: 14,047 x 4
   FireName State  Size ContainDate
   <chr>    <chr> <dbl> <date>
 1 NA       AK    71760 NA
 2 NA       OK    70000 NA
 3 NA       AK    57576 NA
 4 NA       KS    55000 NA
 ...
```

Save the cleaned data

```
saveRDS(fires, file = "../../data/fires_clean.rds")
```

Description

- This analysis focuses on fires that burned 100 or more acres.
- More than a third of the rows for this analysis are missing contain dates and almost a quarter are missing names.

Figure 10-2   Clean the data (part 2)

Prepare the data

Figure 10-3 shows how to prepare the fires data. To start, the first example reads the clean fires data from the saved RDS file.

Add, modify, and select columns

To be able to analyze fires by the month in which they occur, this analysis adds a Month column. The second example in this figure shows how to add this column by using the format() function to extract the month from the DiscoveryDate column. This works because this analysis converted the DiscoveryDate column to the Date type during the cleaning phase.

For this analysis, we decided to analyze fires only by discovery date, not contain date, as the contain date column has many NA values. That means that this analysis can drop the ContainDate column from the data set as shown in the third example. This makes it easier to view the columns when displaying the tibble on the console.

The fourth example modifies the FireName column so the names of the fires use title case instead of uppercase. That's because title case is easier to read than uppercase and looks better when displaying the names of the fires in a table or on a plot.

The fifth example changes the sequence of the columns. Again, this isn't required for analysis, but having the columns in a sequence that makes sense to you can make it easier to work with the data. For instance, this code puts the Year, Month, and DiscoveryDate columns next to each other since they all contain data about the discovery date.

Sort the rows

The fifth example also sorts the rows by size in descending order. This stores the largest fires at the top of the tibble. Since this analysis focuses on analyzing the largest fires, sorting the fires like this may be helpful when viewing the tibble on the console.

The sixth example displays the tibble in its current state, which displays the data for the 10 largest fires. This shows that the largest fire in this data set is the Inowak fire that burned in Alaska in 1997. This fire burned 6.07e5 (607,000) acres. That's a large fire!

Read the cleaned data

```
fires <- readRDS("../../data/fires_clean.rds")
```

Add a column for the discovery month

```
fires <- fires %>%
  mutate(Month = format(DiscoveryDate, "%m"))
```

Drop the column for the contain date

```
fires <- fires %>% select(-ContainDate)
```

Modify the column for the fire name

```
fires <- fires %>% mutate(
  FireName = str_to_title(FireName))
```

Change the sequence of the columns and sort the rows by size

```
fires <- fires %>%
  select(FireName, State, Size, Year, Month,
         DiscoveryDate, Latitude, Longitude) %>%
  arrange(desc(Size))
```

The prepared data

```
fires
# A tibble: 54,093 x 8
   FireName    State  Size  Year Month DiscoveryDate Latitude Longitude
   <chr>       <chr> <dbl> <int> <chr> <date>           <dbl>     <dbl>
 1 Inowak      AK    6.07e5  1997 06    1997-06-25        62.0     -157.
 2 Long Draw   OR    5.58e5  2012 07    2012-07-08        42.4     -118.
 3 Wallow      AZ    5.38e5  2011 05    2011-05-29        33.6     -109.
 4 Boundary    AK    5.38e5  2004 06    2004-06-13        65.3     -147.
 5 Minto Fla~  AK    5.17e5  2009 06    2009-06-21        64.7     -150.
 6 Biscuit     OR    5.00e5  2002 07    2002-07-13        42.0     -124.
 7 Dall City   AK    4.83e5  2004 07    2004-07-06        66.3     -150.
 8 Hwy 152     TX    4.80e5  2006 03    2006-03-12        35.7     -101.
 9 Billy Ck    AK    4.64e5  2004 06    2004-06-18        63.8     -144.
10 Holloway    OR    4.61e5  2012 08    2012-08-05        42.0     -118.
# ... with 54,083 more rows
```

Description

- The data set now contains separate columns for the year and month of discovery.
- The columns are rearranged and sorted for convenience.

Figure 10-3   Prepare the data

Analyze the data

Figure 10-4 shows how to analyze the fires data set. This involves preparing some data sets that summarize data. In addition, it involves plotting the summarized data so you can visualize the relationships between the variables.

Plot the largest fire per year in California

The first example in part 1 starts by getting the largest fire for each year in California. To do that, this code filters the data for only California, groups the data by the fire year, and adds a column that stores the maximum fire size.

After preparing the data set, this example uses a bar plot to plot the largest fire in California by year. This plot shows that the four fires over 200,000 acres are significantly larger than the rest of the fires from 1992 to 2015. This seems to show a trend in the size of largest fire for each year. However, since this plot only shows the largest fire for each year, all of the fire sizes in this plot are outliers. Furthermore, one fire burning many acres is equivalent to several smaller fires burning the same number of acres. Therefore, this plot has limited uses depending on the goals of your analysis.

Plot the largest fire per year in California

Prepare the data

```
fires_ca <- fires %>%
  filter(State == "CA") %>%
  group_by(Year) %>%
  summarize(MaxSize = max(Size))

fires_ca
# A tibble: 24 x 2
    Year MaxSize
   <int>   <dbl>
 1  1992   64000
 2  1993   43201
 3  1994   48851
...
```

Plot the data

```
ggplot(fires_ca, aes(x = Year, y = MaxSize)) +
  geom_col(fill = "darkorange") +
  labs(title = "Largest Fire in California by Year",
       y = "Acres Burned") +
  theme(plot.title = element_text(hjust = 0.5))
```

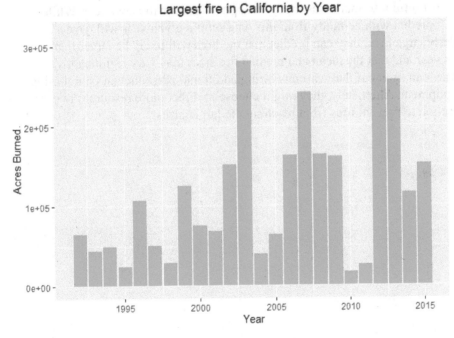

Description

- There have been some very large fires in California over the years, and they seem to be getting larger.

Figure 10-4   Analyze the data (part 1)

Plot the mean and median acres burned in California

Part 2 of figure 10-4 shows how to visualize the mean and median acres burned by a fire per year in California. To start, this example prepares a data set by filtering for California, grouping by year, and adding columns that summarize the mean and median fire size for each year. Then, the code creates long data by pivoting the two summary columns. This is necessary to be able to plot the data set.

After preparing the data, the code displays a line plot with both the mean size and median. This shows that the mean is more sensitive to outliers than the median. For this reason, it's often a good idea to look at the mean and median together when you analyze data. In this plot, there seems to be an upward trend to the mean size of the fires over the years, though that's difficult to determine due to the erratic swings in the line.

This plot also shows that the median size seems rather stable. However, that's because of the differing scales from plotting the median and the mean together. Plotting just the median shows that it also has trended up and with just as many erratic swings as the mean.

This is useful information as firefighters plan ahead to the next year. While they can't predict with certainty from this data set if a given year will have particularly large fires, they can predict that the fires will likely be larger than previous year and that the mean and median fire sizes may vary significantly. Then, they can decide if that warrants hiring additional personnel and purchasing more equipment. Alternately, they might choose to direct more resources to early fire detection to prevent fires from reaching the larger sizes.

Plot the mean and median acres burned in California

Prepare the data

```
fires_ca <- fires %>%
  filter(State == "CA") %>%
  group_by(Year) %>%
  summarize(MeanSize = mean(Size),
            MedianSize = median(Size))

fires_ca <- fires_ca %>%
  pivot_longer(cols = c(MeanSize, MedianSize),
               names_to = "Name", values_to = "Value")

fires_ca
# A tibble: 48 x 3
    Year Name          Value
   <int> <chr>         <dbl>
 1  1992 MeanSize      1078.
 2  1992 MedianSize     266.
 3  1993 MeanSize      1444.
 4  1993 MedianSize     250
...
```

Plot the data

```
ggplot(fires_ca, aes(x = Year, y = Value, color = Name)) +
  geom_line(size = 1) +
  labs(title = "Mean and Median Acres Burned Per Year (CA)",
       y = "Acres Burned") +
  theme(plot.title = element_text(hjust = 0.5),
        legend.position = "bottom")
```

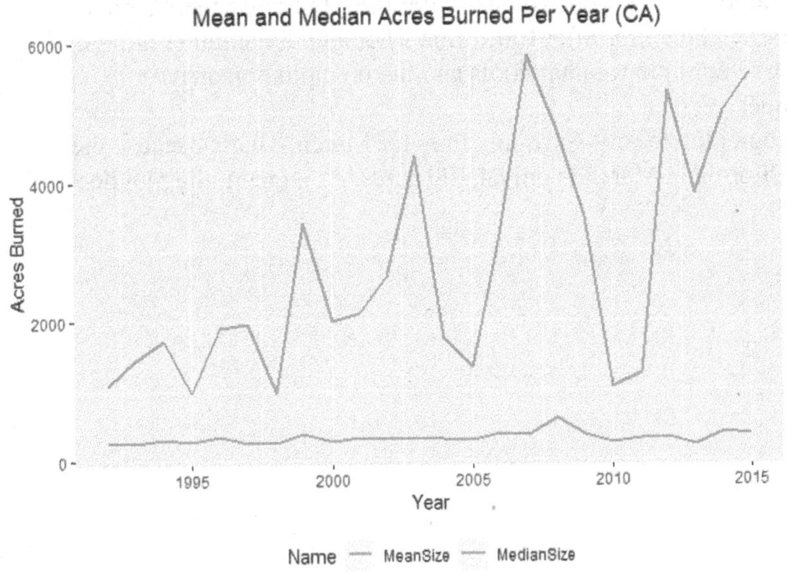

Description

- When the median and mean are far apart, that means there are many outliers in the data set and the median is more useful for determining an "average" value.

Figure 10-4   Analyze the data (part 2)

Plot the fires per month in California

Another way to analyze fires in California is to look at the fires by month. The first example in part 3 of figure 10-4 displays a bar plot of the number of fires for each month. To do that, the code selects the rows for California and uses the Month column to count the number of rows for each month.

The resulting plot shows that the summer months have the most fires. The most likely explanation for this is that California has hot summers with little precipitation, which greatly increases the risk of wildfires. Furthermore, California gets most of its precipitation from November to April, exactly the months with significantly fewer fires.

The second example displays a box plot that shows the sizes of the fires by month for California. To create more interesting quartiles, this code only plots fires that are larger than 10,000 acres. Otherwise, the four quartiles of each box would be clustered closely together for the small fires, and most large fires would be displayed as outliers. This makes sense for this analysis since it focuses on the larger fires. For an analysis of the complete data set, focusing only the very largest fires would distort the analysis.

This plot is revealing because it shows that even for these large fires, 75% (the third quartile) are less than 50,000 acres even in the summer, and the top quartile is less than 100,000 acres. However, the outliers are extreme, with one topping more than 300,000 acres.

Like the bar plot in the first example, this box plot clearly shows that there are many more fires in California in the summer. In addition, this plot shows that the fires tend to be larger in the summer months, and that September and October also tend to have larger fires as well. You can infer this is because there's plenty of dry vegetation to fuel the fires after California's hot and dry summer months and that remains true until the weather cools and the precipitation arrives, usually in November.

Note that the box plot doesn't show any fires for March. That's because the largest fire for California in March is only 4,100 acres. As a result, the plot does not include March.

Plot the number of fires per month in California

```
ggplot(fires %>% filter(State == "CA"),
       aes(x = Month)) +
  geom_bar(fill = "darkorange") +
  labs(title = "Fires By Month (CA)", y = "Count") +
  theme(plot.title = element_text(hjust = 0.5))
```

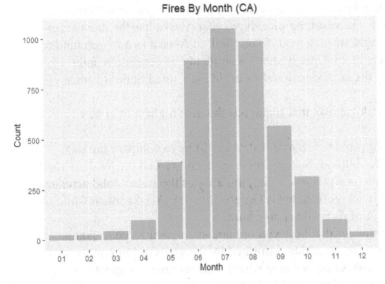

Plot the distribution of large fires per month in California

```
ggplot(filter(fires, State == "CA" & Size >= 10000),
       aes(x = Month, y = Size)) +
  geom_boxplot(fill = "darkorange") +
  theme(plot.title = element_text(hjust = 0.5)) +
  labs(title = "Fires >= 10,000 Acres By Month (CA)",
       x = "Month", y = "Fire Size")
```

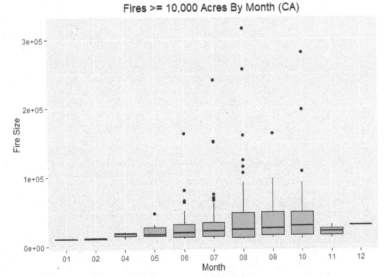

Figure 10-4   Analyze the data (part 3)

Plot the total acres burned for the top 10 states

Part 4 of figure 10-4 shows how to plot the total acres burned for the top 10 states. To start, the first example prepares a summary data set. To do that, it groups the fires by state, creates a column for the total acres burned, sorts the rows by acres burned in descending order, and selects the top 10 rows.

After preparing the data, the second example plots each state by total acres burned with a bar plot. The resulting plot shows that Alaska has the most acres burned by far. This might surprise you if you think of Alaska as a frozen tundra, but it's the largest state in the United States with plenty of vegetation and timberland, including the largest national forest in the United States (Tongass National Forest).

After Alaska, the plot shows that Idaho has the next highest total acres burned. And so on.

A more interesting way to look at this data might be to compare the total size of the state to acres burned. Such a comparison shows that the total number of burned acres in Idaho is equal to a whopping 25% of the state's total acreage. The next closest in this set of 10 states is Oregon, at 13%. Alaska, meanwhile, had an equivalent 7.5% of its total acreage burn.

This is another reminder that how you prepare and visualize your data is crucial to your analysis and should be informed by the goals of your analysis. For example, if your goal is to determine which state requires the most firefighting resources, you might say Alaska based on the total acreage. On the other hand, if your goal is to determine which state needs priority in overhauling firefighting techniques, you might look at the percentage of acres burned and decide to look more closely at Idaho.

Plot the total acres burned for the top 10 states

Prepare the data

```
top_10_fire_states <- fires %>%
  group_by(State) %>%
  summarize(AcresBurned = sum(Size)) %>%
  arrange(desc(AcresBurned)) %>%
  head(10)
```

```
top_10_fire_states
# A tibble: 10 x 2
   State AcresBurned
   <chr>       <dbl>
 1 AK      32189139.
 2 ID      13567949.
 3 CA      12321083.
 4 NV       8956626.
 5 TX       8880251.
 6 OR       8317403.
 7 NM       6239172.
 8 MT       6168346.
 9 AZ       5410534.
10 WA       4708081.
```

Plot the data

```
ggplot(top_10_fire_states,
       aes(x = State, y = AcresBurned, fill = State)) +
  geom_col() + theme(plot.title = element_text(hjust = 0.5)) +
  labs(title = "Acres Burned by State", y = "Acres Burned")
```

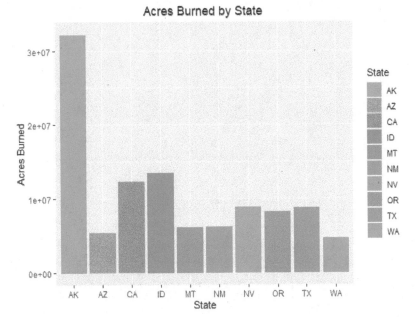

Figure 10-4   Analyze the data (part 4)

Plot the acres burned per year for the top 4 states

Part 5 of figure 10-4 shows how to plot the total acres burned by year for each of the "top" four states. To start, the first example prepares the data. To do that, it groups the fires by state and year, adds a column that holds the total acres burned for each year and state (YearSize), another column that holds the total acres burned for each state (StateSize), and a column that ranks the states by total acres burned (StateRank).

To rank the states, this code uses the dense_rank() function rather than the rank() function. That way, rows with the same values get the same rank. This is necessary because the StateSize column contains the same value for each state.

The second example displays a grid of line plots. Here, each plot in the grid displays the data for one of the states in the prepared data. This shows that each state has years with extreme outliers in the number of acres burned, represented as spikes on the plot. However, Alaska's outliers are the most dramatic.

Plot the acres burned per year for the top 4 states

Prepare the data

```
states_ranked <- fires %>%
  group_by(State, Year) %>%
  summarize(YearSize = sum(Size)) %>%
  mutate(StateSize = sum(YearSize)) %>%
  ungroup() %>%
  mutate(StateRank = dense_rank(desc(StateSize)))

states_ranked
# A tibble: 1,025 x 5
   State  Year YearSize StateSize StateRank
   <chr> <int>    <dbl>     <dbl>     <int>
 1 AK     1992   141207 32189139.         1
 2 AK     1993   684670. 32189139.        1
 3 AK     1994   260102. 32189139.        1
 4 AK     1995    42726 32189139.         1
 5 AK     1996   596706. 32189139.        1
 6 AK     1997  2023370 32189139.         1
 7 AK     1998   119738 32189139.         1
 8 AK     1999  1004150. 32189139.        1
 9 AK     2000   755703. 32189139.        1
10 AK     2001   219392 32189139.         1
```

Plot the data

```
ggplot(states_ranked %>% filter(StateRank < 5),
       aes(x = Year, y = YearSize, color = State)) +
  geom_line() +
  facet_grid(vars(State)) +
  labs(title = "Acres Burned Per Year (Top 4 States)",
       x = "Year", y = "Acres Burned") +
  theme(plot.title = element_text(hjust = 0.5))
```

Figure 10-4   Analyze the data (part 5)

Plot the data on a map

Figure 10-5 shows how to plot the locations of the fires on a map. You can use similar code to plot fire locations for any individual state or even a map of the entire United States. This type of plotting is often helpful to gain geographical insights.

Plot the 20 largest fires in California

Part 1 of figure 10-5 shows how to plot the 20 largest fires in California on a map of the state. To do this, this figure prepares two data sets, one for the locations of the fires and another for the map of the state.

The first example shows how to prepare the data set for the 20 largest fires in California. To start, it selects all fires for California and sorts them by size in descending order. Then, it uses the head() function to select the first 20 rows in the data set. Note that each row has a Latitude and Longitude column that's used to plot the location of each fire.

The second example shows how to get the data to display the map for the state of California. To do that, this example passes arguments of "state" and "california" to the map_data() function. The resulting data set contains columns named long and lat that store the longitude and latitude for the points that make an outline for the map of California.

To create the plot, the third example starts by plotting the map data. To do that, it uses the geom_polygon() function to draw the outline of the state in black with a white background. Then, it adds the fires to the map as red data points. To do that, it calls the geom_point() function and sets the color of each point to red. In addition, it sets the size of each point so that it corresponds to the fire size.

The resulting plot shows the locations and sizes of the 20 largest fires. You might note that none of the largest fires occurred in the southeast, but otherwise this plot doesn't give you many insights.

Plot the 20 largest fires in California

Prepare the fires data

```
ca_top_20 <- fires %>%
  filter(State == "CA") %>%
  arrange(desc(Size)) %>%
  head(20)
```

```
ca_top_20
# A tibble: 20 x 8
  FireName   State   Size  Year Month DiscoveryDate Latitude Longitude
  <chr>      <chr>  <dbl> <int> <chr> <date>           <dbl>     <dbl>
1 Rush       CA    3.16e5  2012 08    2012-08-12        40.6     -120.
2 Cedar      CA    2.80e5  2003 10    2003-10-25        33.0     -117.
3 Rim        CA    2.56e5  2013 08    2013-08-17        37.9     -120.
...
```

Get the map data

```
ca_map <- map_data("state", "california")
```

Plot the fires on the map

```
ggplot() +
  geom_polygon(data = ca_map,
               mapping = aes(x = long, y = lat),
               color = "black", fill = "white") +
  geom_point(data = ca_top_20,
             mapping = aes(x = Longitude, y = Latitude, size = Size),
             color = "red") +
  labs(title = "The 20 Largest Fires in California", x = "", y = "") +
  theme(plot.title = element_text(hjust = 0.5)) +
  coord_quickmap()
```

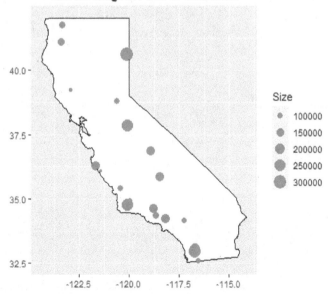

Figure 10-5   Plot the data on a map (part 1)

Plot all fires in California larger than 500 acres

Part 2 of figure 10-5 shows another way to plot fires in California on a map. This time, the plot displays all fires over 500 acres. In addition, it scales the color and transparency of each point so that the color and transparency correspond to the fire size. Here, the smallest fires are displayed in yellow and the largest fires are displayed in dark red.

To create the color scale, this plot uses the scale_color_gradient2() function. This function allows you to specify colors for the low, middle, and high values of the gradient range.

The resulting plot makes some trends in fire locations more apparent. If you're familiar with California's geography, you can see that the fire locations tend to correspond with the forested mountain ranges. Furthermore, all of the biggest fires occurred in mountainous regions.

This plot also shows that there are three regions with a distinct lack of fires. The first is the southeastern region. This region is mostly desert. The second is the northwest coast. This region tends to be wetter than other parts of the state. And the third is the San Joaquin and Sacramento valleys. This region is heavily farmed and irrigated.

Plot all fires in California larger than 500 acres

```
ggplot() +
  geom_polygon(data = ca_map,
               mapping = aes(x= long, y = lat),
               color = "black", fill = "white") +
  geom_point(data = filter(fires, Size > 500 & State == "CA"),
             mapping = aes(x = Longitude, y = Latitude,
                           size = Size, color = Size, alpha = Size)) +
  scale_color_gradient2(low = "yellow", mid = "orange",
                        high = "darkred") +
  labs(title = "Fires > 500 acres in California", x = "", y = "") +
  theme(plot.title = element_text(hjust = 0.5),
        axis.ticks = element_blank(),
        axis.text = element_blank()) +
  coord_quickmap()
```

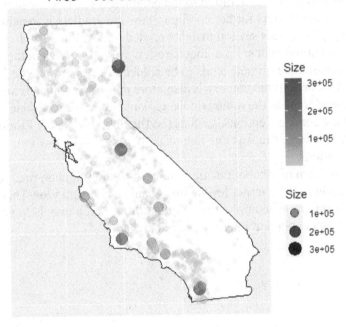

Figure 10-5   Plot the data on a map (part 2)

Plot all fires in the U.S. larger than 100,000 acres

Part 3 of figure 10-5 shows how to plot fires larger than 100,000 acres for the entire continental United States. To start, the first example uses the map_data() function to get the points for the map of all states in the United States.

The second example plots all very large fires in the continental United States. To start, it uses the geom_polygon() function to display the map of the United States with a black outline and a white background. Then, it uses the geom_point() function to display the fires on the map. But first, it selects all fires that are larger than 100,000 acres and not in Alaska or Hawaii. This is necessary because Alaska and Hawaii aren't in the continental United States.

Like the previous plot, this plot uses the scale_color_gradient2() function to set the colors for the plots. However, this function call includes the breaks, labels, and limits parameters in addition to the parameters used by the previous figure. These three parameters further configure how the gradient is generated.

The resulting plot reveals several insights related to the climate and topography of the United States. First, there are few very large fires in the eastern states because their climate tends to be colder and wetter than the western states. That area of the country is also more densely populated. Second, the largest fires tend to be in the mountainous regions of the western states. However, there are some exceptions, such as the Bugaboo fire on the Florida-Georgia border. This large fire was due to unusually high winds, low humidity, and drought conditions.

If you want to learn more about a large fire such as the Bugaboo fire, you can begin by searching the internet for the fire's name, state, and year. This often provides more information about the fire, and that information may help you understand your data better and improve your analysis.

Plot all fires in the U.S. larger than 100,000 acres

Get the map data

```
us_map <- map_data("state")

head(us_map, 3)
        long      lat group order  region  subregion
1 -87.46201 30.38968     1     1 alabama       <NA>
2 -87.48493 30.37249     1     2 alabama       <NA>
3 -87.52503 30.37249     1     3 alabama       <NA>
```

Create the plot

```
ggplot() +
  geom_polygon(
    data = us_map,
    mapping = aes(x = long, y = lat, group = group),
    fill = "white", color = "black") +
  geom_point(
    data = filter(fires, Size > 100000 &
                    State != "AK" & State != "HI"),
    mapping = aes(x = Longitude, y = Latitude,
                  size = Size, color = Size)) +
  scale_color_gradient2(low = "yellow", mid = "orange",
                        high = "darkred",
                        breaks = c(100000, 337500, 575000),
                        labels = c("Low", "Med", "High"),
                        limits = c(100000, 575000)) +
  coord_quickmap() +
  labs(title = "Fires > 100,000 Acres in the U.S.", x = "", y = "") +
  theme(plot.title = element_text(hjust = 0.5))
```

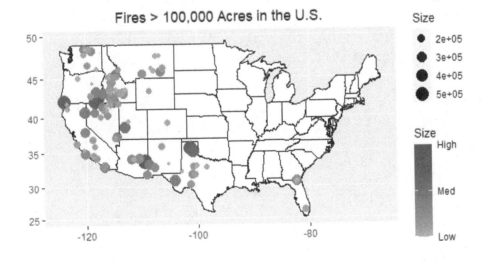

Figure 10-5   Plot the data on a map (part 3)

Perspective

The Fires analysis presented in this chapter showed how you can apply R and tidyverse skills to real-world data. Because the data for this analysis came from a well-maintained database, it didn't require a lot of cleaning and preparation. However, this analysis still illustrates many useful skills such as how to get data from a SQLite database and how to analyze data by plotting it, including how to plot geographic data on a map.

Although this analysis reveals some insights, there are many others that could be gleaned from this data. So, what are some other ways you could analyze this data? You might want to focus your analysis on a single state such as Idaho. That would make it possible to analyze all fires in the state, not just the larger ones. Or, you might want to analyze other properties of the fires such as the length of each fire. To do that, you could calculate the length for every fire that has a contain date by subtracting the contain date from the discovery date. The possibilities are endless. But, if you ask the right questions and look at the data in the right way, you will discover something interesting or useful.

Exercise 10-1   Another analysis of the fires data

This exercise guides you through another analysis of the fires data. Before you do this analysis, it's a good idea to run the example analysis from this chapter and experiment with it to make sure you understand how it works.

Get the data

1. Create a new R script.
2. Load the tidyverse, DBI, and RSQLite packages.
3. If you haven't already done so, unzip the fires zip file included with the download for this book and extract the SQLite database named FPA_FOD_20170508.sqlite.
4. Connect to the database and create a tibble from the fires table containing the following columns: fire_name, fire_size, state, latitude, longitude, fire_year, stat_cause_descr, nwcg_reporting_agency, fire_size_class, source_system_type, and discovery_date.

    To learn about the values for these columns, visit this url:
    https://data.fs.usda.gov/geodata/edw/edw_resources/meta/S_USA.FPA_FOD_4thedition.xml
5. Rename the columns as follows: FireName, Size, State, Latitude, Longitude, Year, Cause, ReportingAgency, SizeClass, SourceType, and DiscoveryDate.
6. Convert the DiscoveryDate column to the Date type, SizeClass to the ordered factor type, and Cause, ReportingAgency, and SourceType to the factor data type.
7. Drop duplicate rows from the tibble.
8. Add a column named Month to store the month each fire was discovered in.

Examine the data

9. Create a bar plot for the Cause column. Are you surprised by which causes are more common? Do you suspect certain causes might be more common at certain times of the year than others?
10. Create another bar plot, this time setting the x axis to the Month column and the fill color to the Cause column. What patterns do you notice?
11. Reverse the plot so that the cause is displayed on the x axis and the month is displayed by the fill color. Is this a better way to visualize the data? Why or why not? Do you see any new patterns now that you didn't notice before?
12. Create a bar plot showing the number of fires caused by fireworks each month. Unsurprisingly, the vast majority of these fires occur in July when the U.S. celebrates its independence from England by shooting off fireworks. However, it might surprise you that June has many more fireworks-caused fires than August. That could be worth deeper analysis.
13. Create a bar plot showing the number of fires caused by arson each month. Do you think there might be more fires caused by arson in March and April because the deadline for filing taxes in the U.S. is in mid April? If so, you could try to confirm this by comparing fire data for countries with different deadlines for filing taxes.

14. Create a bar plot for the SizeClass column. The documentation for this data set says that the size classes A and B indicate fires less than 10 acres and C indicates fires less than 100 acres. As a result, this plot shows that fires of 100 acres or more are outliers in the data set.

15. Create another bar plot for the SizeClass column that sets the fill color to the Cause column. Is there a pattern here, or does cause not have much effect on fire size?

16. Create a bar plot for the ReportingAgency column. An overwhelming majority of fires are reported by state or local organizations (ST/C&L), with the Forest Service (FS) and Bureau of Indian Affairs (BIA) in distant second and third. Why would that be?

17. Create a bar plot for the SourceType column. How does it correlate with the bar plot for the ReportingAgency column?

18. Create a bar plot with the x axis set to the Cause column and the fill color set to the ReportingAgency column. This shows that most fires reported by the Forest Service are caused by lightning, which you might expect, and campfires are common as well.

Deeper analysis

19. Create a new tibble from the fires tibble containing only fires in Idaho (ID).

20. Plot the fires per month in Idaho. This should show that most of the fires take place in the summer, with a high peak in August. Is there anything else to investigate?

21. Plot the cause of Idaho fires by month with fill color set to the Cause column. It seems that lightning causes a large number of fires in August. Why would Idaho have so many more fires caused by lightning in August? You could learn more about weather conditions in Idaho to find out.

22. Plot the cause of Idaho fires again but filter out fires caused by lightning. This shows that August and July have nearly the same number of fires if you exclude fires caused by lightning.

23. Create a tibble that contains the 100 largest fires in Idaho and another tibble that contains the 100 smallest fires.

24. Plot the largest fires on a map of Idaho with the color for each fire set to its cause. This shows that most of the largest fires were caused by lightning and none burned in the upper part of Idaho known as the panhandle.

25. Plot the smallest fires on a map of Idaho with the color of each fire set to cause. This shows that the smallest fires had many different causes and that many of them burned in the panhandle.

26. Plot all fires in Idaho bigger than 500 acres on a map and set the color to reporting agency. Can you tell approximately which parts of Idaho are managed by the Forest Service and which by the Bureau of Land Management?

27. Plot all fires greater than 1,000 acres on a map of the continental US with the color set to reporting agency. Do you see any interesting geographical distributions?

28. Further experiment with plotting the data. What other patterns can you find?

Chapter 11

The Basketball Shots analysis

This chapter presents an analysis of the shots taken by Stephen Curry, a basketball player for the Golden State Warriors of the National Basketball Association (NBA). The data for this analysis comes from the NBA website. It includes the shot locations for every made and missed shot in every game that Curry played from 2009 to 2019.

Get the data

The data for the Basketball Shots analysis is stored in a JSON file that's several levels deep. As a result, you can't read this data directly into a tibble. Instead, you need to read the data from the JSON file into a multidimensional list. Then, you can extract the data you want and use it to build a tibble for this analysis as shown in figure 11-1.

Since chapter 5 describes this process in detail, this figure only provides a review. If you want a detailed explanation of how to get this data, you can refer to chapter 5.

Load the packages

The first example loads the packages needed by this analysis. In addition to the tidyverse package, this analysis uses the RJSONIO package to read JSON data from a file and the ggforce package for some plotting functions such as the facet_zoom() function.

Read the data

The second example reads the data for this analysis. Step 1 downloads the JSON file and saves it in a file named shots.json. Then, step 2 uses the fromJson() function of the RJSONIO package to read the JSON file into a list of lists.

Build the tibble

The third example builds the tibble that stores the data for this analysis. Here, step 1 gets the column names for data set and assigns them to a variable. Then, step 2 gets the rows for the data set and assigns them to another variable. Next, step 3 creates an empty data frame, uses a loop to add each row to this data frame, and adds the column names to it. Finally, step 4 converts the data frame to a tibble. The resulting tibble has 24 columns and 11,846 rows, one for each shot Curry took during the ten seasons that are included in this data set.

Load the packages

```
library("tidyverse")
library("RJSONIO")
library("ggforce")
```

Read the data

Step 1: Download the JSON file

```
url = "https://www.murach.com/python_analysis/shots.json"
download.file(url, "../../data/shots.json")
```

Step 2: Read the JSON file

```
json_data <- fromJSON("../../data/shots.json")
```

Build the tibble

Step 1: Get the column names

```
column_names <- json_data[["resultSets"]][[1]][["headers"]]
```

Step 2: Get the rows

```
rows <- json_data[["resultSets"]][[1]][["rowSet"]]
```

Step 3: Build the data frame

```
shots <- data.frame()
for(row in rows) {
  shots <- rbind(shots, row)
}
names(shots) <- column_names
```

Step 4: Convert the data frame to a tibble

```
shots <- as_tibble(shots)
```

The data set so far

```
shots
# A tibble: 11,846 x 24
    GRID_TYPE        GAME_ID GAME_EVENT_ID PLAYER_ID PLAYER_NAME TEAM_ID
    <chr>            <chr>           <dbl>     <dbl> <chr>          <dbl>
 1 Shot Chart Deta~ 002090~             4    201939 Stephen Cu~   1.61e9
 2 Shot Chart Deta~ 002090~            17    201939 Stephen Cu~   1.61e9
 3 Shot Chart Deta~ 002090~            53    201939 Stephen Cu~   1.61e9
 4 Shot Chart Deta~ 002090~           141    201939 Stephen Cu~   1.61e9
 5 Shot Chart Deta~ 002090~           249    201939 Stephen Cu~   1.61e9
 6 Shot Chart Deta~ 002090~           277    201939 Stephen Cu~   1.61e9
 7 Shot Chart Deta~ 002090~           413    201939 Stephen Cu~   1.61e9
 8 Shot Chart Deta~ 002090~           453    201939 Stephen Cu~   1.61e9
 9 Shot Chart Deta~ 002090~           487    201939 Stephen Cu~   1.61e9
10 Shot Chart Deta~ 002090~           490    201939 Stephen Cu~   1.61e9
# ... with 11,836 more rows, and 18 more variables: TEAM_NAME <chr>,
#   PERIOD <dbl>, MINUTES_REMAINING <dbl>, SECONDS_REMAINING <dbl>,
#   EVENT_TYPE <chr>, ACTION_TYPE <chr>, SHOT_TYPE <chr>,
#   SHOT_ZONE_BASIC <chr>, SHOT_ZONE_AREA <chr>,
#   SHOT_ZONE_RANGE <chr>, SHOT_DISTANCE <dbl>, LOC_X <dbl>,
#   LOC_Y <dbl>, SHOT_ATTEMPTED_FLAG <dbl>, SHOT_MADE_FLAG <dbl>,
#   GAME_DATE <chr>, HTM <chr>, VTM <chr>
```

Figure 11-1   Get the data

Clean the data

Figure 11-2 shows how to clean the data for this analysis. This includes examining the data to determine which columns to select.

Examine the unique values

The first example displays a count of the unique values for each column. This also displays the name of each column in a way that's easy to read. This helps to determine which columns to select and which ones to drop.

To start, you can drop the columns that only contain a single unique value. For example, the PLAYER_NAME column contains a single unique value of "Stephen Curry". Therefore, this column isn't needed for the analysis.

In addition, you can determine other columns that you want to keep or drop by reading their column names and using RStudio to view the data for the tibble. After doing this, you can determine that you don't need the GAME_ID and GAME_EVENT_ID columns for this analysis because you can uniquely identify each game by its date.

For this analysis, we decided to focus on analyzing the number of shots attempted, shots made, and points. To do that, we decided to use the GAME_DATE, SHOT_TYPE, EVENT_TYPE, and SHOT_MADE_FLAG columns. In addition, we decided to analyze the locations of the shots, so we also use the LOC_X, LOC_Y, and SHOT_ZONE_BASIC columns.

The second example displays the unique values for the SHOT_MADE_FLAG column. The result shows that this column uses 1 to indicate that the shot was made and 0 to indicate that the shot was missed.

Select and rename the columns

The third example shows how to select the columns for this analysis. This drops the rest of the columns from this data set. In addition, this example renames these columns to use title case.

Improve the data types for two columns

The fourth example shows how to improve the data types for two columns. To do that, it converts the GameDate and ShotMadeFlag columns to the Date and logical types. At this point, it might make sense to convert the ShotType, EventType, and Zone columns to the factor type. However, for the purposes of this analysis, these columns work fine with the character type. This is another reminder that there are multiple ways to perform most analyses.

Examine the count of unique values for each column

```
apply(X = shots, MARGIN = 2, FUN = unique) %>%
   lapply(FUN = length) %>% str()
List of 24
 $ GRID_TYPE          : int 1
 $ GAME_ID            : int 692
 $ GAME_EVENT_ID      : int 703
 $ PLAYER_ID          : int 1
 $ PLAYER_NAME        : int 1
 $ TEAM_ID            : int 1
 $ TEAM_NAME          : int 1
 $ PERIOD             : int 6
 $ MINUTES_REMAINING  : int 12
 $ SECONDS_REMAINING  : int 60
 $ EVENT_TYPE         : int 2
 $ ACTION_TYPE        : int 51
 $ SHOT_TYPE          : int 2
 $ SHOT_ZONE_BASIC    : int 7
 $ SHOT_ZONE_AREA     : int 6
 $ SHOT_ZONE_RANGE    : int 5
 $ SHOT_DISTANCE      : int 71
 $ LOC_X              : int 489
 $ LOC_Y              : int 438
 $ SHOT_ATTEMPTED_FLAG: int 1
 $ SHOT_MADE_FLAG     : int 2
 $ GAME_DATE          : int 692
 $ HTM                : int 32
 $ VTM                : int 32
```

Examine the unique values for the SHOT_MADE_FLAG column

```
unique(shots$SHOT_MADE_FLAG)
[1] 0 1
```

Select and rename the columns

```
shots <- shots %>%
  select(GAME_DATE, SHOT_TYPE, EVENT_TYPE, SHOT_MADE_FLAG,
         LOC_X, LOC_Y, SHOT_ZONE_BASIC)

names(shots) <- c("GameDate", "ShotType", "EventType", "ShotMadeFlag",
                  "LocX", "LocY", "Zone")
```

Improve the data types for two columns

```
shots <- shots %>% mutate(
  GameDate = as.Date(GameDate, format = "%Y%m%d"),
  ShotMadeFlag = as.logical(ShotMadeFlag))
```

The data so far

```
shots
# A tibble: 11,846 x 7
  GameDate    ShotType        EventType ShotMadeFlag  LocX  LocY Zone
  <date>      <chr>           <chr>     <lgl>        <dbl> <dbl> <chr>
 1 2009-10-28 3PT Field Goal  Missed S~ FALSE           99   249 Abov~
 2 2009-10-28 2PT Field Goal  Made Shot TRUE          -122   145 Mid-~
 3 2009-10-28 2PT Field Goal  Missed S~ FALSE          -60   129 In T~
...
```

Figure 11-2   Clean the data

Prepare the data

One important part of preparing a data set is adding new columns to help you analyze the data. For this analysis, figure 11-3 shows how to add several useful new columns to the data set.

Add a Season column

The first example in this figure defines a function that gets the season for the game based on the GameDate column. This is important because a basketball season starts in fall and goes through spring. As a result, games from a single season take place over two calendar years. So, you can't use the year to select a season. That's why it's helpful to add a season column.

To do that, you can define a get_season() function like the one shown in the first example. Then, you can apply that function to the data set as shown in the second example. To make sure this code works correctly, you can view the GameDate and Season columns as shown in the third example. The resulting tibble shows that Season column corresponds with the GameDate column.

Add a Points column

The fourth example shows that there are two types of shots in the ShotType column. There are three-point field goals, which are shots that are made from behind the three-point line, and there are two-point field goals, which are shots made inside the three-point line. In a regular basketball game, there are also free throws that count for one point, but those aren't included in this data set.

The fifth example shows how you can use the ShotType and ShotMadeFlag columns to calculate the number of points made for each shot and store it in a new column named Points. To do that, the code uses nested ifelse() functions.

The first ifelse() function checks if the shot was made. If so, it continues to the nested ifelse() function. Otherwise, the expression returns 0. Within the nested function, the code checks if the shot was a three-point shot. If so, it returns 3. Otherwise, the shot must have been a two-point shot, so it returns 2.

Add some summary columns

The sixth example shows how to create three summary columns. PointsPerGame stores the number of points per game, AttemptedPerGame stores the number of shots attempted per game, and MadePerGame stores the number of shots made in each game.

To make sure this code adds these summary columns correctly, you can view them as shown in the seventh example. At this point, you can consider adding other summary columns. However, for this analysis, these columns are a good starting place.

Define a function that gets the season

```
get_season <- function(row) {
  month <- as.integer(format(as.Date(row["GameDate"]), "%m"))
  year <- as.integer(format(as.Date(row["GameDate"]), "%Y"))
  season <- ""
  if(month > 6) {
    season <- str_c(year, "-", year+1)
  }
  else {
    season <-str_c(year-1, "-", year)
  }
  return(season)
}
```

Apply the function to add a Season column

```
shots <- shots %>%
  mutate(Season = apply(X = shots, MARGIN = 1, FUN = get_season))
```

View the new column

```
shots %>% select(GameDate, Season) %>% unique()
# A tibble: 692 x 2
   GameDate    Season
   <date>      <chr>
 1 2009-10-28  2009-2010
 2 2009-10-30  2009-2010
 3 2009-11-04  2009-2010
...
```

Examine the shot types

```
unique(shots$ShotType)
[1] "3PT Field Goal" "2PT Field Goal"
```

Add a Points column

```
shots <- shots %>%
  mutate(Points = ifelse(ShotMadeFlag == TRUE,
                         ifelse(ShotType == "3PT Field Goal", 3, 2),
                         0))
```

Create three summary columns for each game

```
shots <- shots %>% group_by(GameDate) %>%
  mutate(PointsPerGame = sum(Points),
         AttemptedPerGame = n(),
         MadePerGame = sum(ShotMadeFlag))
```

View the new columns

```
shots %>% select(GameDate, Points, PointsPerGame,
                 AttemptedPerGame, MadePerGame)
# A tibble: 11,846 x 5
# Groups:   GameDate [692]
   GameDate    Points PointsPerGame AttemptedPerGame MadePerGame
   <date>      <dbl>      <dbl>          <int>          <int>
 1 2009-10-28      0         14             12              7
 2 2009-10-28      2         14             12              7
 3 2009-10-28      0         14             12              7
...
```

Figure 11-3   Prepare the data

Analyze the shot statistics

Now that you've prepared the data, you can begin analyzing it as shown in figure 11-4. As usual, plotting the data is a critical part of analyzing it. In addition, as the analysis progresses, it may be necessary to do some additional preparation.

Plot shots made per game by season

Part 1 shows how to plot the shots made per game for each season. To do this, it creates a box plot for each season to represent the number of shots made in each game. This shows that Stephen Curry had one of his best seasons for scoring in the 2015-2016 season, much better than his first few seasons.

Plot shots made per game by season

```
ggplot(shots, aes(x = Season, y = MadePerGame, fill = Season)) +
  geom_boxplot() +
  labs(title = "Shots Made Per Game", x = "", y = "") +
  theme(plot.title = element_text(hjust = 0.5),
        axis.text.x = element_text(angle = 45, hjust = 1))
```

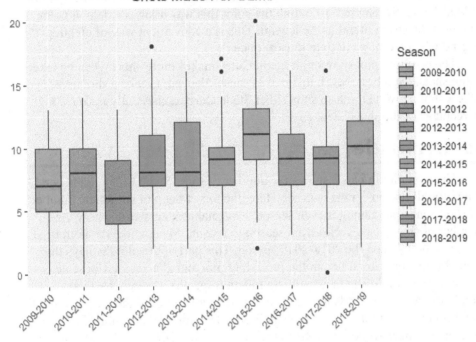

Figure 11-4   Analyze the shot statistics (part 1)

Plot shots attempted vs. made per game

The plot in part 1 of figure 11-4 provided a high-level summary of the shots made by season. Now, part 2 of this figure takes a closer look at the individual games in a single season. To start, it plots the number of shots attempted compared to the number of shots made for the 2018-2019 season.

Here, the color arguments in aes() are set not to columns but to string literals ("Made" and "Attempted"). Coding the color this way assigns a default color and uses the string literal in the legend. This is a convenient way of creating a legend when plotting multiple lines at once.

The resulting plot shows that Steph Curry makes more shots when he takes more shots. On a basic level, this makes sense. You can't make a shot unless you take a shot. However, it also shows that Steph Curry consistently makes a high percentage of the shots he takes.

Plot shots made per game for all seasons

The next plot in part 2 shows the number of shots Steph Curry made per game for all seasons in the data set. Here, the top of the plot shows the number of shots made per game for each season. This makes it easy to see how the number of shots made varies from season to season. Meanwhile, the bottom of the plot zooms in on the 2018-2019 season. This part of the plot displays the same line as the second line in the preceding plot, but at a zoomed-in scale.

In this example, the facet_zoom() function sets the x parameter to a condition. As a result, this function zooms in on all values in the x axis where the condition is true. In this case, the condition is true for all dates in the 2018-2019 season. So, the code zooms in from the lowest value (October 16, 2018) to the highest value (April 9, 2019). By contrast, to use the xlim parameter to zoom in on the same data, you would need to set the xlim parameter to these two values.

Plot shots attempted vs. made per game

```
ggplot(filter(shots, Season == "2018-2019"),
       aes(x = GameDate)) +
  geom_line(aes(y = MadePerGame, color = "Made"), size = 1) +
  geom_line(aes(y = AttemptedPerGame, color = "Attempted"), size = 1) +
  labs(title = "2018-2019 Season", x = "", y = "", color = "Shots") +
  theme(plot.title = element_text(hjust = 0.5))
```

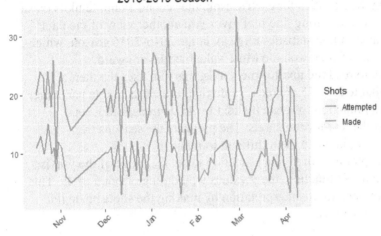

Plot shots made per game with facet zoom

```
ggplot(shots, aes(x = GameDate, y = MadePerGame)) +
  geom_line(size = 1, color = "blue") +
  labs(title = "Shots Made Per Game", x = "", y = "") +
  theme(plot.title = element_text(hjust = 0.5)) +
  facet_zoom(x = Season == "2018-2019")
```

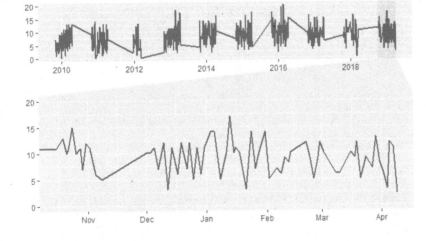

Figure 11-4   Analyze the shot statistics (part 2)

Plot shot statistics by season

Another way to analyze the shots data is to look at trends in the number of shots attempted, shots made, and points by season over time. To do that, part 3 of figure 11-4 begins by summarizing some statistics for each season. In particular, it creates columns that store the season averages for the number of shots made, the number of shots attempted, and the number of points.

After preparing the statistics for each season, the second example plots these statistics over time. The resulting line plot gives you another view of the data. This plot shows that the three statistics all peak in the 2015-2016 season, which was the first season that Curry won the Most Valuable Player award.

To make it easy to read the text for the x axis, this plot uses the theme() function to rotate that text by 45 degrees. To do that, it uses the angle parameter of the element_text() function to rotate the text by 45 degrees, and it sets the resulting text as the text for the x axis. The prevents the seasons from overlapping, which would make them difficult to read.

In addition, this plot uses the ylim() function to set the range of the y-axis from 0 to 30. Without this function, the y values would range from 5 to 25. That would leave the plot open to misinterpretation by making the shots made line seem lower than it actually is.

Plot shot statistics per season

Prepare the data

```
shots_season <- shots %>%
  group_by(Season) %>%
  summarize(MeanMade = mean(MadePerGame),
            MeanAttempted = mean(AttemptedPerGame),
            MeanPoints = mean(PointsPerGame))
```

```
shots_season
# A tibble: 10 x 4
   Season     MeanMade MeanAttempted MeanPoints
   <chr>        <dbl>        <dbl>        <dbl>
 1 2009-2010     7.60         16.4         17.6
 2 2010-2011     7.62         15.7         17.5
 3 2011-2012     6.81         13.5         16.3
 4 2012-2013     8.59         18.9         21.0
 5 2013-2014     9.09         19.2         21.9
 6 2014-2015     8.74         17.8         21.3
 7 2015-2016    10.8          21.3         27.0
 8 2016-2017     8.96         19.1         22.2
 9 2017-2018     8.84         17.8         22.0
10 2018-2019     9.8          20.7         25.1
```

Plot the data

```
ggplot(shots_season, aes(x = Season, group = 1)) +
  geom_line(aes(y = MeanMade, color = "Made"), size = 1) +
  geom_line(aes(y = MeanAttempted, color = "Attempted"), size = 1) +
  geom_line(aes(y = MeanPoints, color = "Points "), size = 1) +
  labs(y = "", color = "") +
  theme(axis.text.x = element_text(angle = 45, hjust = 1)) +
  ylim(0, 30)
```

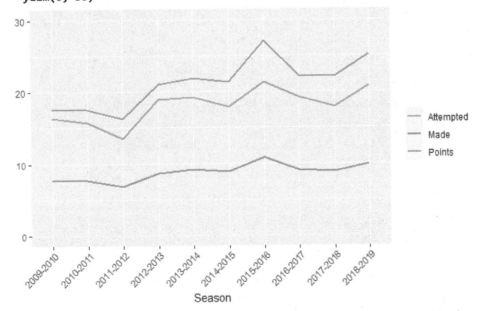

Figure 11-4   Analyze the shot statistics (part 3)

Plot shooting percentages per season

Part 4 of figure 11-4 plots the average shooting percentage by season. To do that, the code prepares the data by calculating the number of shots attempted and made per season. Then, it adds calculated columns for the shooting percentage and cumulative shooting percentage for each season. The resulting tibble shows that Curry's shooting percentage for each season ranges from 45% to 50%, but his cumulative shooting percentage never drops below its starting point of 46%.

After preparing the data, the code plots the shooting percentages per season. The resulting line plot shows how Curry's shooting percentage hits its lowest point in the 2012-2013 season, then climbs to its highest level in the 2015-2016 season. Meanwhile, his cumulative shooting percent dips slightly during the 2012-2013 season but overall trends mostly upward.

Plot shooting percentages per season

Prepare the data

```
shots_pct_season <- shots %>%
  group_by(Season) %>%
  summarize(Made = sum(ShotMadeFlag),
            Attempted = n()) %>%
  mutate(SeasonPct = (Made / Attempted),
         CumulativePct = (cumsum(Made) / cumsum(Attempted)))

shots_pct_season
# A tibble: 10 x 5
   Season    Made Attempted SeasonPct CumulativePct
   <chr>     <int>   <int>     <dbl>      <dbl>
 1 2009-2010  528    1143     0.462      0.462
 2 2010-2011  505    1053     0.480      0.470
 3 2011-2012  145     296     0.490      0.473
 4 2012-2013  626    1388     0.451      0.465
 5 2013-2014  652    1383     0.471      0.467
 6 2014-2015  653    1341     0.487      0.471
 7 2015-2016  804    1596     0.504      0.477
 8 2016-2017  674    1442     0.467      0.476
 9 2017-2018  428     864     0.495      0.477
10 2018-2019  632    1340     0.472      0.477
```

Plot the data

```
ggplot(shots_pct_season, aes(x = Season, group = 1)) +
  geom_line(aes(y = SeasonPct, color = "Percent"), size = 1) +
  geom_line(aes(y = CumulativePct, color = "Cumulative Percent"),
            size = 1) +
  labs(title = "Shooting Percent by Season", x = "", y = "",
       color = "") +
  theme(plot.title = element_text(hjust = 0.5),
        axis.text.x = element_text(angle = 45, hjust = 1))
```

Figure 11-4   Analyze the shot statistics (part 4)

Analyze the shot locations

So far, this analysis has focused on statistics such as shots attempted, shots made, points, and average shooting percentage. Now, it looks at the locations of the shots.

Plot shot locations for two games

Part 1 of figure 11-5 uses a scatter plot to plot the shot locations for two games. But first, it views some select columns from the shots data set so you can familiarize yourself with the data.

Each shot has an X and Y location recorded. The units here are one tenth of a foot, with the center of the court at 0. For example, a shot with a LocX value of 99 and LocY of 249 was made 9.9 feet to the right of the court's center and 24.9 feet above the basket.

Before plotting the data, the code creates a vector to select two specific game dates and plots the shot locations for those games. Then, the code for the plot sets the color parameter to the EventType column so the plot displays made shots in one color and missed shots in another. In addition, it calls the coord_fixed() function so the x and y axes use the same scale.

Although this is a good start, it's hard to interpret this data without seeing the court. That's why the next part of this figure shows how create a function to draw a court on the plot.

Plot shot locations for two games

View the columns

```
select(shots, GameDate, EventType, LocX, LocY)
# A tibble: 11,846 x 4
# Groups:    GameDate [692]
   GameDate    EventType     LocX  LocY
   <date>      <chr>        <dbl> <dbl>
 1 2009-10-28 Missed Shot     99   249
 2 2009-10-28 Made Shot     -122   145
 3 2009-10-28 Missed Shot    -60   129
 4 2009-10-28 Missed Shot   -172    82
 5 2009-10-28 Missed Shot    -68   148
...
```

Plot the data

```
game_dates <- c(as.Date("2019-01-13"), as.Date("2019-02-28"))

ggplot(filter(shots, GameDate %in% game_dates),
       aes(x = LocX, y = LocY, color = EventType)) +
  geom_point() +
  labs(title = "Shot Location by Game", x = "", y = "", color = "") +
  theme(plot.title = element_text(hjust = 0.5)) +
  facet_wrap(facets = vars(GameDate)) +
  coord_fixed()
```

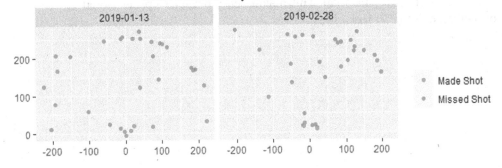

Figure 11-5   Analyze the shot locations (part 1)

Define a function for drawing the court

Part 2 of figure 11-5 defines a function named draw_court() that plots the lines of a basketball court. Although you may sometimes need to define functions like this one, you can often find the code for them by searching the internet. Then, if necessary, you can modify that code to customize it for your purposes.

The draw_court() function returns a vector of function calls. Within this function, the calls to the geom_rect(), geom_segment(), geom_circle(), and geom_arc() functions draw the different components of the court. If you add this vector of function calls to an existing plot, it draws the lines for the court on the existing plot.

Every component in this function sets the data parameter to the data frame defined at the top of the function (dummy_data). This data frame contains only one column and one row. In addition, every component sets the inherit.aes parameter to FALSE. This prevents the code from trying to use data that's set in the ggplot() function when plotting the court, which can cause the draw_court() function to run inefficiently. As a result, no matter how this function is called, it always executes efficiently.

Throughout the draw_court() function, the unit of measurement is 0.1 feet, the same scale as the LocX and LocY columns in the shots data set. For example, the x values for the first call to the geom_rect() function range from 250 to -250, for 500 units total. This makes sense because a professional basketball court is 50 feet wide.

The draw_court() function

```
draw_court <- function() {
  dummy_data <- data.frame(col1 = c(1))
  court <- c(
    coord_fixed(),

    # court outline
    geom_rect(aes(xmin = -250, xmax = 250, ymin = -47.5, ymax = 470),
              fill = NA, color = "black",
              data = dummy_data, inherit.aes = FALSE),

    # backboard
    geom_rect(aes(xmin = -30, ymin = -8.5, xmax = 30, ymax = -7.5),
              fill = NA, color = "black",
              data = dummy_data, inherit.aes = FALSE),

    # outer and inner paint boxes
    geom_rect(aes(xmin = -80, ymin = -47.5, xmax = 80, ymax = 142.5),
              fill = NA, color = "black",
              data = dummy_data, inherit.aes = FALSE),
    geom_rect(aes(xmin = -60, ymin = -47.5, xmax = 60, ymax = 142.5),
              fill = NA, color = "black",
              data = dummy_data, inherit.aes = FALSE),

    # left and right 3pt legs
    geom_segment(aes(x = -220, xend = -220, y = -47.5, yend = 92.5),
                 data = dummy_data, inherit.aes = FALSE),
    geom_segment(aes(x = 220, xend = 220, y = -47.5, yend = 92.5),
                 data = dummy_data, inherit.aes = FALSE),

    # hoop and free throw circles
    geom_circle(aes(x0 = 0, y0 = 0, r = 7.5),
                data = dummy_data, inherit.aes = FALSE),
    geom_circle(aes(x0 = 0, y0 = 142.5, r = 60),
                data = dummy_data, inherit.aes = FALSE),

    # restricted zone arc
    geom_arc(aes(x0 = 0, y0 = 0, r = 40, start = -pi/2, end = pi/2),
             data = dummy_data, inherit.aes = FALSE),

    # inner and outer center court arcs
    geom_arc(aes(x0 = 0, y0 = 470, r = 20, start = pi/2, end = 3*pi/2),
             data = dummy_data, inherit.aes = FALSE),
    geom_arc(aes(x0 = 0, y0 = 470, r = 60, start = pi/2, end = 3*pi/2),
             data = dummy_data, inherit.aes = FALSE),

    # three-point arc
    geom_arc(aes(x0 = 0, y0 = 0, r = 238, start = -1.18, end = 1.18),
             data = dummy_data, inherit.aes = FALSE)
  )
  return(court)
}
```

Figure 11-5   Analyze the shot locations (part 2)

Plot shot locations for two games on a court

Part 3 of figure 11-5 shows how to use the draw_court() function to draw a basketball court. To do that, the code adds the draw_court() function to the ggplot() function. This displays the lines for a basketball court.

The second example shows how to use draw_court() to plot the locations of two games on the court. This code works like the code shown in part 1 of figure 11-5. However, it adds the draw_court() function to the end of the code to draw the court on the scatter plot.

The resulting plot shows that a high percentage of Curry's shots from these games were from just beyond the three-point line. In addition, it shows a cluster of shots in or near the restricted area, which is the small arc that's close to the basket.

Plot the court

```
ggplot() + draw_court()
```

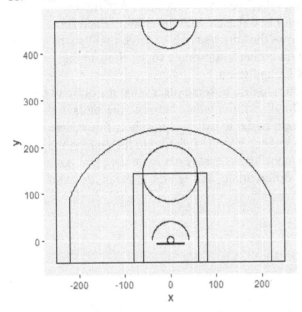

Plot shot locations for two games on the court

```
ggplot(filter(shots, GameDate %in% game_dates),
       aes(x = LocX, y = LocY, color = EventType)) +
geom_point() +
labs(title = "Shot Location by Game", x = "", y = "", color = "") +
theme(plot.title = element_text(hjust = 0.5)) +
facet_wrap(facets = vars(GameDate)) +
draw_court()
```

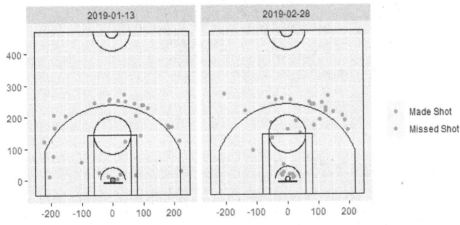

Figure 11-5   Analyze the shot locations (part 3)

Plot shots by zone for one season

To help you visualize the zones on the basketball court, part 4 of figure 11-5 plots the locations of the shots for one season with the color of each shot determined by its zone. This plot shows the various types of three-point shots (above the break, left corner, and right corner), mid-range shots, shots in the paint, shots in the restricted area (RA), and so on.

In the first example, the shots are so dense in some places that it's difficult to interpret them. For example, it's difficult to differentiate between the shots in the paint and the restricted area. To make it easier to examine the shots from these two zones, the second example adds a facet_zoom() function to the code in the first example. This specifies x and y coordinates that zoom in on the paint zone. The resulting plot makes it easier to differentiate between the shots in the paint and shots in the restricted area.

Plot shots by zone for one season

```
ggplot(data = filter(shots, Season == "2009-2010"),
       aes(x = LocX, y = LocY, color = Zone)) +
  geom_point() +
  labs(title = "2009-2010 Season", x = "", y = "") +
  theme(plot.title = element_text(hjust = 0.5)) +
  draw_court()
```

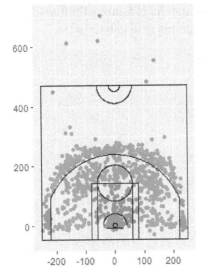

The same plot after adding the following facet_zoom() function

```
facet_zoom(xlim = c(-60, 60), ylim = c(-50, 150), zoom.size = 1)
```

Figure 11-5   Analyze the shot locations (part 4)

Plot shot count by zone

Now that you are familiar with the zones on a basketball court, you might want to look at some statistics for each zone. To start, part 5 of figure 11-5 uses a bar plot to display Steph Curry's shot count by zone for all seasons.

The resulting plot shows that the highest shot count is for a three-point shot from above the break. During the seasons in this data set, Curry attempted more than 4,000 shots from this zone. In fact, as of 2021, he holds the record for the most three-point shots made of all time. Although he is famous for his three-pointers, this bar plot also shows that Curry attempts many shots in the mid-range and restricted area.

Plot shot count by zone

```
ggplot(shots, aes(x = Zone, fill = Zone)) +
  geom_bar() +
  labs(x = "", y = "Count", title = "Shot Count by Zone") +
  theme(plot.title = element_text(hjust = 0.5),
        axis.text.x = element_text(angle = 45, hjust = 1),
        legend.position = "none")
```

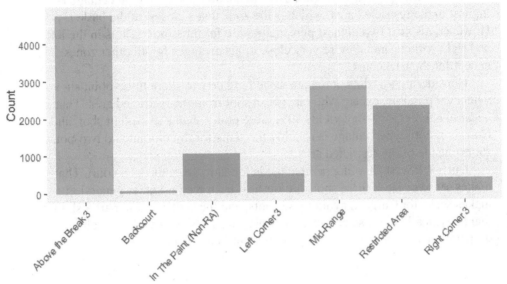

Figure 11-5   Analyze the shot locations (part 5)

Plot shooting percentage by zone

Part 6 of figure 11-5 dives deeper into the data by plotting the shooting percentage by zone. To do that, the code begins by grouping the data by the Zone column and calculating summary columns for the number of shots made and attempted. Then, it adds a column that calculates the shooting percentage for each zone.

The resulting data set and plot show that Curry's shooting percentage is highest in the restricted area, which is the zone that's closest to the basket. However, his next two highest percentages are for three-point shots in the left and right corners, and they're very close to his averages for all other zones except for the backcourt.

Considering this data, it makes sense for Curry to shoot three-point shots whenever possible, except when he gets a shot from the restricted area. That's because a three-point shot yields 50% more points than a two-point shot, and Curry's shooting percentage is roughly the same for three-point and two-point shots outside of the restricted area.

Curry's lowest shooting percentage is for shots from the backcourt. That makes sense because these shots are furthest from the basket. In general, it makes sense for Curry to avoid these shots, and the shot count in part 5 shows that he typically does, as do most other players. These shots are typically desperation shots made as time is running out in a period.

Plot shooting percentage by zone

Prepare the data

```
shots_zone <- shots %>%
  group_by(Zone) %>%
  summarize(Made = sum(ShotMadeFlag),
            Attempted = n()) %>%
  mutate(Percent = (Made / Attempted)) %>%
  arrange(desc(Percent))

shots_zone
# A tibble: 7 x 4
```

	Zone	Made	Attempted	Percent
	<chr>	<int>	<int>	<dbl>
1	Restricted Area	1422	2282	0.623
2	Right Corner 3	181	359	0.504
3	Left Corner 3	249	506	0.492
4	Mid-Range	1305	2830	0.461
5	Above the Break 3	2047	4746	0.431
6	In The Paint (Non-RA)	439	1047	0.419
7	Backcourt	4	76	0.0526

Plot the data

```
ggplot(shots_zone,
       aes(x = Zone, y = Percent, fill = Zone)) +
  geom_col() +
  labs(x = "", title = "Shooting Percentage by Zone") +
  theme(plot.title = element_text(hjust = 0.5),
        axis.text.x = element_text(angle = 45, hjust = 1),
        legend.position = "none")
```

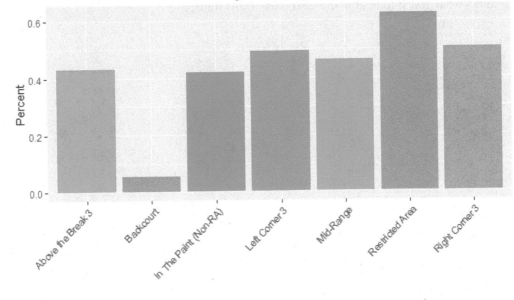

Figure 11-5   Analyze the shot locations (part 6)

Plot shot density

The first example in part 7 of figure 11-5 shows how to plot the made and missed shot densities for the 2009-2010 season. Here, the plot uses the geom_density2d() function to create the density lines. A 2D density plot is similar to a topographic elevation map. If the lines are closer together, the data is denser. If the lines are farther apart, the data is less dense.

The plot for the first example shows five zones that Curry tends to shoot in. The first is the restricted zone that's close to the hoop. The next two are the zones outside the three-point line in the left and right corners of the court. And the last two zones are outside the three-point line to the left and right of the center. While the preference for these two zones is visible in this plot, the differences aren't very distinct. Furthermore, there are some inconsistencies between made and missed shots.

The second example plots the made and missed shot density for all seasons. The code for this plot is similar to the code for the plot in the first example, but it doesn't filter the data set to just one season.

The plot in the second example more clearly shows the five zones that Curry tends to shoot in. It also shows a sixth zone at the top center of the three-point line. In addition, the symmetry between the plots for made and missed shots indicates that there is no obvious zone in which Curry is more or less likely to make the shot aside from the restricted zone.

Overall, the second example makes it easier to spot trends in the shot zones than the first example. When plotting density data, having more data typically makes it easier to see trends in the data. However, it would also be instructive to compare the individual seasons and see when certain trends became clear, so that's what you'll look at in the next part of this figure.

Plot shot density for one season

```
ggplot(shots %>% filter(Season == "2009-2010")) +
  geom_density2d(aes(x = LocX, y = LocY), bins = 40, alpha = 0.9) +
  labs(title = "2009-2010 Season", x = "", y = "") +
  theme(plot.title = element_text(hjust = 0.5)) +
  facet_wrap(facets = vars(EventType)) +
  theme(axis.ticks = element_blank(), axis.text = element_blank()) +
  draw_court()
```

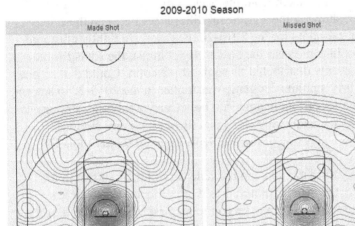

Plot shot density for all seasons

```
ggplot(shots) +
  geom_density2d(aes(x = LocX, y = LocY),  bins = 40, alpha = 0.9) +
  labs(title = "All Shots 2009-2019", x = "", y = "") +
  theme(plot.title = element_text(hjust = 0.5)) +
  facet_wrap(facets = vars(EventType)) +
  theme(axis.ticks = element_blank(), axis.text = element_blank()) +
  draw_court()
```

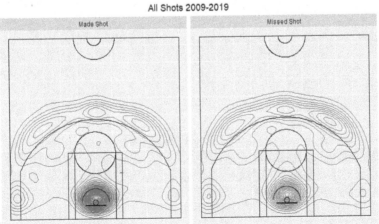

Figure 11-5   Analyze the shot locations (part 7)

Compare shot locations and density for two seasons

Part 8 of figure 11-5 compares the shot locations and density for two important seasons for Stephen Curry, his rookie season (2009-2010) and his first MVP season (2015-2016). The first example uses a scatter plot to plot the locations for every shot that Curry took during those two seasons, with the exception of shots that were taken from the backcourt.

The resulting plot shows some significant differences between these two seasons. In particular, the shots for the 2015-2016 season are tightly clustered behind the three-point line and near the basket. By contrast, the shots for the 2009-2010 are more evenly distributed throughout the court. Could that be one of the reasons that Curry's points per game was higher in the 2015-2016 season?

Although this data reveals an insight into how Curry's shooting strategy changed over time, there are so many shots that there aren't any obvious "hot spots" of activity. That's one of the problems of using scatter plots for high-density data sets.

To further analyze this data, you can use a density plot, as shown in the second example. Here, the density plot makes it easier to see how Curry's shot locations evolved from his rookie season to his first MVP season. For the 2015-2016 season, the density lines outside the three-point circle are much closer together than they were for his rookie season. That's also true for the lines in the restricted area near the basket.

Plot shot locations for two seasons

```
seasons <- c("2009-2010", "2015-2016")
ggplot(filter(shots, Season %in% seasons & LocY < 475),
       aes(x = LocX, y = LocY, color = EventType)) +
  geom_point() +
  labs(title = "Shot Location by Season", x = "", y = "", color = "") +
  theme(plot.title = element_text(hjust = 0.5)) +
  facet_wrap(facets = vars(Season)) +
  theme(axis.ticks = element_blank(), axis.text = element_blank()) +
  draw_court()
```

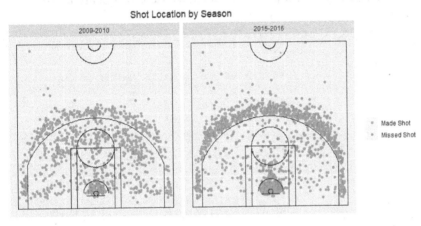

Plot shot density for the same two seasons

```
ggplot(shots %>% filter(Season %in% seasons)) +
  geom_density2d(aes(x = LocX, y = LocY), bins = 40, alpha = 0.9) +
  labs(title = "Shot Density by Season", x = "", y = "") +
  theme(plot.title = element_text(hjust = 0.5)) +
  facet_wrap(facets = vars(Season)) +
  draw_court()
```

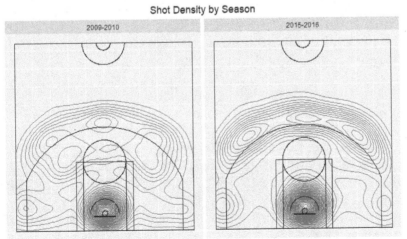

Figure 11-5   Analyze the shot locations (part 8)

Perspective

This chapter shows just one way you can use R and the tidyverse package to analyze sports data. In fact, sports analytics using large data sets have changed the way that many sports are played, coached, and managed. One well-known example is how the Oakland Athletics baseball team used data analysis to assemble a competitive baseball team despite Oakland's small budget. This is documented in Michael Lewis' book *Moneyball* and the movie of the same name from 2011. These days, most professional teams use sports analytics to stay competitive.

Exercise 11-1   More analysis of the Curry data

This exercise guides you through more analysis of the data for Stephen Curry's shots.

Get the data

1. Open the script named exercise_11-1.R that's in this folder:
    exercises/ch11

2. Set your working directory to the source file location.

3. Review the code in the script as you run it.

4. Note that most of the code works like the code presented in this chapter. However, it also adds some new columns such as the TotalSeconds and BasicActionType columns.

5. Note that the code that adds the TotalSeconds column uses the Period, MinutesLeft, and SecondsLeft columns to calculate the total elapsed time since the game began. However, there's a special case for games with a fifth, overtime period, which is only 5 minutes (300 seconds) long, not the usual 12 minutes (720 seconds).

6. Note that the code that adds the BasicActionType column uses the word() function to get the second-to-last word from the ActionType column.

Explore the TotalSeconds column

7. To confirm that the math creating the TotalSeconds column is correct, generate a box plot with the Period column on the x axis and the TotalSeconds column on the y axis. Since the top of each whisker reaches the bottom of the next box's whisker, the total seconds are calculated correctly.

    Testing the math also shows some interesting data. Periods 1 and 3 have medians close to the center of their boxes, but periods 2 and 4 do not. Why is that?

8. Create a histogram that has 60 bins with the x axis set to the TotalSeconds column, and the fill color set to the Period column. This shows that Steph Curry attempts fewer shots at the beginning of the second and fourth periods. Perhaps his team is resting him after the first and third periods? It's hard to say why, but the pattern is clearly visible.

9. Create another histogram that uses facet_wrap() to display a separate plot for each season. This shows that the pattern of Curry attempting fewer shots at the beginning of the second and fourth periods was already evident in his second season, and became prominent by the 2013-2014 season.

    As it happens, this grid of plots also shows that Curry had an unusually low number of shots in 2011-2012 and 2017-2018. Research indicates that the 2011-2012 season was shortened due to the NBA lockout and that Curry was injured for part of the 2017-2018 season. This is a good example of a pattern in the data giving you a direction for more research.

Explore the ActionType and BasicActionType columns

10. Use the table() function to view the count of each value in the ActionType column. Note that some of these actions are very common, but some only have one instance in the entire data set.

11. Use the table() function to view the count of each value in the BasicAction-Type column.

12. Generate a bar plot of the basic action types. This shows that jump shots overwhelmingly outnumber all other types of shots.

13. Create a bar plot of all action types that plots only shots with "Jump" as the basic action type. This shows that most jump shots have an action type of "Jump Shot", not one of the more specific types such as "Turnaround Jump Shot".

14. Use the table() function to view which zone each type of shot is made in, like this:

```
table(shots[c("Zone","BasicActionType")])
```

This shows that dunks, for example, are only made from the restricted zone, but this is a little hard to visualize from a table of numbers.

15. Plot the shots on the court and set the color of each shot to the BasicAc-tionType column. To do that, use the draw_court() function provided by this script.

16. Plot the shots on the court but filter out the jump shots. This makes it easier to see patterns in the other types of shots.

17. Plot just the jump shots on the court setting the color of each shot to the ActionType column. What new patterns do you see?

18. Plot just the jump shots on the court setting the color of each shot to the ActionType column, but filter out the "Jump Shot" values. This makes the patterns easier to see.

19. Plot just dunk shots on the court. Since dunks only happen near the basket, use facet_zoom() to zoom in on the basket. Do you see any patterns? If not, is this because dunks are randomly distributed? Or could it be because there are too few dunks in the data set to draw conclusions?

20. Continue experimenting with the data and look for other patterns and correlations. If you ask the right questions and keep trying different approaches, you're bound to find something interesting!

Section 4

An introduction to data modeling

So far, this book has shown you how to use R for descriptive analysis. Now, this section introduces a solid set of skills for using R for predictive analysis. All three chapters in this section show how to use the tidymodels package to create statistical models that you can use to make predictions.

To start, chapter 12 shows how to use a linear regression model to make predictions based on a single variable. Then, chapter 13 shows how to expand on those skills by using a linear regression model to make predictions based on multiple variables. Finally, to complete this introduction to data modelling, chapter 14 shows how to use a classification model to make predictions that categorize data.

Chapter 12

How to work with simple regression models

This chapter shows how to use the tidymodels package to create a simple regression model that makes predictions based on a single variable. In addition, it presents many skills for identifying correlations between variables and understanding the formulas created by the models. But first, it introduces some concepts and terms that apply to predictive analysis.

Introduction to predictive analysis

So far, the analyses in this book have focused on *descriptive analysis*. Descriptive analysis analyzes historical data to understand the past, identify patterns, and gain insights. However, it isn't designed to predict future or unknown values. That's the job of *predictive analysis*.

Types of predictive models

The first table in figure 12-1 presents five types of *predictive models*. This isn't a complete list, but it gives you an idea of what predictive analysis can be used for. This book focuses on the forecast and classification models. These models are used in a wide variety of industries and are relatively easy to understand, giving you a solid foundation in predictive analysis.

Introduction to regression analysis

In a *regression analysis*, you use one or more *independent variables* to predict the value of another variable, called the *dependent variable*. Regression analysis is one of the main tools used in predictive analysis. It involves creating and using *regression models* to make predictions.

A *linear regression model* is a mathematical model that predicts the value of a dependent variable based on the values of one or more independent variables. This figure lists several examples of how linear regression models are used in the real world. These listings use DV to identify the dependent variable and IV to identify the independent variables. Note that some of these examples are much more complicated than others.

For instance, if you want to predict the odds that a patient will be readmitted to the hospital following release (the dependent variable), you may focus on data such as the patient's current condition, blood pressure, or hemoglobin count (independent variables). You may also want to consider the patient's status at admission, such as whether surgery was involved and whether the surgery was planned or an emergency. Generally, but not always, the more independent variables taken into account, the better the prediction.

When a regression model uses only one independent variable to predict the value of a dependent variable, it is called a *simple linear regression*. When a regression model uses more than one independent variable, it's called a *multiple linear regression*. This chapter shows how to work with simple regression models. Then, the next chapter shows how to work with multiple regression models.

Some types of predictive models

Model	Description
Forecast model	Makes predictions about numeric data based on historical data.
Time series model	Makes predictions about numeric data based on historical data with time as a factor.
Classification model	Categorizes data into predefined groups based on historical data.
Clustering model	Identifies similarities in data and creates groups based on them.
Outliers model	Identifies anomalies relative to historical data.

Real-world applications of predictive analysis

Industry	Usage
Finance	Credit scores, loan applications, risk analysis, fraud detection.
Healthcare	Personalized healthcare, patient deterioration detection, readmission prevention.
Manufacturing	Maintenance prediction, product quality prediction.
Marketing	Product recommendation, customer satisfaction analysis.
Sports	Individual performance prediction, team performance prediction.
Entertainment	Content curation, content adoption prediction.

An introduction to regression analysis

- A *linear regression model* is a mathematical model that uses an equation to predict an unknown value called a *dependent variable* (DV) based on one or more known values called *independent variables* (IVs).
- A *simple linear regression* uses only one independent variable. A *multiple linear regression* uses two or more independent variables.

Real-world linear regression examples

- Predict revenue (DV) based on advertising budget (IV).
- Predict patient readmission rate (DV) given factors like hemoglobin count (IV), prior admissions (IV), blood pressure (IV), current medications (IV), and age (IV).
- Predict household energy consumption (DV) given factors like the number of people living there (IV), the size of the house (IV), the energy consumption of surrounding houses (IV), and past energy consumption (IV).
- Predict the odds of a sports team winning a game (DV) given factors like who they are playing (IV), the current season record (IV), any injured players (IV), and where the game is played (IV).

Figure 12-1   Introduction to predictive analysis

The diamonds data set

In general, predictive models work best when they work with large data sets. That's why this chapter uses the diamonds data set that contains data about more than 50,000 diamonds. In this chapter, the examples use the price column as the dependent variable. As a result, the other columns in this data set are all potential independent variables. In other words, these examples attempt to predict the price of a diamond based on its other characteristics.

How to get the data

Figure 12-2 presents the diamonds data set. This data set is available from the ggplot2 package. As a result, you can access it after you load the tidyverse package as shown in the first example.

The second example displays the data set. This shows that the data set is stored as a tibble and that the carat, depth, table, price, x, y, and z columns store continuous numeric data while the cut, color, and clarity columns store categorical data. Since numeric data works best for simple regressions, this chapter focuses on the numeric columns.

The third example shows how to view the documentation for the diamonds data set. After you do that, take a moment to read through it to familiarize yourself with each column. For example, carat is a unit of weight while x, y, and z describe the diamond's size. Similarly, depth is related to the way a diamond has been cut, and table refers to the diamond's biggest facet. This is important because understanding the data is crucial to choosing good independent variables for your model.

Load the package that contains the data set

```
library("tidyverse")
```

View the data set

```
diamonds
# A tibble: 53,940 x 10
   carat cut       color clarity depth table price    x    y    z
   <dbl> <ord>     <ord> <ord>   <dbl> <dbl> <int> <dbl> <dbl> <dbl>
 1  0.23 Ideal     E     SI2      61.5    55   326  3.95  3.98  2.43
 2  0.21 Premium   E     SI1      59.8    61   326  3.89  3.84  2.31
 3  0.23 Good      E     VS1      56.9    65   327  4.05  4.07  2.31
 4  0.29 Premium   I     VS2      62.4    58   334  4.2   4.23  2.63
 5  0.31 Good      J     SI2      63.3    58   335  4.34  4.35  2.75
 6  0.24 Very Good J     VVS2     62.8    57   336  3.94  3.96  2.48
...
```

View the documentation for the data set

```
? diamonds
```

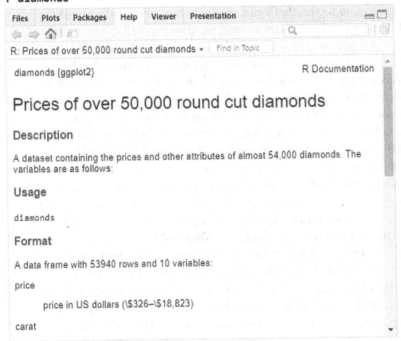

Description

- The diamonds data set contains attributes for more than 50,000 diamonds.

Figure 12-2   How to get the data

How to examine and clean the data

When creating a linear regression model, it's important to clean your data. Since the diamonds data set is an example data set that's intended for learning, it's already pretty clean. However, before you use this data set, you should confirm that by examining the data and cleaning it based on your findings as shown in figure 12-3.

The first example renames the columns in the data set to use title case. This makes it easier to identify the column names in the code.

The second example checks for NA values. This shows that there are no NA values in the set. As a result, you don't need to drop any rows that contain NA values.

After checking for NA values, you can check for invalid values. One easy way to do this is to view the tibble in RStudio and sort each numeric column in both ascending and descending order. When you sort the X, Y, and Z columns, you'll find that they contain some values of 0. Since these columns are a measurement of a diamond's size, that's not a valid value. The third example displays some of these invalid values by sorting the data set by the X column in ascending order.

The fourth example shows how to drop the rows that have invalid values in the X, Y, and Z columns. To do that, this example uses the filter() function to return all rows where X, Y, and Z are not equal to 0.

When you examine your data, it's important to get a sense of how many outliers you have in the data. This is important because regression models rely heavily on calculations like the mean and median. Because of that, outliers can throw off the model's calculations. If 99 people eat 1 scoop of ice cream a day and Ice Cream George eats 1,000 scoops, the mean number of scoops consumed per person each day is 11. However, a model that predicts that any given person would eat 11 scoops a day would be very inaccurate. Ice Cream George is a clear outlier and should be dropped from the data set before making a model.

To do a more in-depth check for outliers, you can use a histogram or KDE plot to view the overall distribution of the values for the numeric columns in your data set. The fifth example in this figure shows a KDE plot for each numeric column in the diamonds data set. In addition to the KDE curve, each plot shows the upper and lower fences as vertical red lines. This helps show which columns likely contain outliers. Although it isn't shown in this figure, the code for generating these plots is included in the example script for this chapter.

Rename columns to use title case

```
names(diamonds) <- names(diamonds) %>% str_to_title()
```

Check for NA values

```
apply(X = diamonds, MARGIN = 2, FUN = function(col) sum(is.na(col)))
  Carat    Cut   Color Clarity   Depth   Table   Price       X       Y       Z
      0      0       0       0       0       0       0       0       0       0
```

Check for invalid values

```
diamonds %>% arrange(X)
# A tibble: 53,940 × 10
    Carat Cut         Color Clarity Depth Table Price     X     Y     Z
    <dbl> <ord>       <ord> <ord>   <dbl> <dbl> <int> <dbl> <dbl> <dbl>
  1  1.07 Ideal       F     SI2      61.6    56  4954     0  6.62     0
  2  1    Very Good   H     VS2      63.3    53  5139     0     0     0
  3  1.14 Fair        G     VS1      57.5    67  6381     0     0     0
```

Drop rows that have invalid values

```
diamonds <- diamonds %>% filter(X != 0 & Y != 0 & Z != 0)
```

Generate KDE plots for the numeric columns

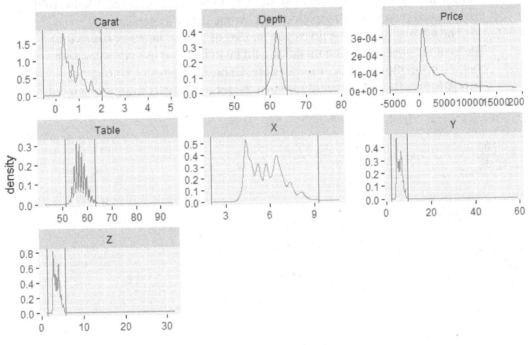

Description

- There are multiple ways to examine the data prior to cleaning, including checking for NA values, sorting the tibble, and creating histograms to see value distribution.

Figure 12-3   How to examine and clean the data

How to find correlations

If two or more variables are related in a linear way, they are said to be linearly correlated. You can measure how close the correlation is with a *correlation coefficient,* or *r-value*. There are different methods of calculating the r-value, but this book uses the Pearson method.

How to interpret correlation coefficients

A correlation coefficient can range from 1 to -1. A coefficient of 1 indicates a *perfect positive relationship*. In other words, given a value for x you can perfectly predict the value of y, and as x increases y also increases. A coefficient of -1 is a *perfect negative relationship* where you can perfectly predict the value of y, but as x increases y decreases. If there is no correlation between the values at all, the coefficient is 0.

When using real-world data, you typically don't come across perfect correlations. If you do it's probably because one variable was calculated using the other. A coefficient of exactly 0 is also extremely unlikely, as even random data usually isn't perfectly scattered. In fact, if you come across an r-value of 0 that's a good reason to be suspicious of how the data was generated.

The table in figure 12-4 gives you an idea of how to interpret the r-value. The goal for your regression model should be to find and use variables with coefficients as close to 1 or -1 as possible. Choosing variables with strong correlation coefficients generally leads to better predictions from your model.

How to interpret a correlation coefficient

r-value	Correlation	r-value	Correlation
1.00	Perfect positive	-1.00	Perfect negative
.70 to .99	Strong positive	-.70 to -.99	Strong negative
.50 to .69	Moderate positive	-.50 to -.69	Moderate negative
.30 to .49	Weak positive	-.30 to -.49	Weak negative
.01 to .30	Negligible positive	-.01 to -.30	Negligible negative
0.00	No correlation	0.00	No correlation

Scatter plots with correlations of various strengths

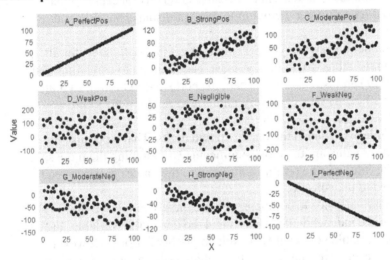

Description

- The *Pearson correlation coefficient* (or *r-value*) is a measure of the linear relationship between two variables.
- An r-value of 1 is a *perfect positive relationship*. That means given a value for x you can perfectly predict the value of y, and as x increases so does y.
- An r-value of -1 is a *perfect negative relationship*. That means given a value for x you can perfectly predict y, but as x increases y decreases.
- An r-value of 0 indicates no relationship between the variables at all, and it is impossible to predict y based on x.
- The specific math for calculating a Pearson correlation coefficient relies on the standard deviations for both x and y.

Figure 12-4   How to interpret a correlation coefficient

How to identify correlations with r-values

To calculate the Pearson correlation coefficient, you can use the cor() function. In some cases, you just want to calculate the coefficient for two variables as shown in the first example in figure 12-5. This example passes the Price and Carat columns from the diamonds data set to the cor() function, and the function returns the calculated r-value.

In other cases, you may want to compare one column with all other numeric columns in the data set as shown in the second example. This example passes the Price column as the first argument and all numeric columns as the second argument. To do that, this code uses the select_if() and is.numeric() functions to only pass numeric columns to cor(). That's because cor() only accepts numeric columns. If you pass a non-numeric column to cor(), it displays an error and doesn't return any r-values.

There may even be times when you will want to get all coefficients for all numeric columns in the table. To do that, you can pass a tibble to cor() after first passing it to select_if() and is.numeric(). Then, the cor() method returns a matrix that contains the coefficients between every numeric variable in the tibble as shown in the third example.

The third example also rounds the results to two digits. That's because the first two digits of the r-value are the most significant. So, rounding them doesn't affect your ability to interpret the r-value and it makes the values easier to read when they're printed to the console.

When you choose variables to compare, you are mostly interested in the correlation between the independent variables and the dependent variable. In this case, the dependent variable you want to be able to predict is Price. However, viewing the table of results provides a way to double-check the integrity of your data set and to make sure you understand the data overall.

For example, the Carat column is very strongly correlated with the X, Y, and Z columns. That makes sense because carat is a unit of weight and x, y, and z describe the diamond's size. Meanwhile, Depth is not strongly correlated with anything, and is only weakly correlated with Table. That makes sense because depth is related to the way a diamond has been cut, a purely aesthetic choice. Table, meanwhile, refers to the diamond's biggest facet, again, an aesthetic choice that doesn't correlate with the overall price of a diamond.

However, r-values only describe relationships that can be described with a straight line. If the relationship forms a curved line, the r-value may be low. That's why it often makes sense to identify correlations visually as well as with r-values.

The cor() function

Function	Description
cor(x, y)	Returns the Pearson correlation coefficient (r-value) for the specified variables. The y argument is optional if x is a tibble.

How to get the coefficient between two variables

```
cor(x = diamonds$Price, y = diamonds$Carat)
[1] 0.9215921
```

How to get all coefficients for a variable

```
cor(diamonds$Price, select_if(diamonds, is.numeric))
          Carat       Depth    Table Price        X         Y         Z
[1,] 0.9215921 -0.01072892 0.1272453     1 0.8872314 0.8678642 0.8682064
```

How to get all coefficients for all numeric variables

```
select_if(diamonds, is.numeric) %>% cor() %>% round(2)
      Carat Depth Table Price     X     Y    Z
Carat  1.00  0.03  0.18  0.92  0.98  0.95 0.96
Depth  0.03  1.00 -0.30 -0.01 -0.03 -0.03 0.10
Table  0.18 -0.30  1.00  0.13  0.20  0.18 0.15
Price  0.92 -0.01  0.13  1.00  0.89  0.87 0.87
X      0.98 -0.03  0.20  0.89  1.00  0.97 0.98
Y      0.95 -0.03  0.18  0.87  0.97  1.00 0.96
Z      0.96  0.10  0.15  0.87  0.98  0.96 1.00
```

Description

- The cor() function is designed to work with numeric data.
- Since r-values are an approximation of the relationship between variables, using more than two significant digits is unnecessary.
- Viewing all of the Pearson coefficients for your data set is a good way to double-check the data set's integrity and make sure you understand the data.
- R-values only describe relationships that form a straight line. If the relationship forms a curved line, the r-value may be low despite a well-defined relationship.

Figure 12-5   How to identify correlations with r-values

How to identify correlations visually

Since an r-value is only sensitive to linear relationships, it doesn't reflect relationships that are non-linear such as exponential, polynomial, or logarithmic relationships. As a result, it's a good practice to attempt to identify correlations between variables visually.

In part 1 of figure 12-6, the first example shows how to visualize the correlation between the Price and Carat variables. These variables have a strong positive correlation that can be described well by a straight line. That's why the r-value for these variables is so high (approximately .92).

The second example shows how to visualize the correlation between the Price and Z variables. These variables also have a strong positive correlation and a high r-value (approximately .86). However, the relationship between these variables could probably be better described by a curved line than a straight one. That can be useful information when creating a predictive model.

A strong positive correlation

```
ggplot(diamonds, aes(x = Carat, y = Price)) +
  geom_point()
```

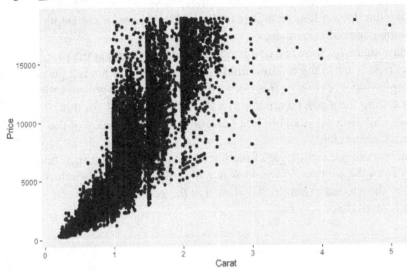

A strong positive correlation with a slight curve

```
ggplot(diamonds, aes(x = Z, y = Price)) +
  geom_point() +
  xlim(2.25, 6)
```

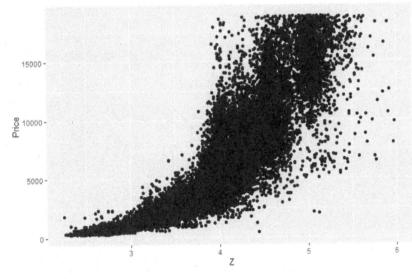

Description

- It's important to identify correlations visually because the r-value only represents the relationship between two variables for a straight line.

Figure 12-6   How to identify correlations visually (part 1)

In part 2 of figure 12-6, both examples show how to visualize the correlation between the Price and Depth variables. These variables have a weak correlation and a low r-value (approximately .01). However, the diamonds data set has more than 50 thousand rows, and when you plot all of these rows, the scatter plot makes it look like the Price might be higher in the middle ranges of the Depth variable than at the edges of this range.

That's because there are so many points on top of each other that the plot hides how many points are in the bottom middle part of the plot. This is a good example of overplotting. Previously, this book showed you how to account for overplotting by setting the alpha parameter to a low number, like .1. In this instance, however, the data set is so big that adjusting the transparency of the data points doesn't help much.

Instead, you can use the sample_n() function to randomly select a specified number of rows from the data set. If you look at a plot of 500 randomly selected rows as shown by the second example, it's clear that the correlation between depth and price is negligible.

A negligible correlation (all rows)

```
ggplot(diamonds, aes(x = Depth, y = Price)) +
  geom_point()
```

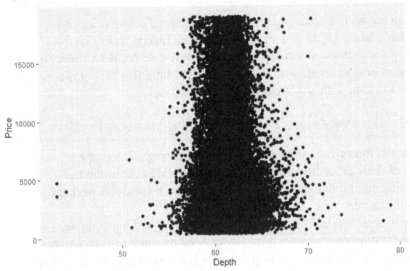

A negligible correlation (500 randomly selected rows)

```
ggplot(sample_n(diamonds, 500), aes(x = Depth, y = Price)) +
  geom_point()
```

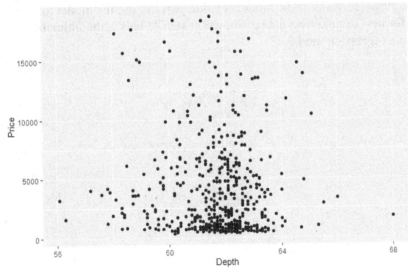

Description

- With a large data set, a scatter plot can be misleading because overplotting can make it difficult to tell where most of the values are.

- You can use the sample_n() function to get a random sample of rows from a data set.

Figure 12-6   How to identify correlations visually (part 2)

How to create a model that uses a straight line

A regression model is a conceptual model used to predict the value of a dependent variable based on the values of one or more independent variables. The simplest type of regression model uses a straight line to predict values, so this chapter shows how to create a model for a straight line. But first, it presents a procedure for working with any type of regression model.

A procedure for working with a regression model

The diagram in figure 12-7 shows the steps for creating and using a regression model. This procedure provides the conceptual background that you need for using any modeling package, including the tidymodels package presented in this chapter.

In step 1, you split the data set into a *training data set* and a *testing data set*. The training and testing data sets are typically created by randomly assigning rows to each group. Incidentally, splitting the data is a part of creating any type of predictive model, not just regression.

In step 2, you use the training data to create and train the regression model. In step 3, you use the testing data set to check how well the model works. If the model predicts values that are close to the actual values in the test data set, the model can be considered valid. If the model is valid, you can use the model to predict values for new or unknown data as shown in step 4. This is the ultimate goal of creating a regression model.

The procedure for creating, testing, and using a regression model

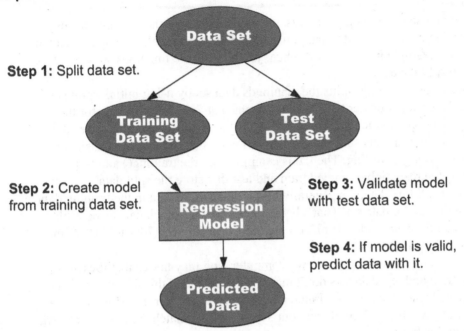

Step 1: Split data set.

Step 2: Create model from training data set.

Step 3: Validate model with test data set.

Step 4: If model is valid, predict data with it.

Description

- The procedure above is used by most R packages that support regression models, including the tidymodels package presented in this chapter.

- After you use the *training data set* to train the model and the *testing data set* to validate the model, you can use the model to predict the value of a dependent variable based on the value of an independent variable.

Figure 12-7   A procedure for working with a regression model

How to split the data

The first step to creating a regression model is to split the data into testing and training data sets. To do that, you can load the tidymodels package as shown in the first example in figure 12-8. Then, you can use the functions shown in this figure to split the data.

The second example splits the diamonds data set by using initial_split() and setting the prop parameter to 0.75. This means that 75% of the data is for the training data set and the remaining 25% is for the testing data set. It stores the result in an interim data set named diamonds_split that has each row assigned to either training or testing. Then, this example uses the training() and testing() functions to create tibbles named train and test that store the split data.

When splitting data, it's generally best to have more data in the training data set than the testing data set. That's because you want as much data as possible when creating your model. Typical regressions use between 70% to 80% of the total data for training.

If you run this code on your own computer, you may notice that the rows in your train and test data sets don't match the example in this figure. That's to be expected since the rows are randomly assigned to each group. However, the number of rows for each tibble on your system should match the number of rows shown in this figure.

The package for creating regression models

```
library("tidymodels")
```

Functions for splitting the data

Function	Description
initial_split(data, prop)	Splits the data into testing and training data sets. The prop parameter specifies the proportion of the data that goes to the training set.
training(x)	Returns the training data from split data (x).
testing(x)	Returns the testing data from split data (x).

How to split the data

```
diamonds_split <- initial_split(diamonds, prop = 0.75)
train <- training(diamonds_split)      # 75% of data set
test <- testing(diamonds_split)        # 25% of data set
```

The training data

```
train
# A tibble: 40,440 × 10
  Carat Cut       Color Clarity Depth Table Price    X     Y     Z
  <dbl> <ord>     <ord> <ord>   <dbl> <dbl> <int> <dbl> <dbl> <dbl>
1  0.7  Ideal     E     VS1      62.6    55  3167  5.66  5.69  3.55
2  0.3  Ideal     E     VVS2     60.2    57   789  4.37  4.4   2.64
3  1.09 Ideal     E     VS2      60.6    58  7796  6.66  6.7   4.05
4  0.38 Ideal     G     SI1      61.5    55   695  4.66  4.68  2.87
5  0.9  Good      D     SI2      64      59  4078  6.04  6.09  3.88
6  0.51 Good      G     VVS2     63.4    57  1800  5.06  5.09  3.22
7  1.73 Premium   I     VS1      62.2    59 12674  7.65  7.62  4.75
...
```

The testing data

```
test
# A tibble: 13,480 × 10
  Carat Cut       Color Clarity Depth Table Price    X     Y     Z
  <dbl> <ord>     <ord> <ord>   <dbl> <dbl> <int> <dbl> <dbl> <dbl>
1  0.23 Good      E     VS1      56.9    65   327  4.05  4.07  2.31
2  0.29 Premium   I     VS2      62.4    58   334  4.2   4.23  2.63
3  0.31 Good      J     SI2      63.3    58   335  4.34  4.35  2.75
4  0.24 Very Good J     VVS2     62.8    57   336  3.94  3.96  2.48
5  0.3  Good      J     SI1      63.8    56   351  4.23  4.26  2.71
6  0.3  Good      I     SI2      63.3    56   351  4.26  4.3   2.71
7  0.23 Very Good H     VS1      61      57   353  3.94  3.96  2.41
...
```

Description

- The rows for the train and test tibbles are randomly assigned to each group.

Figure 12-8   How to split the data

How to drop outliers from the training data set

Now that you've split your data, you can drop outliers from the training data. This typically improves the accuracy of your model. However, since the data that you make predictions for might include outliers, you typically don't want to drop outliers from the test data. If you do, you won't be able to evaluate how well your model handles outliers during testing.

Part 1 of figure 12-9 shows how to drop the outliers for six of the numeric columns in the diamonds data set. To do this, the first example defines functions named get_upper_fence() and get_lower_fence() that get the upper and lower fences for the outliers for a column. Then, this example uses the filter_at() function to filter six numeric columns (Carat, Depth, Table, X, Y, and Z) to select only values that lie within the fences for each variable.

Note that this code doesn't filter the Price column. In a real analysis, it would be ok to drop outliers for your dependent variable. But to keep things simple, this analysis just focuses on the independent variables.

The filter_at() function works similarly to the filter() function, but it can be used on multiple columns at once. To do that, it provides an extra parameter named .vars that specifies which columns should be filtered. Then, the conditional expression (or predicate) that's used to filter the values can use the all_vars() function to only select rows where the value lies within the upper and lower fences.

In the predicate, this example uses a period (.) to specify the current column. This placeholder is necessary to allow you to apply the same predicate to multiple columns.

A simple filter like the one shown in this figure can drop the rows for most of the outliers, but there are often a few that make it past the filter. To check for these outliers, you can plot the training data set to identify any remaining outliers that you might want to remove from the training data set as shown in the second part of this figure.

Some functions for filtering

Function	Description
filter_at(.tbl, .vars, .predicate)	Returns the rows for the columns specified by .vars that meet the conditions specified by .predicate.
all_vars(expr)	Returns the values for which the expression is true.

A placeholder symbol

Symbol	Description
.	Refers to the current variable.

How to drop outliers from the training data set

```
# Custom functions for getting the upper and lower fences
get_upper_fence <- function(x) {
  quantile(x, 0.75) + (1.5 * IQR(x))
}

get_lower_fence <- function(x) {
  quantile(x, 0.25) - (1.5 * IQR(x))
}

# Returns only rows with values between the fences
train <- train %>% filter_at(vars(Carat, Depth, Table, X, Y, Z),
          all_vars(. > get_lower_fence(.) & . < get_upper_fence(.)))

train
# A tibble: 36,814 x 10
   Carat Cut        Color Clarity Depth Table Price     X     Y     Z
   <dbl> <ord>      <ord> <ord>   <dbl> <dbl> <int> <dbl> <dbl> <dbl>
 1  0.9  Very Good  H     SI2      59.5    63  3160  6.26  6.31  3.74
 2  0.3  Very Good  G     VS2      61.7    56   541  4.32  4.35  2.67
 3  1.51 Very Good  D     VS2      63.2    57 12311  7.27  7.25  4.59
 4  1.74 Premium    J     VS1      62.5    58 11050  7.67  7.65  4.79
 5  1.13 Ideal      F     VVS2     62.1    54 10742  6.67  6.71  4.16
 6  0.4  Ideal      E     VS2      60.9    56   904  4.78  4.81  2.92
 7  1.05 Very Good  H     VS2      63.3    56  5614  6.48  6.41  4.08
 8  0.3  Ideal      F     VVS2     61.8    55   694  4.33  4.35  2.68
 9  1    Premium    E     SI2      61      62  4200  6.41  6.34  3.89
10  1.21 Ideal      F     VS2      61.2    57  8452  6.87  6.82  4.19
# ... with 36,804 more rows
```

Description

- The get_upper_fence() and get_lower_fence() functions calculate the upper and lower fences for an outlier using the same formula that's used by a box plot.
- The filter_at() and all_vars() functions can be used to apply the same expression or condition to several columns at once.

Figure 12-9   How to drop outliers from the training data set (part 1)

The first example in part 2 of figure 12-9 displays a scatter plot for the Price and Z columns. This plot shows that there are still a few obvious outliers in the Z column even though part 1 used fences to drop most outliers. In particular, this plot shows some outliers that have values less than 2.25 that might cause the model to be less accurate. That's why the second example drops the rows that contain these values from the training data set.

The third example shows the same plot after removing these outliers. This shows that there are still some values that stick out a little bit, but the most obvious ones are gone. For a data set this size, that's often good enough.

The Z column after the outliers have been dropped

```
ggplot(data = train) +
  geom_point(aes(x = Z, y = Price))
```

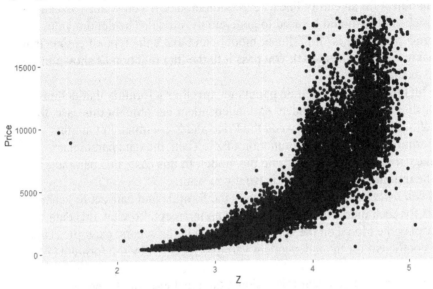

How to drop some of the remaining outliers

```
train <- train %>% filter(Z > 2.25)
```

The Z column after some more outliers have been dropped

```
ggplot(data = train) +
  geom_point(aes(x = Z, y = Price))
```

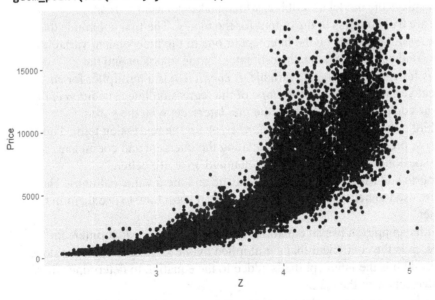

Figure 12-9   How to drop outliers from the training data set (part 2)

How to create a model

The first example in figure 12-10 shows how to use the linear_reg() function to create an object for an empty linear regression model. If you want, you can specify which engine should be used to generate the model. The default value for the engine parameter is "lm" (linear model), and that's the type of engine that we'll be using. To train the model, you pass it to the fit() function as shown in the second example.

In the fit() function, the formula parameter specifies a formula that defines the relationship between a dependent and independent variable. In this case, the formula defines a relationship between the Price and Z columns. Put another way, this formula says, "Price is a function of Z". Then, the data parameter specifies the data set that's used to train the model. In this case, this parameter specifies the training data set, which is what you want.

When you *train* a model, the model uses the formula and data set to generate coefficients for each independent variable and an intercept. To view this data, you can display the model on the console as shown in the second example. This shows the coefficient for the independent variable specified by the formula (Z) and the intercept.

If you want to create a linear regression model that uses the lm engine, you can alternately use the lm() function as shown in the third example. This function provides an easy way to create a linear model object and fit it. However, it doesn't let you specify other engines like the linear_reg() and fit() functions do.

Once you've fit the model, you can pass it to the tidy() function to view more information about the model as shown in the fourth example. The tidy() function displays the model as a tibble, which makes it easier to read.

There are five columns in the output for the model. The first column is the *regression term*, which can be the intercept or one of the independent variables.

The second column contains the estimates for the intercept and the coefficients for each independent variable. A *coefficient* is a multiplier for an independent variable that forms the *slope* of the regression line. The *intercept* is added to the equation to adjust where the line intersects with the y axis.

The third column contains the *standard error* for the regression term. This tells you how precisely the model is estimating the intercept and coefficient. Generally speaking, lower values for the standard error are better.

The fourth and fifth columns are the statistic and the p.value columns. These columns are used to help judge a regression. You'll learn how to use them in the next chapter.

The fifth example shows an equation for a straight line in slope-intercept form. Here, m is the coefficient that's multiplied by the x value to form the slope of the line, and b is the intercept that's added to the equation to determine where the line intersects with the y axis.

The sixth example shows the equation that the model generated based on the formula and the training data. This shows that the model calculates the Price variable by multiplying the Z variable by 4642 and then subtracting 12617.

Some functions for creating and fitting a linear model

Function	Description
`linear_reg(engine)`	Creates an object to hold the model. Valid values for the engine parameter include "lm" (linear model), "glm" (generalized linear model), "keras", "stan", and others. Default is "lm".
`fit(object, formula, data)`	Calculates the coefficients and intercept for the model based on the model's engine, formula, and training data.
`lm(formula, data)`	Creates a linear regression model that uses the lm engine and calculates the coefficients and intercept for the model based on the model's formula and training data.
`tidy(object)`	Shows additional data about a model.

Two ways to create a linear regression model

```
linear_reg()
linear_reg(engine = "lm")
```

How to fit a model

```
model <- fit(object = linear_reg(), formula = Price ~ Z, data = train)
model
Coefficients:
(Intercept)              Z
    -12617           4642
```

A quick way to create a linear regression model that uses the lm engine

```
model <- lm(formula = Price ~ Z, data = train)
```

How to view more information about the model

```
model %>% tidy()
# A tibble: 2 × 5
  term          estimate std.error statistic p.value
  <chr>            <dbl>     <dbl>     <dbl>   <dbl>
1 (Intercept)   -12617.       48.2     -262.       0
2 Z               4642.       13.7      339.       0
```

An equation in slope-intercept form

```
y = m * x + b                # m is coefficient, b is intercept
```

The equation generated by this model

```
Price = 4642 * Z - 12617
```

Description

- An equation in slope-intercept form uses a *coefficient* to adjust the slope of the line and an *intercept* to adjust where the line intersects with the y axis.
- When using R to specify a formula, you use a tilde (~) to separate the left and right sides of an equation, and you don't specify the coefficients or the intercept.
- The linear_reg() function generates a model object that can be *trained* with fit().

Figure 12-10   How to create a model

How to use a model to make predictions

Now that you've created a model, you can make some predictions by passing the trained model and a data set to the predict() function. Typically, you begin by passing the testing data set to the predict() function as shown in the first example of figure 12-11. In this example, the function returns a tibble with a single column named .pred that contains the predicted values for the price. The resulting values show that some of these predicted prices are negative, which is clearly incorrect. As a result, the model definitely has room for improvement.

After you get the predicted values, you can add them to the testing tibble and store them in a new variable named model_results as shown in the second example. Then, you can plot the actual and predicted values as shown in the third example. This shows that the predicted values form a straight line that's calculated by the equation described in the previous figure.

The predict() function

Function	Description
`predict(object, new_data)`	Returns a data object of predicted values given a model object and a data set to make predictions for. The new data set must have the same column names as the data set used to train the model.

How to view the predictions

```
predict(model, new_data = test)
# A tibble: 13,480 × 1
     .pred
     <dbl>
1 -1895.
2  -409.
3   148.
4 -1105.
5   -37.8
...
```

How to add the predictions to the testing data set

```
model_results <- test %>% mutate(predict(model, new_data = test))
```

How to plot the predictions

```
ggplot(data = model_results) +
  geom_point(aes(x = Z, y = Price)) +
  geom_point(aes(x = Z, y = .pred), color = "red")
```

Description

- If you use the fit() function to train your model, the predict() function returns the predictions in a tibble.
- If you use the lm() function to train your model, the predict() function returns the predictions in a vector.

Figure 12-11 How to use a model to make predictions

How to work with formulas

So far, this chapter has shown how to use a simple formula to create a model that's based on a straight line. Now, you're ready to learn how to use more complex formulas to create models for curved lines.

How to plot an equation

To plot an equation without a data set, you can use geom_function(). This function takes a function as an argument. As a result, you can use it to plot an equation that's defined by a custom function or a lambda expression.

The first example in figure 12-12 shows how to create a function named cube_x() that accepts a variable, cubes it, and returns the result. Then, the second example shows how to use geom_function() to plot this custom function. In this example, the code uses xlim() to specify that the plot should range from 0 to 10 along the x axis. Without it, the plot would default to a range of 0 to 1.

The third example works like the second example. However, it uses a lambda expression instead of a custom function to achieve the same result.

A function for plotting equations

Function	Description
geom_function(fun)	Plots the given function.

A custom function that defines an equation

```
cube_x <- function(x) {
  return(x^3)                # defines y = x^3
}
```

How to plot the custom function

```
ggplot() +                            # No data set
  geom_function(fun = cube_x) +       # Plot the cube_x function
  xlim(0, 10)                         # Limit x axis from 0 to 10
```

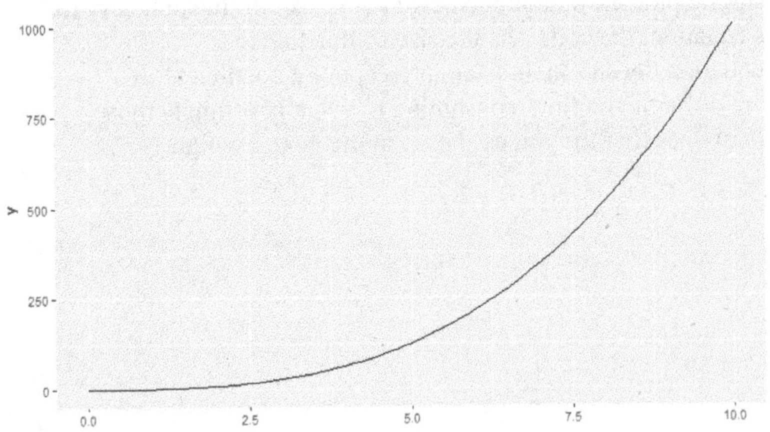

How to use a lambda expression instead of a custom function

```
ggplot() +
  geom_function(fun = function(x) x^3) +
  xlim(0, 10)
```

Description

- You can use geom_function() to plot a function that defines an equation.

Figure 12-12   How to plot an equation

How to plot an equation on a scatter plot

Figure 12-13 shows how to use geom_function() to plot an equation on a scatter plot. The first example shows how to plot the equation for the straight line that's used by the regression model presented earlier in this chapter. To start, it displays a scatter plot for the Price and Z variables from the testing data set. Then, it plots the equation for the straight line. To do that, this code uses the intercept and coefficient calculated by the model. This results in a plot like one shown in figure 12-11.

The second example shows how to use geom_function() to visually estimate an equation for a model that displays a curved, exponential line. To do that, this code uses trial and error. It begins by displaying a scatter plot for the Price and Z variables for the training data set. Then, it plots a series of equations from $y = x$ to $y = x^7$. The resulting plot shows that the best fit is $y = x^6$. To make it easy to see the line for this equation, the code sets the size of this line to 2.

Note that this equation doesn't include an intercept or a coefficient. It would be much more difficult and time-consuming to visually estimate those. Fortunately, the model does that for you as shown in the next few figures.

How to plot a straight line with an intercept and coefficient

```
ggplot(data = test) +
  geom_point(aes(x = Z, y = Price)) +
  geom_function(fun = function(x) 4642 * x - 12617,
                color="red", size = 2)
```

How to plot an exponential line without an intercept and coefficient

```
ggplot(data = train) +
  geom_point(aes(x = Z, y = Price)) +
  geom_function(fun = function(x) x,    color="red") +
  geom_function(fun = function(x) x^2, color="orange") +
  geom_function(fun = function(x) x^3, color="yellow") +
  geom_function(fun = function(x) x^4, color="green") +
  geom_function(fun = function(x) x^5, color="blue") +
  geom_function(fun = function(x) x^6, color="purple", size = 2) +
  geom_function(fun = function(x) x^7, color="violet") +
  ylim(0, 20000)
```

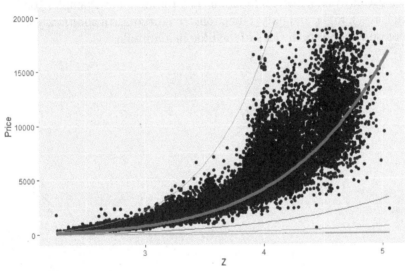

An equation for Price and Z

```
Price = Z^6
```

Description

- You can plot a function on a scatter plot to experiment with formulas that might work for a model.

Figure 12-13   How to plot an equation on a scatter plot

How to code formulas

A *formula* is a special construct in R for specifying a relationship between variables. Once you determine an equation for a line, you can convert it to a formula so R can work with it. To do that, R formulas use a tilde (~) instead of the equals sign (=) that's used by an equation. This tells R that the expression is a formula rather than an assignment.

R formulas have special meanings for certain operators like * and ^. To indicate that you want the arithmetic meaning, you can use I() as shown in figure 12-14. I() tells R that you don't want it to convert what's inside the parentheses to any other type. Instead, you want the object to be used "as is".

For instance, if you want to code a formula where y is a function of x times 3, you can code y ~ I(x*3). Or, if you want to code a formula where y is a function of x squared, you can code y ~ I(x^2). In each case, using I() prevents R from misinterpreting your arithmetic operators.

You can also use any mathematical functions in a formula. For example, you can use the sqrt(), abs(), log(), and poly() functions in a formula. In addition, you can use functions on either or both sides of the tilde in a formula.

An operator and a function for coding formulas

Operator	Description
~	Separates the left and right sides in a formula. Similar to an equals sign (=).

Function	Description
I(x)	Prevents the conversion of the object inside to another type. Means "is" or "as is".

How to code formulas

Price is a function of Z
```
Price ~ Z
```

Price is a function of X times Z
```
Price ~ I(X * Z)          # I() protects * from being misinterpreted
```

Price is a function of Z raised to the 6th power
```
Price ~ I(Z ^ 6)          # I() protects ^ from being misinterpreted
```

Price is a function of the square root of Table
```
Price ~ sqrt(Table)
Price ~ I(sqrt(Table))    # I() is unnecessary but doesn't cause any errors
```

The log of Price is a function of the log of Z
```
log(Price) ~ log(Z)       # Formula operators can be used on either side of ~
```

Description

- A *formula* is a special construct in R for specifying the relationship between variables.
- To use arithmetic operators like * and ^ in a formula, you must enclose them within the I() function.

Figure 12-14   How to code formulas

How to plot a formula on a scatter plot

If you want to quickly plot a formula on a scatter plot, you can use the geom_smooth() function to do that as shown in figure 12-15. The example in this figure begins by displaying a scatter plot for the Price and Z variables of the training data. Then, it uses the geom_smooth() function to plot the line defined by the formula y ~ I(x^6). Because x is set to Z and y is set to Price, this is the same as Price ~ I(Z^6).

Since this code sets the method to lm, the geom_smooth() function uses a linear model to calculate a coefficient and intercept for the line. As a result, it fits the training data a little better than the purple line displayed in figure 12-13.

The geom_smooth() function

Function	Description
geom_smooth(data)	Generates a smooth line for the given data that can be used to quickly visualize the correlation coefficient between two variables.

Some parameters of this function

Function	Description
method	The method to use to calculate the line. Valid methods include lm (linear model), loess (locally weighted polynomial regression), or gam (generalized additive model).
formula	The formula used to calculate the line in terms of x and y.

How to plot a formula

```
ggplot(data = train, aes(x = Z, y = Price)) +
  geom_point() +
  geom_smooth(method = lm, formula = y ~ I(x^6),
              color="purple", size = 2)
```

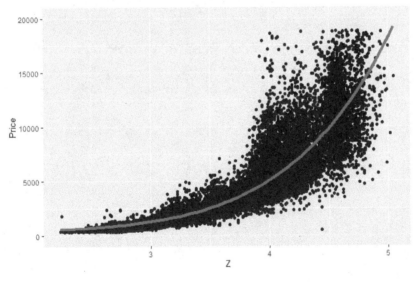

Description

- When you use the geom_smooth() function to plot a formula, it generates the coefficients and intercept for the specified formula.

Figure 12-15   How to plot a formula on a scatterplot

How to create a model for a curved line

Figure 12-16 shows how to create a model for the curved line defined by the formula of `Price ~ I(Z^6)`. The predictions made by this model don't include any negative prices, which is good. However, the predictions for the smallest and largest diamonds in the data set appear to be off by significant amounts. Still, the standard error for the independent variable is much lower than it was in the previous model.

How to create and view the model

```
model_exp <- fit(object = linear_reg(),
                 formula = Price ~ I(Z^6),
                 data = train)
model_exp %>% tidy()
# A tibble: 2 × 5
  term          estimate std.error statistic p.value
  <chr>            <dbl>     <dbl>     <dbl>   <dbl>
1 (Intercept)     435.      10.3       42.4       0
2 I(Z^6)            1.13     0.00269   420.        0
```

The equation generated by the model

```
Price = 1.13 * Z^6 + 435
```

How to view the predictions

```
predict(model_exp, new_data = test)
# A tibble: 13,480 × 1
   .pred
   <dbl>
1  606.
2  606.
3  923.
4  691.
5  731.
...
```

How to add the predictions to the testing data set

```
model_exp_results <- test %>% mutate(predict(model_exp, new_data = test))
```

How to plot the new predictions

```
ggplot(data = model_exp_results) +
  geom_point(aes(x = Z, y = Price)) +
  geom_point(aes(x = Z, y = .pred), color = "red")
```

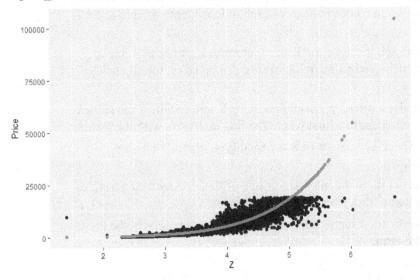

Figure 12-16   How to create a model for a curved line

Perspective

Now that you've completed this chapter, you should be able to create and use a simple regression model. More specifically, you should know how to select an independent variable, create formulas to define a relationship between two variables, and improve your models by adjusting this formula.

Summary

- *Descriptive analysis* uses historical data to provide insights about the past that can be used to make decisions. *Predictive analysis* attempts to predict future or unknown values.

- A *linear regression model* is a conceptual model that uses an equation to predict an unknown value called a *dependent variable* based on one or more known values called *independent variables*.

- A *simple linear regression* uses only one independent variable, and a *multiple linear regression model* uses two or more independent variables.

- The *Pearson correlation coefficient* (or *r-value*) is a number between 1 and -1 that measures the type and strength of the linear correlation between the two variables.

- An r-value of 1 is a *perfect positive relationship*. That means given a value for x you can perfectly predict the value of y, and as x increases so does y.

- An r-value of -1 is a *perfect negative relationship*. That means given a value for x you can perfectly predict y, but as x increases y decreases.

- When you create a predictive model, you split your data set into a *training data set* and a *testing data set* by randomly assigning rows to each group.

- When you *train* a model (or *fit* a model), the model uses the given formula and data set to generate coefficients for each independent variable and an intercept.

- With a large data set, it can be difficult to interpret a scatter plot due to *overplotting*, which is when many points are displayed on top of each other.

- An equation in slope-intercept form uses a *coefficient* to adjust the slope of the line and an *intercept* to adjust where the line intersects with the y axis.

- A *formula* is a special construct in R for specifying the relationship between variables.

- A *regression term* can be the intercept or one of the independent variables used by the equation for a linear regression model.

- The *standard error* for a regression term tells you how precisely the model is estimating the term.

Exercise 12-1   Create a simple regression model

In this exercise, you'll create a simple regression model that predicts the price of a house based on its building area.

Get, clean, and examine the data

1. Create a new R script.

2. Load the tidyverse and tidymodels packages.

3. Read the housing data from the melbourne_housing.csv file that's in the murach_r/data directory. Ignore any warnings about parsing problems.

4. Examine the data for missing values.

5. Drop all rows that contain missing values.

6. Examine the r-values between the Price column and the other numeric columns.

7. Display a scatter plot that shows the relationship between the Price and BuildingArea variables.

Prepare the data

8. Split the data into testing and training data sets with 75% of the data in the training data set.

9. Create functions for calculating the upper and lower fences for outliers.

10. In the training data, use the fences to drop rows that contain outliers from the Price and BuildingArea columns.

11. Visually identify any additional outliers in the training data by displaying a scatter plot for the Price and BuildingArea columns.

12. Drop any other rows that contain additional outliers such as rows where the BuildingArea is very low but the Price is high.

Create a model

13. Create a linear regression model where the Price variable is a function of the BuildingArea variable.

14. Use the model to make predictions for the price in the testing data set and add those predictions to the testing data set.

15. Display a scatter plot that shows the actual and predicted values in the testing data set.

Plot equations and formulas

16. Display the intercept and coefficient for the equation that's used by the model.

17. Use the geom_function() function to plot the equation that's used by the model over a scatter plot of the test data set.

18. Use the geom_smooth() function plot the formula that's used by the model over a scatter plot of the test data set.

Chapter 13

How to work with multiple regression models

In the previous chapter, you learned how to make predictions with a regression model that uses a single independent variable to predict the dependent variable. However, you can usually predict the dependent variable more accurately by using a regression model that uses multiple independent variables. This chapter shows how to use several different techniques to create, fit, and judge multiple regression models.

How to work with multiple variables

This chapter begins by showing a basic way to add multiple independent variables to a regression model. This type of model is known as a *multiple regression model*.

How to create and fit the model

Before showing how to create and fit the model, the first example in figure 13-1 shows how to load the packages for this chapter. Most of these packages are described in the previous chapter. The only new package is the dotwhisker package that contains the dwplot() function described later in this chapter.

The second example reads the cleaned testing and training data sets used in this chapter. These data sets are the same ones used in the previous chapter.

The third example shows how to create and fit a model that uses the + operator to add multiple independent variables to a formula. Using this operator in a formula isn't the same as using it as an arithmetic operator. That's why you need to code arithmetic operators within the I() function as shown in the previous chapter if you want to use them in formulas.

Using the + operator states that each variable is independent from the others. In other words, the formula in this example states that a diamond's width (X) has no relation to its length (Y) or depth (Z).

The resulting model contains a list of terms for the model. To view these terms and their values, you can pass the model to the tidy() function as shown in the fourth example. This shows the intercept for the equation as well as the coefficients for each independent variable. Using the values shown in the fourth example, you can determine that the model uses the equation shown in the fifth example.

The sixth example shows how to use the model to make predictions. It adds a column named Predictions to the testing data set and stores the new data set in a variable named model_results. To get this to work, the code passes the model and the testing data set to the predict() function. Since this returns a tibble with the predictions stored in a column named .pred, the code accesses this column to return its vector of values.

Load the packages for this chapter

```
library("tidyverse")
library("tidymodels")
library("ggforce")
library("ggpubr")
library("dotwhisker")
```

Get the cleaned testing and training data sets for this chapter

```
train <- readRDS("../../data/train_clean.rds")
test <- readRDS("../../data/test_clean.rds")
```

A formula operator for adding variables to the model

Operator	Description
+	Add a variable to the model.

How to create and fit the model

```
model <- fit(linear_reg(), Price ~ X + Y + Z, data = train)
```

How to view the regression terms

```
model %>% tidy()
# A tibble: 4 × 5
  term        estimate std.error statistic  p.value
  <chr>          <dbl>     <dbl>     <dbl>    <dbl>
1 (Intercept)  -14160.      47.6    -297.   0
2 X              2852.      43.4      65.7  0
3 Y               145.      31.6       4.58 0.00000467
4 Z               257.      53.7       4.79 0.00000171
```

The equation for the model

```
Price = (2852 * X) + (145 * Y) + (257 * Z) - 14160
```

How to use the model to make predictions

```
model_results <- test %>% mutate(
  Predictions = predict(model, new_data = test)$.pred)
```

Description

- A *multiple regression model* uses more than one independent variable to predict a dependent variable.

- The dotwhisker package contains the dwplot() function.

Figure 13-1   How to create and fit a multiple regression model

How to judge the model by its R^2 value

Once you've generated a multiple regression model, you'll want to know if it's accurate. However, visualizing multiple regression models is difficult because you can't plot multiple variables on a two-dimensional plot. So how can you tell if your model is accurate?

One way is to use the rsq() function to calculate the *coefficient of determination* (or *R^2 value*) as shown in the first example in figure 13-2. The rsq() function accepts a data frame and parameters specifying the actual and predicted values of the dependent variable. Then, it uses these values to calculate the coefficient of determination.

The rsq() function returns a value between 0 and 1. An R^2 value of 1 means the model works perfectly, with every predicted value matching the actual value. An R^2 of 0 means there is no correlation at all between the predicted values and actual values. What counts as a "good" R^2 value varies by field. Some types of data, like weather systems, are extraordinarily difficult to model accurately.

The first example shows that the model returns an R^2 value of .788. That means that the model is able to explain 78.8% of the variance in the price data.

How to judge the model by its residuals

Another way to judge a model is to plot its residuals. A *residual* provides a way to measure how far the prediction is from the actual value. To start, you can calculate the residuals by subtracting the predicted value for each price from the actual value as shown in the second example in figure 13-2.

The third example plots the residuals to help you visualize how well the model performed. In this plot, anything above the red line indicates that the prediction was too low and anything below the red line indicates that the prediction was too high.

Residual plots can tell you a lot about your model. In general, if the model is working well, the residual plot should show a balanced cloud of points. Any trend in the data indicates that the model could be improved.

The plot in this figure shows that the data is unbalanced on the y axis and trends down on that axis. In addition, this plot has *heteroscedasticity*. This means that the residuals get larger as predictions get larger, creating a cone shape in the data. Both of these trends indicate that the model can be improved.

The plot in this figure shows that this model is not very accurate, especially for diamonds with very low or very high actual prices. In fact, many of the predictions are negative, which is obviously wrong. This doesn't mean X, Y, and Z aren't important independent variables for predicting Price. It just means the model isn't using the right formula yet.

The rsq() function

Function	Description
rsq(data)	Calculate the coefficient of determination (R^2).
Parameter	Description
truth	The actual values of the dependent variable.
estimate	The predicted values of the dependent variable.

How to check the R^2 value for a model

```
model_results %>% rsq(truth = Price, estimate = Predictions)
# A tibble: 1 × 3
  .metric .estimator .estimate
  <chr>   <chr>          <dbl>
1 rsq     standard       0.788
```

How to add a column for the residuals

```
model_results <- model_results %>% mutate(
  Residuals = Price - Predictions)
```

How to plot the residuals

```
ggplot(data = model_results) +
  geom_point(aes(x = Predictions, y = Residuals), alpha = .2) +
  geom_hline(yintercept = 0, color = "red")
```

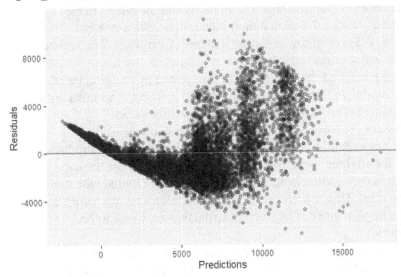

Description

- You can use the *coefficient of determination* (or *R^2 value*) to judge a model.
- *Residuals* are calculated by subtracting the predicted value for each price from the actual value. They measure how far off the predicted value is from the actual value.

Figure 13-2   How to judge a model by its R^2 value and its residuals

How to work with variable interactions

Sometimes the interaction between variables can be used to build a better model. For example, it's unlikely for a commercially sold diamond to have a short width and a long length. Another example of related variables in the diamonds data set is Carat, X, Y, and Z. That's because Carat is the weight of a diamond and X, Y, and Z are its length, width, and depth. Similarly, Table and X are related because the table of a diamond is the width of its top relative to its widest point.

More formula operators

To reflect these interactions in a model, you can use *interaction terms*. An interaction term is a measure of the effect of two or more independent variables considered together independent of their individual effects. Figure 13-3 presents some operators you can use to create formulas that use interaction terms.

An individual interaction term is indicated with the interaction operator (`:`). If a formula is `c ~ a:b`, that means c is a function of the effects of a and b taken together (independent of their individual effects).

If you want to use multiple interaction terms in your formula, writing each one out can be tedious and error prone. That's why there are additional formula operators to make writing formulas easier.

The cross operator (`*`) indicates that a formula should include both the independent variables being crossed and their interactions. So, `a * b` is equivalent to `a + b + a:b`. You can cross any number of variables. The second example shows how this works.

Sometimes, you may find that one of the terms created by crossing variables isn't useful to the model. In that case, you can remove it from the formula with the remove operator (`-`). The third example shows how this works.

The carat operator (`^`) crosses variables but limits the degree of the interaction terms. For example, crossing X, Y, and Z would normally produce interaction terms for all three variables up to X:Y:Z. But, if you use the carat operator as shown in the fourth example, the formula doesn't include the third degree interaction term. This is useful when you want to cross many variables without creating a huge number of interaction terms that are likely to be insignificant anyway.

The examples in this figure show that there are multiple ways to write formulas and get the same results. You can confirm this yourself by passing these formulas to the lm() function. Then, if two formulas generate the same coefficients and intercept, they both get the same results.

More formula operators

Operator	Description
:	The interaction between two variables.
*	Cross two variables.
-	Remove a variable.
^	Cross variables to the specified degree.

How to write formulas for interactions

Price is a function of the interaction between X and Y
```
Price ~ X:Y
Price ~ I(X * Y)
```

Price is a function of X, Y, and the interaction between X and Y
```
Price ~ X + Y + X:Y
Price ~ X * Y
```

Price is a function of X and the interaction between X and Y
```
Price ~ X + X:Y
Price ~ (X * Y) - Y
```

Price is a function of X, Y, Z and their second-degree interactions
```
Price ~ X + Y + Z + X:Y + X:Z + Y:Z
Price ~ X * Y * Z - X:Y:Z
Price ~ (X + Y + Z)^2
```

Price is a function of X, Y, Z, and their interactions
```
Price ~ X + Y + Z + X:Y + X:Z + Y:Z + X:Y:Z
Price ~ X * Y * Z
Price ~ (X + Y + Z)^3
```

Description

- An *interaction term* is created by multiplying two or more independent variables together. In R formulas, the interaction operator is a colon (:).

- Formula operators use the same symbols as arithmetic operators. That's why you need to use I() to specify arithmetic operators in a formula.

- There are multiple ways to write formulas that get the same results.

Figure 13-3   More formula operators

How to create and fit the model

Now that you understand the basics for creating formulas with several interaction terms, you can create a model for the diamonds data set that uses interaction terms. For instance, the first example in figure 13-4 creates and fits a linear regression model that uses interaction terms by crossing the X, Y, Z, and Table variables. Then, it assigns that model to a variable named model_int.

How to view the model's terms

Once you've generated your model, you can use the tidy() function to view the *terms* of the model. A term can be either an intercept, an independent variable taken on its own, or an interaction of multiple independent variables taken together. Furthermore, the model gives an estimate, standard error, statistic, and p-value for each term.

The *estimate* is the model's calculation for the term's coefficient. In the formula for a straight line (`y = mx + b`), m is the coefficient for x. If an estimate is close to zero, that means it's not significant to the model. It's quite literally not adding much to the formula.

In a regression model, the *standard error* is the average distance between the observed values and the regression line. The smaller the error, the more precise the estimate is.

The *statistic*, which is also called the *t-statistic* or *t-value*, is the estimate divided by the standard error. For our purposes, it is mostly notable as a step used for calculating the p-value.

The *p-value* is the probability of the null hypothesis being true given the data used to train the model. The *null hypothesis* states that there is no relationship between the independent variables and the dependent variables. So, if the p-value is high, that means the null hypothesis is likely true and there is no linear relationship between the independent variable and the dependent variable. Generally, a p-value of 0.05 or lower is statistically significant.

How to create and fit the model

```
model_int <- fit(linear_reg(), Price ~ X * Y * Z * Table, data = train)
```

Parts of a model

Part	Description
term	A variable, interaction of variables, or intercept.
estimate	The estimated value (coefficient) of the regression term.
std.error	The standard error for the regression term.
statistic	The estimate divided by the standard error. Also called a t-value or t-statistic.
p.value	The probability of getting the observed results assuming that the null hypothesis is true for the regression term.

How to view the terms

```
model_int %>% tidy() %>% mutate(p.value = round(p.value, 3))
# A tibble: 16 × 5
```

	term	estimate	std.error	statistic	p.value
	<chr>	<dbl>	<dbl>	<dbl>	<dbl>
1	(Intercept)	61312.	22845.	2.68	0.007
2	X	-165193.	12335.	-13.4	0
3	Y	111384.	12785.	8.71	0
4	Z	26720.	9145.	2.92	0.003
5	Table	-209.	396.	-0.529	0.597
6	X:Y	9130.	923.	9.89	0
7	X:Z	33629.	3161.	10.6	0
8	Y:Z	-37225.	2780.	-13.4	0
9	X:Table	2657.	214.	12.4	0
10	Y:Table	-2061.	222.	-9.28	0
11	Z:Table	-610.	160.	-3.81	0
12	X:Y:Z	-587.	163.	-3.59	0
13	X:Y:Table	-125.	16.1	-7.78	0
14	X:Z:Table	-553.	55.0	-10.1	0
15	Y:Z:Table	669.	48.5	13.8	0
16	X:Y:Z:Table	5.67	2.83	2.01	0.045

Description

- A model's terms may or may not be useful. You can use tidy() to view each term's *estimate* (or coefficient), *standard error*, *t-value* (or *t-statistic*), and *p-value*.

- The *null hypothesis* states that there is no relationship between the independent and dependent variables.

Figure 13-4   How to create and fit a model that uses variable interactions

How to remove insignificant terms

Figure 13-5 shows how to judge a model's terms by looking at the p-value for each term. In most scientific contexts, a p-value of .05 or less is considered significant. Another way to say that is that 0.05 is the *significance level*, which is sometimes also called *alpha* because it is represented in formulas by α.

The first example in this figure shows how to filter for terms with a p-value greater than or equal to 0.05. This shows that only one term in our current model meets that condition. Then, the second example shows how to use the - operator to remove this term from the model. In addition, it stores the new model in a variable named model_int2.

After you drop insignificant terms, the new model may identify additional insignificant terms. If so, you can optionally repeat this process until you have a model with no p-value above 0.05.

To compare the first interaction model with the second, the third example uses the predict() function to create predictions for both models. Then, the fourth example, checks the R^2 value for both models. The results show that the two models both have an R^2 value of 0.874. This makes sense as dropping a single insignificant term isn't likely to have much of an impact on a model. As a result, this technique is typically more valuable when a model contains multiple terms with insignificant p-values.

So why not just cross all variables and remove the insignificant terms? Because it's important to avoid *overfitting* your model. Overfitting occurs when your model fits your training data set too closely and doesn't work as well on your testing data set. In other words, the model's predictions for the training data set are extremely accurate, but its predictions for the testing data set aren't as accurate.

Overfitting is commonly caused by including more independent variables and interactions than you need in a model. By contrast, *underfitting* occurs when you don't use enough variables and your model doesn't fit the training data set very well.

In general, the goal of data modelling it to create a model that selects variables that capture the *signal* (actual trend) in the data without capturing the *noise* (change in the data due to random or unpredictable factors). To do that well, you need to understand your data and carefully select your variables. If you succeed, your model will perform well when you use it to predict values from data sets other than your training and testing data sets.

How to interpret the p-value

p-value	Description
p > 0.05	No statistically significant evidence against the null hypothesis.
p <= 0.05	Statistically significant evidence against the null hypothesis.

How to view coefficients that are not statistically significant

```
model_int %>% tidy() %>% filter(p.value > 0.05)
# A tibble: 1 × 5
  term  estimate std.error statistic p.value
  <chr>    <dbl>     <dbl>     <dbl>   <dbl>
1 Table    -209.      396.    -0.529   0.597
```

How to recreate the model without the insignificant terms

```
model_int2 <- fit(linear_reg(),
                  Price ~ X * Y * Z * Table - Table,
                  data = train)
```

How to create the predictions for both models

```
model_int_results <- test %>%
  mutate(Predictions = predict(model_int, new_data = test)$.pred,
         Predictions2 = predict(model_int2, new_data = test)$.pred)
```

Check the R² value for the model

```
model_int_results %>% rsq(truth = Price, estimate = Predictions)
# A tibble: 1 × 3
  .metric .estimator .estimate
  <chr>   <chr>          <dbl>
1 rsq     standard       0.874
```

```
model_int_results %>% rsq(truth = Price, estimate = Predictions2)
# A tibble: 1 × 3
  .metric .estimator .estimate
  <chr>   <chr>          <dbl>
1 rsq     standard       0.874
```

Description

- You can use the p-value to determine if a regression term is statistically significant or not. If it isn't, you can remove it from the model.

- Using too many terms can cause your model to *overfit* your data set. *Underfitting*, by contrast, occurs when you don't have enough variables.

Figure 13-5   How to remove insignificant terms

How to plot regression coefficients

When you first generate a model that has many terms, it can be difficult to grasp the scale of the coefficients or how confident the model is about their accuracy. To quickly visualize the coefficients and standard error for each term, you can use a *dot and whisker plot*. This plots the estimated coefficient as a dot and a 95% confidence interval as the whiskers.

To generate dot and whisker plots, you can use the dwplot() function as shown in the examples in figure 13-6. This function is part of the dotwhisker package and accepts a model as its first parameter. That's why both examples pass the model as the first parameter. In addition, they both make the plot easier to read by setting the dot_args parameter to a list of parameters that set the size and color of the dot. Finally, the code adds a blue vertical line where x is zero.

The first example shows that the X, Y, Z, X:Z, and Y:Z terms have estimated coefficients that are the furthest away from zero, which is represented by the vertical blue line. However, the coefficients for X, Y, and Z have wide confidence intervals. In fact, they're so wide that the confidence intervals for other terms appear quite small on this plot.

That's why the second example zooms in on the values between -1000 and 1000. This code uses coord_cartesian(), not xlim(), to make sure that this plot includes all data.

What does the second plot show? It shows that X:Y:Z:Table has a coefficient that's very close to zero. As a result, it doesn't have much impact on the model. It also shows that the confidence interval for the Table coefficient crosses zero. This means that it's possible that the Table coefficient could be zero, which means that it would have no impact on the model.

How to plot the regression coefficients and their confidence intervals

```
dwplot(model_int, dot_args = list(size = 2, color = "black")) +
  geom_vline(xintercept = 0, color = "blue")
```

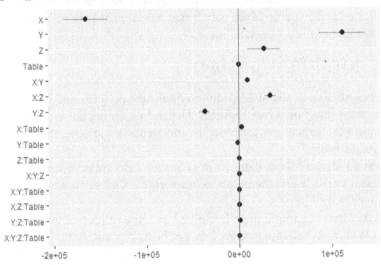

How to zoom in

```
dwplot(model_int, dot_args = list(size = 2, color = "black")) +
  coord_cartesian(xlim = c(-1000,1000)) +
  geom_vline(xintercept = 0, color = "blue")
```

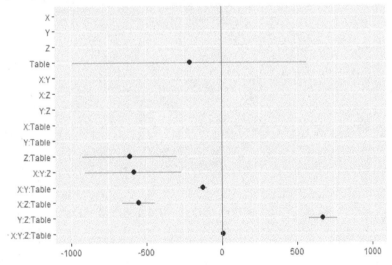

Description

- A *dot and whisker plot* plots the estimated coefficient as a dot with "whiskers" on each side of the dot that provide a confidence interval for the coefficient.

Figure 13-6   How to plot regression coefficients

How to work with nonlinear patterns

So far, you've learned how to create simple and multiple regression models with linear data. But what do you do if the data forms a nonlinear pattern? You use *data transformations* to make the data linear.

Five common nonlinear patterns

Figure 13-7 presents five of the most common nonlinear patterns and how to transform these patterns into linear relationships. Depending on the pattern, you may need to transform the dependent variable, the independent variable, or both to create a linear relationship.

The easiest way to identify these patterns in your data is to create a scatter plot for the dependent variable and the independent variable that you want to test and see what the data looks like.

Keep in mind that these example plots just give you a rough idea of what the pattern looks like. With real data, the pattern may not be as obvious.

Common nonlinear patterns and how to transform them

Name	Pattern	Transformation
Power	$y = x^b$	$\log(y) = \log(x)$
Exponential	$y = b^x$	$\log(y) = x$
Logarithmic	$y = \log(x^b)$	$y = \log(x)$
Reciprocal	$y = 1/(a + bx)$	$1/y = x$
Square root	$y = a + b * \mathrm{sqrt}(x)$	$y = \mathrm{sqrt}(x)$

Examples of these patterns given a = 1 and b = 2

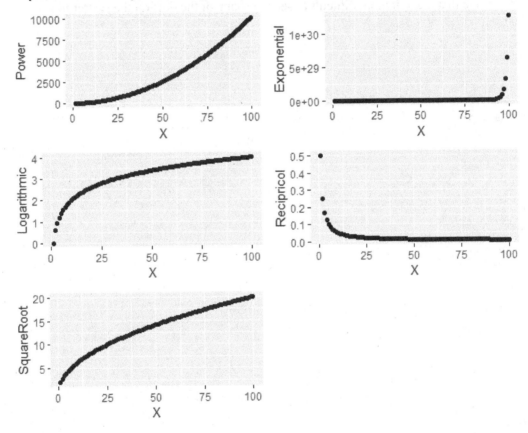

Description

- If the points on a scatter plot curve, the relationship between the variables is nonlinear.
- When a dependent and independent variable have a nonlinear relationship, you can *transform the data* to be linear.

Figure 13-7   Five common nonlinear patterns

How to transform variables

Part 1 of figure 13-8 shows how to use a transformation to make the data for the Z variable linear with Price. To do this, it takes the log of the Price variable to implement the exponential transformation. The first plot shows the data and correlation coefficient (r-value) before the transformation, and the second plot shows the data and r-value after the transformation. As you can see, the relationship is clearly more linear after the transformation.

Note that both of these plots use the xlim() function to limit the range of the x axis. That's because the training data set contains some outliers outside of this range that make it more difficult to see the shape of the dots on the scatter plot.

The Z and Price variables

```
ggplot(data = train) +
  geom_point(aes(x = Z, y = Price)) +
  labs(title = paste("r-value:", cor(x = train$Z, y = train$Price))) +
  xlim(c(2,6))
```

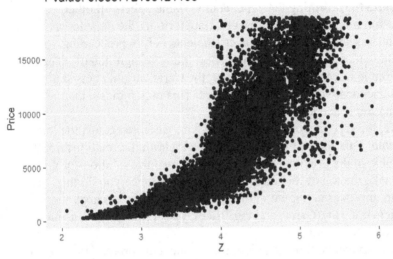

The Z and log(Price) variables

```
ggplot(data = train) +
  geom_point(aes(x = Z, y = log(Price))) +
  labs(title = paste("r-value:",
                     cor(x = train$Z, y = log(train$Price)))) +
  xlim(c(2,6))
```

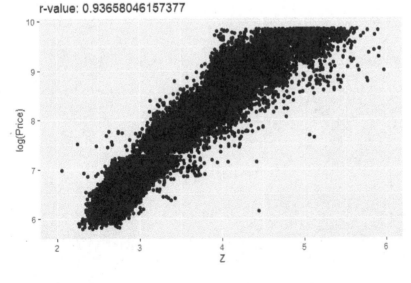

Figure 13-8   How to transform variables (part 1)

When you transform a dependent variable, it affects its relationship with all other variables in the model. That's because you make transformations prior to creating the model.

Part 2 of figure 13-8 begins by adding a PriceLog column that stores the value of the log(Price) transformation. Then, it calculates the r-values for each numeric column and displays them for the Price and PriceLog columns. This makes sense because transforming the dependent variable affects all other variables. As a result, it makes sense to start by transforming the dependent variable and evaluating the new correlation coefficients before proceeding.

The results show that the X, Y, and Z variables improve significantly when the dependent variable is transformed. However, the Carat variable gets slightly worse when the dependent variable is transformed. This may indicate that another transformation is needed.

To investigate, the second example plots the Carat variable against the log of the Price variable. This shows that this data has a logarithmic relationship. If you refer to the table in the previous figure, you'll see that the solution for this relationship is to take the log of the independent variable. Once you do this, the linear relationship improves as shown in the third example. If you check the correlation coefficient for log(Caret) and log(Price), you'll find that it's as high as the coefficients for X, Y, and Z.

Nonlinear patterns can be time consuming to locate and correct. However, the improvement in your regression results is often well worth the effort. For instance, you might want to check whether taking the log of X, Y, and Z improves their r-values relative to the log of Price. As it turns out, this does improve their r-values slightly, and it does seem to improve the accuracy of the model. However, to keep things simple, the formulas presented in this chapter don't take the log of these variables.

The r-values for the transformed data

```
train_log <- train %>% mutate(PriceLog = log(Price)) %>%
  select_if(is.numeric) %>%
  cor() %>%
  round(3)

train_log[, c("Price","PriceLog")]
          Price PriceLog
Carat     0.921    0.920
Depth    -0.008    0.001
Table     0.126    0.159
Price     1.000    0.896
X         0.887    0.961
Y         0.861    0.931
Z         0.864    0.937
PriceLog  0.896    1.000
```

Investigate a variable that didn't improve as much

```
ggplot(data = train) +
  geom_point(aes(x = Carat, y = log(Price)))
```

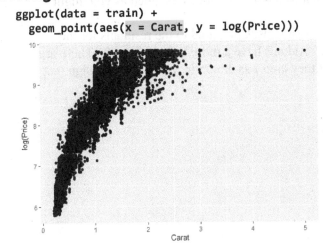

Compare with a transformation to the variable

```
ggplot(data = train) +
  geom_point(aes(x = log(Carat), y = log(Price)))
```

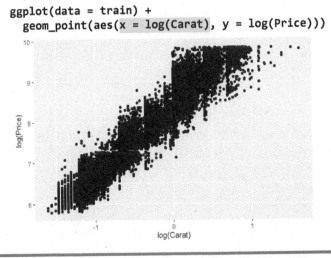

Figure 13-8   How to transform variables (part 2)

How to create, fit, and judge the model

Figure 13-9 begins by showing how to create and fit a model that uses transformations. Here, the formula transforms the dependent variable (Price) as well as one of the independent variables (Carat). In addition, it crosses the X, Y, Z, and Table variables as in the previous model.

After creating and fitting the model, this figure judges the model. To start, the second example gets the predictions for the test data. To do that, it adds a column named Predictions that stores the results of the predict() function.

However, since the formula uses the log of Price, the code needs to undo that transformation before it can compare the predicted prices to the actual ones. To do that, this example uses the exp() function to undo the log() function. Then, it adds the Residuals column by subtracting the predicted prices from the actual prices.

The third example calculates the R^2 value for the model. This value is a slight improvement over the previous model, but not by as much as you might want.

The fourth example plots the residuals. Compared to the residuals for the first model in this chapter, these residuals have a more balanced shape. There are still some trends in the data, but they aren't as dramatic. That's another sign that the model is improving.

How to create and fit the model

```
model_trans <- fit(linear_reg(),
                   log(Price) ~ log(Carat) + X * Y * Z * Table,
                   data = train)
```

How to calculate predictions and residuals

```
model_trans_results <- test %>% mutate(
  Predictions = predict(model_trans, new_data = test)$.pred,
  Predictions = exp(Predictions),  # Undo the log operation
  Residuals = Price - Predictions)
```

How to check the R² value

```
model_trans_results %>% rsq(truth = Price, estimate = Predictions)
# A tibble: 1 × 3
  .metric .estimator .estimate
  <chr>   <chr>          <dbl>
1 rsq     standard       0.879
```

How to plot the residuals

```
ggplot(data = model_trans_results) +
  geom_point(aes(x = Predictions, y = Residuals), alpha = .2) +
  geom_hline(yintercept = 0, color = "red")
```

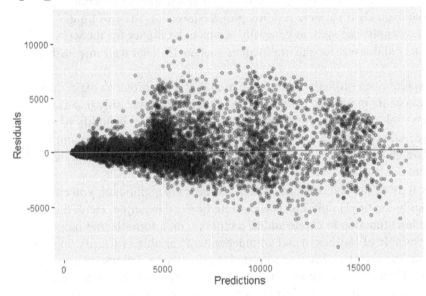

Description

- If you transform the dependent variable in the model, you need to undo the transformation for the predictions. Otherwise, the predictions are for the transformed dependent variable, not the untransformed independent variable.

Figure 13-9   How to create, fit, and judge a model that uses transformations

How to work with ordinal variables

So far, the regression models in this chapter have only used numeric data. But, you can also use some categorical data in your regression models. Remember that categorical data can only take on a set number of values. If the values can't be ordered from good to bad (such as up, down, left, right), it makes sense to store them using the factor type. However, if they can be ordered (such as bad, moderate, good, excellent), it makes sense to store them as the ordinal factor type.

How to examine ordinal variables

If your categorical variable is ordinal, you can include it in your model like any numeric variable. In the diamonds data set, the clarity is stored as an ordinal variable with the following possible values: I1 (worst), SI2, SI1, VS2, VS1, VVS2, VVS1, IF (best).

When you choose ordinal variables to include in your regression models, the r-value doesn't indicate if there's a relationship between the variables. So, you have to rely on plots to check for relationships.

The first plot that you should use to test for a relationship between an ordinal and the dependent variable is a box plot as shown in part 1 of figure 13-10. In this example, the median generally seems to be higher for the worse values such as I1 and lower for the high values such as IF. That warrants further investigation.

Before including an ordinal variable in a model, it's important to make sure that each level occurs in enough rows to be significant. If a level only has a small number of rows relative to the number of rows in the data set, it is highly likely that levels are outliers or that the level is not significant. In the second example, the results show that each level occurs in 500 or more rows. For this analysis, that's sufficient.

To check if all of the levels for an ordinal variable are significant, you can pass the model to the tidy() function as shown in the third example. Here, the code uses the lm() function to create and fit a model with a formula that has a dependent variable of log(Price) and an independent variable of Clarity. In addition, it uses the mutate() function to round the p.value to 3 digits.

The output for the third example shows all of the terms for the model. This model doesn't show a single term for the Clarity variable as you might expect. Instead, it generates as series of terms that have letters or powers to show their place in the equation (L for linear, Q for quadratic, C for cubic). Due to the way the model generates these terms, there's one less term than the number of levels for the ordinal variable. However, the important part is the p-values of these terms.

If any level of an ordinal variable isn't statistically significant ($p > .05$), it doesn't make sense to use that variable in a model. That's because you can't drop the terms that are generated for the individual levels of the ordinal variable. However, it still might make sense to use the ordinal variable as part of an interaction term.

How to plot an ordinal variable to look for a trend

```
ggplot(data = train) +
  geom_boxplot(aes(x = Clarity, y = log(Price)))
```

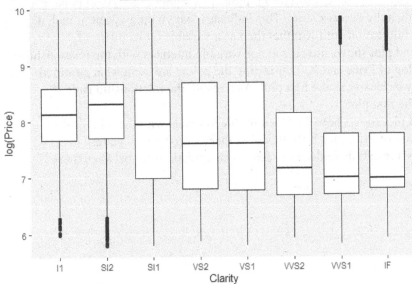

How to check if an ordinal variable has enough values of each level

```
train$Clarity %>% table()
  I1  SI2  SI1  VS2  VS1 VVS2 VVS1   IF
 564 6988 9756 9127 6131 3813 2725 1336
```

How to check if all terms are significant

```
lm(formula = log(Price) ~ Clarity, data = train) %>%
  tidy() %>%
  mutate(p.value = round(p.value, 3))
# A tibble: 8 × 5
  term         estimate std.error statistic p.value
  <chr>           <dbl>     <dbl>     <dbl>   <dbl>
1 (Intercept)   7.72      0.00748   1032.      0
2 Clarity.L    -0.746     0.0285     -26.2     0
3 Clarity.Q     0.0105    0.0279       0.376   0.707
4 Clarity.C     0.130     0.0241       5.41    0
5 Clarity^4     0.00312   0.0193       0.162   0.872
6 Clarity^5     0.179     0.0158      11.3     0
7 Clarity^6    -0.0666    0.0138      -4.82    0
8 Clarity^7    -0.00706   0.0122      -0.579   0.563
```

Description

- Before you include an ordinal variable in a model, you should make sure it has enough values in each level to prevent outliers from distorting the model.

- With an ordinal variable, you can't drop terms individually. If an ordinal variable generates terms that aren't significant, it might still make sense to use the ordinal variable as an interaction term if it has high collinearity with another variable.

Figure 13-10 How to examine ordinal variables (part 1)

Part 2 of figure 13-10 shows how to examine potential interaction terms by plotting them. Here, first plot shows how the Clarity variable interacts with the relationship between the log of Price and X. In this plot, there's clearly some stratification as the lighter colors (which indicate a higher grade) tend to have a higher price than the darker colors. This indicates that you may want to include Clarity as an interaction with X rather than as a variable on its own.

The second plot shows how the Color variable interacts with the relationship between the log of Price and X. In this plot, the colors are somewhat stratified, but not nearly as clearly as the first plot. As a result, the relationship isn't as strong as in the first plot.

The third plot shows how the Cut variable interacts with the relationship between the log of Price and X. In this plot, there's no obvious pattern to the colors. As a result, there doesn't seem to be a relationship between these variables.

Plotting X and Clarity against log(Price)

```
ggplot(data = train) +
  geom_point(aes(x = X, y = log(Price), color = Clarity), alpha = 0.5)
```

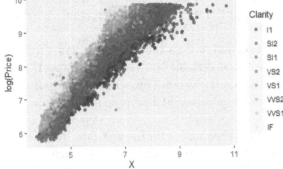

Create a similar plot with the Color variable

```
ggplot(data = train) +
  geom_point(aes(x = X, y = log(Price), color = Color), alpha = 0.5)
```

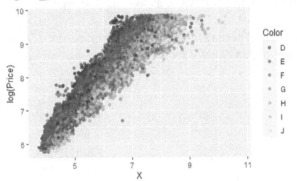

Create a similar plot with the Cut variable

```
ggplot(data = train) +
  geom_point(aes(x = X, y = log(Price), color = Cut), alpha = 0.5)
```

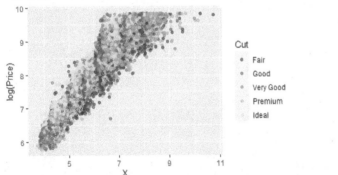

Figure 13-10   How to examine the ordinal variables (part 2)

How to create, fit, and judge the model

Figure 13-11 shows how to create, fit, and judge a model that uses ordinals. To start, the first example creates and fits the model with a formula that states that the log of Price is a function of the log of Carat, X, and the interaction between X and Clarity.

The second example uses the tidy() function to view the regression terms for this model. This shows that the log of the Carat variable has the biggest impact on the model and that the X:Clarity interaction terms don't have as much impact. Still, these interaction terms are statistically significant and seem to improve the model's performance.

The third example gets the predictions and residuals for the test data. To do that, it adds a column named Predictions that stores the results of the predict() function, and it adds a column named Residuals that stores the residuals. As with the previous model presented in this chapter, this code uses the exp() function to undo the log transformation for the predicted values.

The fourth example calculates the R^2 value for the model. This value is higher than the previous model, which is what you want, as long as you aren't overfitting the model.

The fifth example plots the residuals. These residuals show a noticeable downward trend as the predictions get higher. That's a sign that there's still room for improvement in the model.

How to create and fit the model

```
model_ord <- fit(linear_reg(),
                 log(Price) ~ log(Carat) + X + X:Clarity,
                 data = train)
```

How to view the terms

```
model_ord %>% tidy() %>% mutate(p.value = round(p.value, 3))
# A tibble: 10 × 5
   term         estimate std.error statistic p.value
   <chr>           <dbl>     <dbl>     <dbl>   <dbl>
 1 (Intercept)   7.80     0.0458     170.      0
 2 log(Carat)    1.59     0.0135     118.      0
 3 X             0.113    0.00705     16.0     0
 4 X:Clarity.L   0.152    0.000941   161.      0
 5 X:Clarity.Q  -0.0321   0.000877   -36.6     0
 6 X:Clarity.C   0.0206   0.000770    26.7     0
 7 X:Clarity^4  -0.0145   0.000640   -22.7     0
 8 X:Clarity^5   0.00624  0.000544    11.5     0
 9 X:Clarity^6   0.00156  0.000481     3.24    0.001
10 X:Clarity^7   0.00892  0.000415    21.5     0
```

How to get the predictions and the residuals

```
model_ord_results <- test %>% mutate(
  Predicted = predict(model_ord, test)$.pred,
  Predicted = exp(Predicted),              # undo log operation
  Residuals = Price - Predicted)
```

How to get the R^2 value

```
model_ord_results %>% rsq(truth = Price, estimate = Predicted)
# A tibble: 1 × 3
  .metric .estimator .estimate
  <chr>   <chr>          <dbl>
1 rsq     standard       0.905
```

How to plot the residuals

```
ggplot(data = model_ord_results) +
  geom_point(aes(x = Predicted, y = Residuals), alpha = .2) +
  geom_hline(yintercept = 0, color = "red")
```

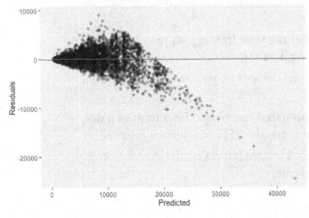

Figure 13-11   How to create, fit, and judge a model that uses ordinals

Perspective

Now that you've completed this chapter, you should have a solid foundation for working with multiple regression models. However, regression analysis is a large and complex subject. In fact, entire books have been written on regression analysis. In addition, the tools and best practices for working with regression analysis are constantly changing, so there's always more to learn about it.

Summary

- A model that uses more than one independent variable to predict a dependent variable is called a *multiple regression model*.
- The *coefficient of determination* (or R^2 *value*) represents how much *variance* is accounted for by the model.
- *Residuals* are calculated by subtracting the predicted value from the actual value. This measures how far off the predictions are from the actual values.
- *Heteroscedasticity* means that the residuals get larger as predictions get larger.
- An *interaction term* is created by multiplying two or more independent variables together.
- A *model term* is either an intercept, an independent variable, or an interaction term.
- The *estimate* is the model's calculation for the term's coefficient.
- The *statistic*, which is also called the *t-statistic* or *t-value*, is the estimate for a coefficient divided by the standard error.
- The *p-value* is the probability of getting the given data set if the null hypothesis is true.
- The *null hypothesis* is that there is no relationship between the independent and dependent variables.
- The cutoff for considering a p-value significant is called the *significance level* (or α for *alpha)*.
- *Overfitting* occurs when your model fits your training set too closely and doesn't work as well on your testing data set.
- *Underfitting* occurs when your model doesn't fit your training data set very well.
- A *dot and whisker plot* plots the estimated coefficient for a term as a dot with a "whisker" on each side that spans the 95% confidence interval.
- When a dependent and independent variable have a nonlinear relationship, you can *transform the data* to be linear.

Exercise 13-1   Create a multiple regression model

In this exercise, you'll create a multiple regression model to predict the prices of cars.

Get and clean the data

1. Create a new R script.
2. Load the tidyverse, tidymodels, and dotwhisker packages.
3. Read the cars data from the cars.csv file that's in the murach_r/data directory.

Prepare the data

4. Split the data into testing and training data sets. Since this splits the data randomly, the rows in each data set will be different every time you run the script.
5. Check the correlations that the numeric variables in the training data set have with the price variable.
6. Plot the seven numeric columns that have the highest correlations with the price column. To do that, use a scatter plot.
7. Create functions for computing the upper and lower fences of a column.
8. Use the functions for the upper and lower fences to remove outliers for the seven numeric columns that have the highest correlation with the price column.

Create and judge a model

9. Create a multiple regression model with the seven independent variables that have the highest correlation with the price column.
10. Use the tidy() function to view the terms of the model.
11. Use the model to add the predicted price and the residuals for the test data set.
12. Judge the model by checking its R^2 value.
13. Judge the model by plotting its residuals.
14. Check the model for statistically insignificant variables.
15. Plot the model with a dot and whisker plot.

Create an improved model and judge it

16. Create a second model that doesn't use the insignificant variables.
17. Use the second model to add the predictions and residuals to the test data set.
18. Judge the second model by calculating its R^2 value. Is this value higher or lower than the first model? If the second model has a lower value, is that OK given how much simpler the second model is?
19. Judge the second model by plotting its residuals.

Chapter 14

How to work with classification models

If you understand the last two chapters, you have a pretty good handle on how regression models work. But, that's only one type of predictive model. In this chapter, you'll learn about another type of predictive model known as a classification model.

Introduction to classification models

Where a regression model attempts to predict a continuous value, a *classification model* attempts to predict a discrete category based on the values of one or more independent variables. In other words, a classification model attempts to classify data by putting it in a category. At first, this may seem like it would be easier than creating a good regression model. However, there are plenty of challenges to creating a good classification model as well.

Introduction to classification analysis

Within classification analysis, there are several different types of tasks that you may face. Some of these types of tasks are listed in the first table in figure 14-1. The first of these is *binary classification*. A binary classifier classifies data into one of two categories, or classes. For example, binary classifiers are commonly used in medical diagnostics to confirm the presence of a disease.

Similarly, *multi-class classification* places data into one of three or more classes. The rest of this chapter works with multi-class classification, but you should know that many of the models that work with multi-class classification also work with binary classification.

Two special types of classification tasks are multi-label classification and imbalanced classification. *Multi-label classification* assigns one or more labels to each observation rather than just a single label. An example of this would be a classifier that labels movies with one or more genres.

In an *imbalanced classification*, there are significantly more of one class than the others. For example, in email spam detection, 85% of all emails are spam. Imbalanced classification tasks can make it more difficult to train an accurate model.

There are many different types of classification models to choose from, and each model has its advantages and disadvantages. The rest of this chapter focuses on *decision trees*. Decision trees are like a series of if-else statements that are used to classify data. Decision trees are not as powerful as some other models, but they are easy to understand and they can still achieve high levels of accuracy.

Types of classification tasks

Type	Description
Binary	There are only two possible classes.
Multi-class	There are more than two possible classes.
Multi-label	Each row can be multiple classes at the same time.
Imbalanced	The instances of one class significantly outweigh others.

Types of classification models

Model	Description
Logistic regression	Models the probability of each row falling into a binary class.
Decision tree	Uses a tree of if-else statements to classify each row.
Random forest	Uses several decision trees to classify each row.
Support vector machine	Separates data into classes by drawing hyperplanes between the classes.
Multilayer perceptron	Uses a neural network to classify each row.

Real-world applications for classification models

Industry	Usage
Tech	Spam detection, optical character recognition, facial recognition, image labeling.
Healthcare	Medical diagnostics, scan interpretation.
Business	Customer churn prediction, automatic inventory labeling, fraud detection, credit scores.

Real-world classification examples

- Predict the presence of cancer given an MRI image (binary classification).
- Predict the species of a plant given information about the plant (multi-class classification).
- Label an image with the species of the birds in the image (multi-class classification).
- Label a movie with all applicable genres (multi-label classification).
- Predict if an email is spam based on keywords, domain name, and email history (imbalanced classification).

Figure 14-1   Introduction to classification analysis

How to get the data for this chapter

Figure 14-2 begins by listing the packages used by this chapter. In addition to familiar packages like tidyverse and tidymodels, this chapter introduces two new packages: rpart.plot and caret. The rpart.plot package contains functions for plotting a decision tree like rpart.plot(), and the caret (Classification And REgression Training) package contains the confusionMatrix() function that's useful for judging a classification model.

The model presented in this chapter uses the irises data set. In this figure, the second example shows how to get that data and store it as a tibble. Then, the third example shows that the Species column contains three unique values, making it a good choice for the dependent variable in a classification model. Since there are three values (more than two), this is a multi-class classification.

As with all predictive models, the first step is to split the data into training and testing data sets. To do this for a classification model, you can use the same functions presented in chapter 12 as shown in the fourth example.

Load the packages for this chapter

```
library("tidyverse")
library("tidymodels")
library("ggforce")
library("rpart.plot")
library("caret")
```

Get the irises data

```
irises <- as_tibble(iris)
irises
# A tibble: 150 x 5
   Sepal.Length Sepal.Width Petal.Length Petal.Width Species
          <dbl>       <dbl>        <dbl>       <dbl> <fct>
 1          5.1         3.5          1.4         0.2 setosa
 2          4.9         3            1.4         0.2 setosa
 3          4.7         3.2          1.3         0.2 setosa
 4          4.6         3.1          1.5         0.2 setosa
 5          5           3.6          1.4         0.2 setosa
 6          5.4         3.9          1.7         0.4 setosa
 7          4.6         3.4          1.4         0.3 setosa
 8          5           3.4          1.5         0.2 setosa
 9          4.4         2.9          1.4         0.2 setosa
10          4.9         3.1          1.5         0.1 setosa
# ... with 140 more rows
```

View the unique values for the independent variable

```
irises$Species %>% unique()
[1] setosa    virginica  versicolor
Levels: setosa versicolor virginica
```

Split the data

```
iris_split <- initial_split(irises, prop = 0.75)
train <- training(iris_split)
test <- testing(iris_split)
```

Description

- The irises data set contains one categorical column (Species) that can be used as the dependent variable in a classification model.
- The irises data set contains four columns (Sepal.Length, Sepal.Width, Petal.Length, Petal.Width) that can be used as independent variables in a classification model.
- Since there three possible classes for the dependent variable, classifying the species is a multi-class classification.

Figure 14-2   How to get the data for this chapter

How to visually investigate the data

Before creating your model, it's worthwhile to visually investigate the data. You may spot an obvious variable for the decision tree to use, or note a cluster of data points that's effectively impossible to classify with 100% accuracy.

It's important to understand that a decision tree is only capable of separating data by drawing straight lines for boundaries. In other words, decision trees can only separate the data with straight vertical and horizontal lines. They can draw zig zags to approximate diagonal lines, but every new direction change requires another node in the decision tree.

That's why it's important to choose variables that are easy to separate. If you don't, you'll need a much more complex model to get good results. Keep in mind that selecting variables visually is not always the best choice because it doesn't work well if you want to build a model with more than two variables. However, plotting your variables can still help you build an initial model. In addition, it can help you recognize if your model is accurate.

You can start by displaying a pairwise grid of plots. This may help you choose a combination of variables that are easy to divide. For example, the first plot in figure 14-3 displays a pairwise grid of plots with scatter plots in the lower left corner, KDE plots on the diagonal, and box plots in the upper right corner. This plot is explained in detail at the end of chapter 8, so you can refer to figure 8-22 if you need a refresher on how it works.

At any rate, this pairwise grid of plots shows that the Petal.Length and Petal.Width variables separate the data better than the Sepal.Length and Sepal. Width variables. In fact, either the Petal.Length or Petal.Width variable makes classifying setosa irises trivial. These variables also make clearer distinctions between versicolor and virginica irises, but there is still overlap.

The second example uses a scatter plot to take a closer look at the interaction between petal length and petal width. This plot shows that the majority of the flowers are fairly easy to group and classify. However, a few flowers in the middle are impossible to distinguish without some other characteristics. In other words, regardless of how the model groups the data, it's likely to misclassify at least a couple flowers.

To make it easier to compare the second plot with the first plot, the code for second plot uses the hex values for the colors. That way, these colors match the colors in the first plot.

How to display a pairwise grid for each independent variable

```
ggplot(irises, aes(x = .panel_x, y = .panel_y)) +
  geom_point(aes(color = Species)) +
  geom_autodensity(aes(fill = Species)) +
  geom_boxplot(aes(fill = Species)) +
  facet_matrix(vars(Sepal.Length:Petal.Width),
               layer.lower = 1,
               layer.diag = 2,
               layer.upper = 3)
```

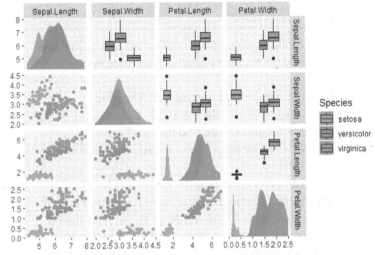

How to display a scatter plot for two overlapping variables

```
ggplot(filter(train, Species != "setosa")) +
  geom_point(aes(x = Petal.Length, y = Petal.Width, color = Species),
             size = 3) +
  scale_color_manual(values = c("#00BA38","#619CFF"))
```

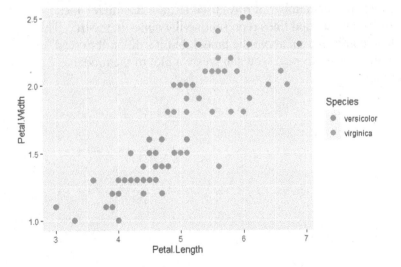

Figure 14-3   How to visually investigate the data

How to create and judge a classification model

The process for working with classification models is similar to the process for working with regression models. You start by cleaning and splitting the data. Then, you create and fit a model. Last, you make predictions, judge those predictions, and revise the model until it's ready for production.

How to create a decision tree

This chapter focuses on a type of classification model called a decision tree. Figure 14-4 shows how to use the decision_tree() function to create and fit a decision tree model.

The decision_tree() function defines several parameters. The mode parameter specifies whether the decision tree is for a classification or regression model. Then, the engine parameter specifies the engine that performs the computations and builds the model. For our purposes, the default value of rpart works perfectly well.

The min_n and tree_depth parameters configure the *nodes* and *depth* of the tree. A node is similar to an if-else statement, and the depth of a tree is how many levels of nodes it has. Both are covered in more detail in the next figure.

The first example in this figure shows how to create a blank decision tree model. Then, the second example shows how to fit the model. To do that, you use a formula to specify the relationship between the variables in the model. In this case, the right side of the formula uses the dot operator (.) to tell the formula to use all variables not already specified. So, in this example, the dot operator is shorthand for Sepal.Length + Sepal.Width + Petal.Length + Petal.Width.

The third example shows how to print the model to the console. When you do that, the console displays the number of rows used to train the model, the format for the rest of the output, and lines representing the nodes in the tree. This output contains valuable information about the model, but it's difficult to read. To make this information easier to read, you can create a plot of the model as shown in the next figure.

The parameters for the decision_tree() function

Function	Description
`decision_tree()`	Creates a blank decision tree model.

Parameter	Description
`mode`	Sets the type of task the model will be used for. Possible values are "unknown", "regression", or "classification". Default is "unknown".
`engine`	Sets the engine to use when fitting the model. Possible values are "rpart", "C5.0", "party", or "spark". Default is "rpart".
`min_n`	Sets the minimum number of data points in a node for the node to be split.
`tree_depth`	Sets the maximum depth of the tree.

How to create a decision tree model

```
dtree_mod <- decision_tree(mode = "classification", min_n = 4, tree_depth = 3)
```

How to fit the model

```
dtree_mod_fit <- fit(object = dtree_mod, formula = Species ~ ., data = train)
```

How to display information about the model

```
dtree_mod_fit
parsnip model object

n= 112

node), split, n, loss, yval, (yprob)
      * denotes terminal node

 1) root 112 69 setosa (0.38392857 0.30357143 0.31250000)
   2) Petal.Length< 2.45 43  0 setosa (1.00000000 0.00000000 0.00000000) *
   3) Petal.Length>=2.45 69 34 virginica (0.00000000 0.49275362 0.50724638)
     6) Petal.Width< 1.7 37  3 versicolor (0.00000000 0.91891892 0.08108108)
      12) Petal.Length< 4.95 33  0 versicolor (0.00000000 1.00000000 0.00000000) *
      13) Petal.Length>=4.95 4  1 virginica (0.00000000 0.25000000 0.75000000) *
     7) Petal.Width>=1.7 32  0 virginica (0.00000000 0.00000000 1.00000000) *
```

Description

- The fit() function works with a classification model created by decision_tree() function just as it does with a linear model created by linear_reg().

- A *node* is like an if-else statement. When you display the information about a decision tree model, each line represents a node.

- The *depth* of a tree is how many layers of nodes it has.

- In a formula, the dot operator (.) indicates that the formula should add all variables not already specified in the formula.

Figure 14-4   How to create a decision tree

How to plot a decision tree

A fitted decision tree model contains two objects: an rpart object and a model_fit object. However, to plot a decision tree, you only need to access the rpart object. To do that, you can pass the fitted model to the extract_fit_engine() function as shown in the first example in figure 14-5. Then, you can pass the rpart object to the rpart.plot() function. This displays a plot of the decision tree like the one shown in this figure.

The plot of a decision tree displays boxes and lines, similar to a flow chart. Each box is called a *node*. Each node contains a name, three decimals, and a percentage.

The name at the top of the node shows the class with the highest proportion at that point in the tree. In this figure, for example, the first node (also called the *root node* because it is at the "root" of the tree) is labeled setosa because there are more setosa irises in this node than the other classes.

After the name, each decimal represents the percent of that class in that node. So, the root node is 38% setosa irises, 30% versicolor, and 31% virginica. The percentage shown at the bottom of the node shows the percentage of the total data represented by the node. The root node will always be 100%.

Beneath each node, the plot displays a conditional expression like Petal.Length < 2.5. The model uses this expression to split the rows in each node into either another node or their final category.

To give you an example of how this decision tree works, imagine that you have a row where the Petal.Length is 4.9 and Petal.Width is 1.5. The root node begins by checking if the Petal.Length is less than 2.5. Since it isn't, the decision tree follows the path on the right ("no"). This node checks if Petal.Width is less than 1.7. Since it is, the tree follows the path to the left ("yes"). This node checks if Petal.Length is less than 5. Since it is, the model classifies the species as versicolor.

When a node has other nodes below it, it can be referred to as a *parent node* and the nodes beneath as *child nodes*. These child nodes can also be parent nodes, just like a family tree. By contrast, *leaf nodes* have no children. Leaf nodes are special because they are used to make the final predictions in a decision tree. They are also called *end nodes*.

The depth of a tree may also be described in terms of *levels*. The root node is the first level, its children are on the second level, the children's children are on the third level, and so on. There is always one more level than the depth of the tree, to account for the root node. The tree in this example has seven nodes, four levels, and a depth of three. Four of its nodes are leaf nodes.

When running rpart.plot(), you may receive a warning message that says "Cannot retrieve the data used to build the model." You can safely ignore this warning message. Or, if you prefer, you can turn this warning off by setting the roundint parameter to FALSE like this:

```
rpart.plot(roundint = FALSE)
```

The functions for plotting a decision tree

Function	Description
extract_fit_engine(x)	Extracts the rpart object from the parsnip object.
rpart.plot(x)	Plots the decision tree for the rpart object.

How to plot the decision tree

```
dtree_mod_fit %>% extract_fit_engine() %>% rpart.plot()
```

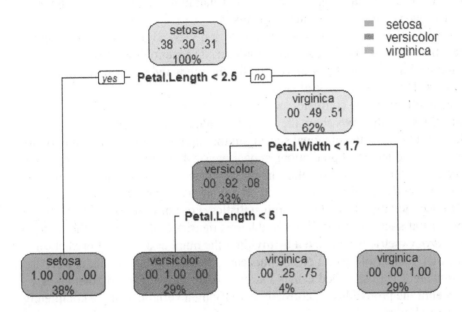

Description

- The *root node* is the node at the top of the tree.
- A *parent node* is a node with *child nodes* below it. A node can be both a parent and a child.
- A *leaf node* (also called an *end node*) has no child nodes.

Figure 14-5   How to plot a decision tree

How to judge a model with a confusion matrix

The first example in figure 14-6 shows how to use the model to make predictions. To do that, it passes the model and the test data set to the predict() function. This returns a tibble that stores the predictions in a column named .pred_class.

After you use a model to make predictions, you can judge the model. One of the most common ways to judge a classification model is with a *confusion matrix*. A confusion matrix is a table of values that makes it easy to see where your model gets confused and makes mistakes.

The second example shows how to create and view a confusion matrix and its statistics. To start, it calls the confusionMatrix() function that's available from the caret package. Then, it passes vectors for the predicted and actual classes as arguments. This displays several tables.

The first table is the confusion matrix. The columns of this matrix show the actual values and the rows show the predicted values. In this matrix, correct predictions appear in the diagonal starting in the top left of the table going to the lower right, and incorrect predictions appear everywhere else. For example, the versicolor column shows two values in the virginica row. This means that the model incorrectly classified two versicolor species as virginica.

The second table shows the overall statistics. For our purposes, the most important statistics are the accuracy and kappa metrics highlighted in the figure. The *accuracy* metric is calculated by dividing the number of correct predictions by the total number of predictions. In this case, that's (7+14+14) / 38, or 92.1%. The *kappa* metric (also called *Cohen's kappa)* is an accuracy metric that takes into account the probability of correctly classifying a variable using a completely random predictor.

This kappa metric is important because it gives a better representation of how the model performs relative to the difficulty of the classification problem. For example, a model that performs with 60% accuracy on a binary classification problem is much less impressive than a model that performs with 60% accuracy on a multi-class classification problem with 100 classes. This is because you would expect random guessing to yield a 50% accuracy on the binary problem but only a 1% accuracy with the multi-class problem. Similarly, if you have an imbalanced binary classification model that is accurate 60% of the time but one of the classes represents 70% of the data, your model isn't performing well.

The third table breaks down statistics by class. These statistics allow you to compare how your model performs for each individual class.

When using the confusionMatrix() function, you may encounter some issues. If it returns a tibble of numbers, the caret package didn't load correctly, and R is using the function from the ModelMetrics package instead. Or, you may get a message that says:

```
cannot open file '…/Documents/R/win-library/4.1/proxy/R/proxy.rdb'
```

This means the named package (in this case, proxy) isn't installed. To fix this, you can use the install.packages() function to install the package.

The confusionMatrix() function

Function	Description
confusionMatrix(data, reference)	Calculates a cross-tabulation of the predicted and actual classes specified by the data and reference parameters.

How to use the model to make predictions

```
predictions <- predict(dtree_mod_fit, new_data = test)
```

How to view the confusion matrix and statistics

```
confusionMatrix(data = predictions$.pred_class,
                reference = test$Species)
Confusion Matrix and Statistics

          Reference
Prediction  setosa versicolor virginica
  setosa       7        0         0
  versicolor   0       14         1
  virginica    0        2        14

Overall Statistics

               Accuracy : 0.9211
                 95% CI : (0.7862, 0.9834)
    No Information Rate : 0.4211
    P-Value [Acc > NIR] : 1.237e-10

                  Kappa : 0.8754

 Mcnemar's Test P-Value : NA

Statistics by Class:

                     Class: setosa Class: versicolor Class: virginica
Sensitivity                 1.0000            0.8750           0.9333
Specificity                 1.0000            0.9545           0.9130
Pos Pred Value              1.0000            0.9333           0.8750
Neg Pred Value              1.0000            0.9130           0.9545
Prevalence                  0.1842            0.4211           0.3947
Detection Rate              0.1842            0.3684           0.3684
Detection Prevalence        0.1842            0.3947           0.4211
Balanced Accuracy           1.0000            0.9148           0.9232
```

Description

- A *confusion matrix* shows where a classification model gets confused and makes predictions that are incorrect.
- The *accuracy* metric is calculated by dividing the number of correct predictions by the total number of predictions.
- The *kappa*, or *Cohen's kappa*, is an accuracy metric that takes into account the probability of correctly classifying a variable using a completely random predictor.

Figure 14-6   How to judge a model with a confusion matrix

How to tune a classification model

Now that you know how to create and judge a decision tree, you're ready to learn how to tune a decision tree. There are two main aspects to tuning a decision tree. The first is *variable selection*, which is selecting the smallest set of variables that gives you an accurate model.

The second is *hyperparameter tuning*. A *hyperparameter* is passed to your model when you create it. Hyperparameters affect how the model is generated and have nothing to do with the data.

How to use variable importance to select variables

There are several ways to select variables for a decision tree. The easiest is to use all of the independent variables available and let decision_tree() choose as shown by the example presented earlier in this chapter. However, this becomes less feasible for data sets that have many variables to choose from.

Another method is to visually investigate the variables with plots and choose the ones that can be separated easily. For instance, the plots in figure 14-3 show that Petal.Length and Petal.Width separate the data better than Sepal.Length and Sepal.Width. However, this method is limited to individual variables or pairs of variables that can be plotted together.

You can also choose independent variables by using *variable importance*, which is a measure of how important each variable is to the model's predictions. To do that, you can use the varImp() function to compare the variables. This function takes an rpart object as an argument and outputs the calculated importance for each variable.

The first example in figure 14-7 shows the output for the varImp() function. This shows that the Petal.Length and Petal.Width variables are much more important than the Sepal.Length or the Sepal.Width variables. This confirms what's shown by the plots in figure 14-3.

Given this information, you may want to drop the Sepal variables to simplify the model. To do that, you can specify a formula like the one shown in the second example in this figure. As it happens, this results in the same decision tree for this particular data set as the formula used earlier in this chapter. However, with more variables and a less clearly separated data set, changing the formula typically results in a different decision tree that may perform better. In addition, this relationship can change as you adjust the values for your hyperparameters. As a general rule, you want your model to be both as accurate and simple as possible.

The parameters for the varImp() function

Function	Description
varImp(object)	Calculates variable importance for the rpart object of a fitted model.

How to calculate variable importance

```
dtree_mod_fit %>% extract_fit_engine() %>% varImp()
             Overall
Petal.Length 71.06617
Petal.Width  69.29097
Sepal.Length 37.47872
Sepal.Width  20.92350
```

How to create a model with only the top two variables

```
dtree_mod_fit2 <- dtree_mod %>%
  fit(Species ~ Petal.Length + Petal.Width, data = train)
```

Description

- *Variable importance* measures how important each variable is to the model's predictions.

Figure 14-7   How to use variable importance to select variables

How to adjust the hyperparameters

Hyperparameters are passed to the model to control aspects of how the model is created. The table at the top of figure 14-8 lists some available hyperparameters for the decision tree model. This chapter has already presented the tree_depth and min_n parameters. As a result, you should already understand how they work.

However, a model can also use the cost_complexity parameter to specify the minimum improvement at each node in the tree. This can help avoid overfitting. The cost_complexity parameter should be set to a number between 0 and 1. The default value of 0 doesn't limit complexity at all and results in the largest, most complex tree possible. By contrast, setting this parameter to 1 removes all complexity and results in a tree with a single root node.

In the first example, the first statement creates a model that adjusts the hyperparameters. These hyperparameters increase the tree depth to 4, decrease the minimum number of data points required to split a node to 1, and leave the cost_complexity parameter at its default value of 0. Then, the second statement fits the new model using a formula that only specifies two independent variables.

The third example plots the decision tree. This shows that the tree now has a new level and that the number of nodes has increased from seven to nine.

The hyperparameters for the decision tree

Parameter	Description
tree_depth	The max depth of the tree.
min_n	The minimum number of required data points for a node to be split.
cost_complexity	A number between 0 and 1 representing a penalty for increased complexity. 0 results in the biggest possible tree, 1 results in a tree with only a root node. Default is 0.

How to create a classification model with different hyperparameters

```
dtree_mod_hyper <- decision_tree(mode = "classification",
                                 min_n = 1, tree_depth = 4,
                                 cost_complexity = 0)
dtree_mod_hyper_fit <- dtree_mod_hyper %>%
                  fit(Species ~ Petal.Length + Petal.Width, data = train)
```

The decision tree

```
dtree_mod_hyper_fit %>% extract_fit_engine() %>% rpart.plot()
```

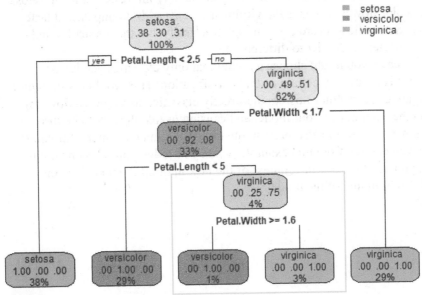

Description

- *Hyperparameters* control how a model is created.
- Tuning the hyperparameters can increase the performance of a model.
- Tuning the hyperparameters can lead to overfitting.
- To reduce overfitting, you can adjust the cost_complexity hyperparameter to reduce overall *complexity* (how long the model takes to run).

Figure 14-8   How to adjust the hyperparameters

How to compare decision trees

Once you've fitted the hyperparameter model, you can generate a new set of predictions for that model and use them to compare this model to the previous model. To do that, you can display the confusion matrixes for each model as shown in figure 14-9. The confusion matrices show a slight difference in the predictions, with the hyper-fit model classifying one more versicolor correctly but one less virginica correctly. But when you compare the accuracy and kappa for each model, the values are nearly identical.

However, judging classification models is not always as straightforward as judging regression models. With regression models, you just want the predicted values to be as close to the true values as possible. With classification models, the consequences for misclassifying a value can vary depending on the task. There are two types of misclassifications. With a *false positive*, the classifier says that a value is a certain class but it isn't. With a *false negative*, the classifier says that a value isn't a certain class but it is.

For example, let's say that your company wants to use these irises to make teas. Furthermore, let's assume your company is only interested in using setosa and versicolor flowers because the virginica species adds an unpleasant taste to the tea. Setosa flowers are easy enough to classify, but the versicolor and virginica species are harder to differentiate.

In this case, you might choose a model that only classifies versicolor flowers if it is absolutely certain of that classification. This would result in more false negatives (versicolor flowers incorrectly classified as virginica flowers), reducing the accuracy of your model and causing usable flowers to be thrown away. However, it would also prevent quality issues and customer complaints later. As you can see from this example, when you judge your classification models, you need to keep the context of the task in mind to make decisions about the final tuning of the model.

How to interpret Cohen's kappa

Kappa	Reliability	Kappa	Reliability
<= 0.2	None	0.21 to 0.39	Minimal
0.40 to 0.59	Weak	0.60 to 0.79	Moderate
0.80 to 0.90	Strong	> 0.90	Nearly Perfect

How to compare the accuracy of each model

```
predictions <- predict(dtree_mod_fit, new_data = test)
confusionMatrix(data = predictions$.pred_class, reference = test$Species)
...
Prediction   setosa versicolor virginica
  setosa        7          0          0
  versicolor    0         14          1
  virginica     0          2         14
...

             Accuracy : 0.9211
               95% CI : (0.7862, 0.9834)
  No Information Rate : 0.4211
  P-Value [Acc > NIR] : 1.237e-10

                Kappa : 0.8754
...
predictions_hyper <- dtree_mod_hyper_fit %>% predict(test)
confusionMatrix(data = predictions_hyper$.pred_class, reference = test$Species)
...
Prediction   setosa versicolor virginica
  setosa        7          0          0
  versicolor    0         15          2
  virginica     0          1         13
...

             Accuracy : 0.9211
               95% CI : (0.7862, 0.9834)
  No Information Rate : 0.4211
  P-Value [Acc > NIR] : 1.237e-10

                Kappa : 0.8751
...
```

Description

- You can judge models by comparing their accuracy and kappa values.
- Unlike regression models, classification models must take into account the consequences of misclassifying in addition to accuracy.
- A *false positive* is when the model says that a value is a certain class but it isn't.
- A *false negative* is when the model says that a value isn't a certain class but it is.

Figure 14-9   How to compare decision trees

How to cross validate a model

Cross validation is an important technique for judging any model. It's used to confirm the metrics used to judge a model and to make sure that the calculated metrics aren't being affected by outliers in the testing data set or by overfitting. The procedure and graphic shown in part 1 of figure 14-10 show how *v-fold* (or *k-fold*) *cross validation* works.

First, cross validation uses only the training data. The training data is split into a number of groups, called *folds*. In this example, there are five folds. However, with larger data sets ten folds are more common.

Second, the model is trained and tested the same number of times as there are folds. Each time, the model's formula and hyperparameters stay the same, but a different one of the folds is used as the testing data set and the metric for evaluating the model is calculated. In this example, that's the accuracy, but you could also choose the kappa. For a regression model, you would choose the R^2 value.

Third, the evaluation metric for each fold is averaged to get the overall evaluation metric (accuracy, kappa, R^2, etc.) for the model. The fold used for testing each time is said to be *held out*.

To split the data into folds, you use the vfold_cv() function as shown by the example in part 1 of this figure. Because this code sets v to 5, it splits the training data into five folds. If you display the folds on the console, the numbers in the brackets indicate how many values are in the regular and held out groups for each fold.

Why would you want to only split the training data set into folds? Even though testing is being performed at each step, the model is still being trained on the data set overall, just with a different group held out each time. To properly test the model afterwards, you need a data set that was not used to train the model at any time.

The steps for performing cross validation

1. Split the training data into v groups, or *folds* (v = 10 is common)
2. Create v sub-models. Use the same formula and hyperparameters for each submodel, but use a different fold for the test data.
3. Calculate the average accuracy for the sub-models.

How cross validation works

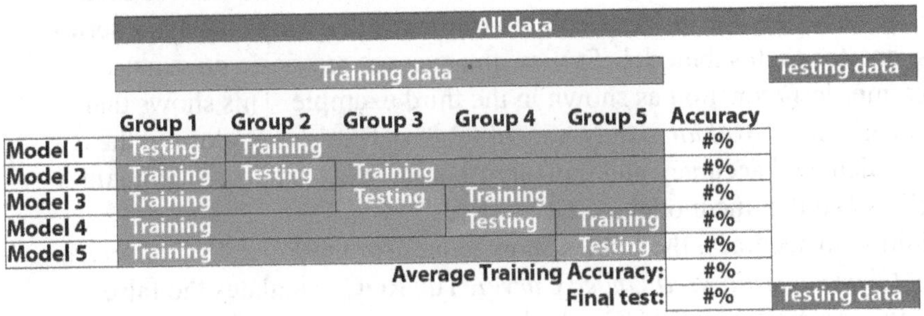

All data						
Training data					Testing data	
	Group 1	Group 2	Group 3	Group 4	Group 5	Accuracy
Model 1	Testing	Training				#%
Model 2	Training	Testing	Training			#%
Model 3	Training		Testing	Training		#%
Model 4	Training			Testing	Training	#%
Model 5	Training				Testing	#%
				Average Training Accuracy:		#%
				Final test:		#%	Testing data

A function for creating folds

Function	Description
vfold_cv(data, v)	Randomly splits the data into v groups of roughly equal size.

How to split the training data set into folds

```
folds <- vfold_cv(train, v = 5)
folds
#  5-fold cross-validation
# A tibble: 5 x 2
  splits          id
  <list>          <chr>
1 <split [89/23]> Fold1
2 <split [89/23]> Fold2
3 <split [90/22]> Fold3
4 <split [90/22]> Fold4
5 <split [90/22]> Fold5
```

Description

- *Cross validation* is used to confirm the accuracy metrics of a model by accounting for possible outliers or overfitting.
- *V-fold cross validation* can be used with many kinds of models as long as you adjust the metric the model is judged by accordingly.
- V-fold cross validation is also called *k-fold cross validation*.
- When creating folds, the testing data for each group is *held out*.
- The original test data set should never be included in the folds. That way, the final model is tested on data that it was never trained on.

Figure 14-10   How to cross validate a model (part 1)

Part 2 of figure 14-10 shows how to use the tune_grid() function to perform cross validation. The tune_grid() function takes an empty model as the first argument. Then, the preprocessor argument specifies the formula for the model, and the resamples argument specifies the folds to use.

When you run this function as shown in the first example, R displays a warning message that indicates that no tuning parameters have been detected. For now, that's fine.

After running the tune_grid() function, you can view the metrics for each submodel by accessing the .metrics column of the results. This displays a series of tibbles, one for each submodel. To view the average metrics, you can use the collect_metrics() function as shown in the third example. This shows that our model has a *cross-validated accuracy* of 93.8%. That's very close to the previously calculated accuracy and indicates that this data set doesn't contain many outliers and that the model isn't overfitted.

In addition to accuracy, the tibbles show values for the *ROC AUC* (*Receiver Operating Characteristic Area Under Curve*). The ROC calculates the false positive rate for a model, and AUC calculates the percentage of the area under the curve represented by ROC on a graph.

If the area is 1, 100% of the data fits within the curve. In other words, the model is perfect. If the area is .5 with two classes, the model is no better than chance. If the area is 0, the model is perfect in reverse. In other words, it's always wrong, so you can make it right by taking the opposite classification! However, the opposite (for example, "not setosa") becomes less useful the more classes you have, so it's better to aim for an ROC AUC closer to 1 than 0.

If you're following along with the examples in RStudio, your metrics may be slightly different. That's because the vfold() function randomly splits each row in the data into the folds. However, your average metrics should be fairly close to those shown in this figure.

Functions for calculating and viewing cross-validation metrics

Function	Description
`tune_grid(object, preprocessor, resamples)`	Calculates performance metrics for the specified model object with the formula specified by the preprocessor parameter and the folds specified by the resamples parameters.
`collect_metrics(x)`	Display the average metrics for a tuned grid.

How to tune the grid

```
dtree_mod <- decision_tree(mode = "classification")     # an empty model
tuned <- tune_grid(dtree_mod, preprocessor = Species ~ ., resamples = folds)
Warning message:
No tuning parameters have been detected, performance will be evaluated using
the resamples with no tuning. Did you want to [tune()] parameters?
```

How to view the metrics

```
tuned$.metrics
[[1]]
# A tibble: 2 x 4
  .metric  .estimator .estimate .config
  <chr>    <chr>          <dbl> <chr>
1 accuracy multiclass     0.870 Preprocessor1_Model1
2 roc_auc  hand_till      0.892 Preprocessor1_Model1

[[2]]
# A tibble: 2 x 4
  .metric  .estimator .estimate .config
  <chr>    <chr>          <dbl> <chr>
1 accuracy multiclass     0.957 Preprocessor1_Model1
2 roc_auc  hand_till      0.981 Preprocessor1_Model1

# plus three more tibbles for the other three submodels
```

How to view the average metrics

```
collect_metrics(tuned)
# A tibble: 2 x 6
  .metric  .estimator  mean     n std_err .config
  <chr>    <chr>      <dbl> <int>   <dbl> <chr>
1 accuracy multiclass 0.938     5  0.0223 Preprocessor1_Model1
2 roc_auc  hand_till  0.962     5  0.0187 Preprocessor1_Model1
```

Description

- The *ROC AUC* (*Receiver Operating Characteristic Area Under Curve*) calculates how many data points are false positives.
- If the ROC AUC is 1, your model is perfect. If the area is .5 with two classes, the model is no better than chance.
- The average accuracy is also called the *cross-validated accuracy*.

Figure 14-10 How to cross validate a model (part 2)

How to tune hyperparameters with a grid search

In the previous figure, tune_grid() generated a warning that no tuning parameters were detected. Now, figure 14-11 shows how to add tuning parameters to the tune_grid() function. Then, you can perform a *grid search* to find the best values for your hyperparameters. This type of search uses a grid of hyperparameter value combinations to find the best values for your hyperparameters.

Before you can choose the parameters to tune, you need to know which hyperparameters are tunable for your model. To find out, you can pass the empty model to the tunable() function as shown in the first example. This returns a tibble listing all of the tunable parameters. In this case, these parameters are the same three shown in 14-8.

To make the hyperparameters tunable in a grid search, you have to mark the parameter as tunable by setting the parameter to a call to the tune() function as shown in the second example. This function call tells the model that you will fill in the values of the hyperparameter later. Those values are specified via a table of all the combinations of values you want to try for your model. So, if you have 3 values you want to try for tree_depth, 2 values for min_n, and 4 values for cost_complexity, you would need a grid with 24 rows for the 24 possible combinations (3*2*4).

There are multiple ways to create a tuning grid. One way is to automatically generate appropriate values for each parameter by calling the grid_regular() function as shown in the third example. Then, you pass the names of the hyperparameters that you want to generate values for as function calls.

Although not shown in this figure, you can use the levels parameter with grid_regular() to specify a certain number of values to generate for each hyperparameter, like this:

```
grid_regular(tree_depth(), min_n(), levels = 3)
```

Another way to create a tuning grid is to manually create a grid of values using the expand.grid() function as shown in the fourth example. Then, you set each hyperparameter to a vector of values. This creates a grid of each possible combination formed by the vectors.

The functions for working with a grid search

Function	Description
`tunable(x)`	Returns a tibble of the tunable hyperparameters for the model.
`tune()`	Used to mark a hyperparameter as tunable in the model declaration.
`grid_regular(x)`	Automatically generates a grid of hyperparameter values.
`expand.grid(...)`	Creates a grid from one or more vectors.

How to find the tunable parameters for the model

```
tunable(dtree_mod)
# A tibble: 3 x 5
  name           call_info            source     component     component_id
  <chr>          <list>               <chr>      <chr>         <chr>
1 tree_depth     <named list [2]>     model_spec decision_tree main
2 min_n          <named list [2]>     model_spec decision_tree main
3 cost_complexity <named list [2]>    model_spec decision_tree main
```

How to create a tunable model

```
dtree_mod_tune <- decision_tree(mode = "classification", tree_depth = tune(),
                            min_n = tune(), cost_complexity = tune())
```

How to create a grid of hyperparameter values automatically

```
tree_grid_auto <- grid_regular(tree_depth(), min_n(), cost_complexity())
tree_grid_auto
# A tibble: 27 x 3
   tree_depth min_n cost_complexity
        <int> <int>            <dbl>
1           1     2      0.0000000001
2           8     2      0.0000000001
3          15     2      0.0000000001
4           1    21      0.0000000001
5           8    21      0.0000000001
...
9          15    40      0.0000000001
10          1     2      0.00000316
# ... with 17 more rows
```

How to create a grid of hyperparameter values manually

```
tree_grid_manual <- expand.grid(tree_depth = 2:5, min_n = seq(1, 51, 10),
                   cost_complexity = c(0,.1,.2,.3)) %>% as_tibble()

tree_grid_manual
# A tibble: 96 x 3
   tree_depth min_n cost_complexity
        <int> <dbl>            <dbl>
1           2     1                0
2           3     1                0
3           4     1                0
4           5     1                0
5           2    11                0
...
```

Figure 14-11   How to tune hyperparameters with a grid search (part 1)

The first example in part 2 of figure 14-11 shows how to add the grid of hyperparameter values to the tune_grid() function. When you run this statement, it may take a little while to execute if you have many value combinations to test.

After running tune_grid(), you can pass the results to the collect_metrics() function to get the cross-validated accuracy and ROC AUC for each combination of hyperparameter values as shown in the second example.

To only view the best performing combination, you can pass the tuning data to the select_best() function with a specified metric. The third example shows that the best combination was 0.0000000001 (effectively 0) for cost_complexity, 8 for tree_depth, and 2 for min_n.

For more information, you can use show_best() to display the top five combinations and accompanying statistics. This shows that the best hyperparameters resulted in average accuracy of 95.5%! However, you should be aware that "best" in this context means the best from the grid of value combinations that you passed to tune_grid(). It's possible there are better or less complex values that you could use for your model.

To find potential better values, you can narrow your search by considering the range of the values you passed. For example, the grid used in this figure contains values of 1, 8, and 15 for tree_depth. The results show that 1 isn't the best choice. That makes sense because a tree with a depth of 1 is a small tree. However, 8 and 15 gave the same results. That means there is probably a number between 1 and 8 that would work just as well if not better. To test this, you could use expand.grid() to manually create a grid of values that sets tree_depth to the numbers from 2 to 7.

You don't always want the smallest number possible for a hyperparameter. For example, a higher min_n reduces complexity in a decision tree. In this example, a min_n of 2 had the best result and a value of 21 performed a little worse. That means there might be a number between 2 and 21 that might work better. To test this, you could try values from 3 to 20 for min_n in a new test, perhaps counting by 2 or 5 instead of trying all 18 values in the same grid.

Sometimes, no matter how much you adjust the hyperparameters, you can't get measurable improvement. It's important to be able to recognize that point.

A function for judging a grid search

Function	Description
select_best(x, metric)	Returns the most accurate submodel within one standard deviation. If there is a tie, returns the least complex, most accurate submodel.
show_best(x, metric)	Returns the top five most accurate submodels and statistics.

How to run the grid search

```
tuning_data <- tune_grid(dtree_mod_tune, preprocessor = Species ~ .,
                         resamples = folds, grid = tree_grid_auto)
```

How to show the cross-validated metrics for each model

```
tuning_data %>% collect_metrics()
```
```
# A tibble: 54 x 9
   cost_complexity tree_depth min_n .metric  .estimator   mean     n std_err .c~
             <dbl>      <int> <int> <chr>    <chr>       <dbl> <int>   <dbl> <chr>
 1   0.0000000001          1     2 accuracy multiclass  0.679     5  0.0369 Pr~
 2   0.0000000001          1     2 roc_auc  hand_till   0.833     5  0      Pr~
 3   0.0000000001          8     2 accuracy multiclass  0.955     5  0.0144 Pr~
 4   0.0000000001          8     2 roc_auc  hand_till   0.966     5  0.0102 Pr~
 5   0.0000000001         15     2 accuracy multiclass  0.955     5  0.0144 Pr~
...
```

How to display the best values for the hyperparameters

```
tuning_data %>% select_best(metric = "accuracy")
```
```
# A tibble: 1 x 4
  cost_complexity tree_depth min_n .config
            <dbl>      <int> <int> <chr>
1    0.0000000001          8     2 Preprocessor1_Model02
```

How to display the top five values for the hyperparameters

```
tuning_data %>% show_best(metric = "accuracy")
```
```
# A tibble: 5 x 9
   cost_complexity tree_depth min_n .metric  .estimator   mean     n std_err .con-
fig
             <dbl>      <int> <int> <chr>    <chr>       <dbl> <int>   <dbl> <chr>
 1   0.0000000001          8     2 accuracy multiclass  0.955     5  0.0144 Pr~
 2   0.0000000001         15     2 accuracy multiclass  0.955     5  0.0144 Pr~
 3   0.00000316            8     2 accuracy multiclass  0.955     5  0.0144 Pr~
 4   0.00000316           15     2 accuracy multiclass  0.955     5  0.0144 Pr~
 5   0.0000000001          8    21 accuracy multiclass  0.938     5  0.0223 Pr~
```

Description

- The "best" values are the best from the values listed in the grid passed to tune_grid().
- Viewing the top five submodels can help show which parameters might be worth additional testing.

Figure 14-11   How to tune the hyperparameters with grid search (part 2)

Perspective

Now that you've completed this chapter, you should be familiar with decision trees and have a solid foundation for working with classification models. In addition, you should understand cross-validation and hyperparameter searches. As with regression models, classification analysis is a large and complex subject that's constantly changing. As a result, there's always more to learn about it.

Summary

- *Classification models* are different than regression models because they predict categorical values (*classes*) rather than continuous values.
- *Binary classification* assigns data to one of two classes while *multi-class classification* assigns data to one of three or more classes.
- *Multi-label classification* assigns one or more labels to each observation.
- *Imbalanced classification* occurs when there are significantly more instances of one class than the others.
- *Decision trees* are like a series of if-else statements used to classify data.
- A decision tree *node* is like an if-else statement.
- The first node in a decision tree is called the *root node*.
- A *parent node* has one or more *child nodes* below it. By contrast, a *leaf node* (or *end node*) doesn't have any child nodes.
- The *depth* of a tree is how many layers of nodes it has. Each layer is called a *level*. To account for the root node, there is always one more level than the depth of the tree.
- A *confusion matrix* is a table of values that shows how many values your model correctly classified.
- The *accuracy* metric is calculated by dividing the number of correct predictions by the total number of predictions.
- The *kappa* metric, or *Cohen's kappa,* is an accuracy metric that takes into account the probability of correctly classifying a variable using a completely random predictor.
- *Variable selection* refers to selecting the smallest set of variables that gives you an accurate model.
- *Hyperparameters* control how a model is created. These parameters can be tuned to produce better models.
- *Variable importance* measures how important each variable is to the model's predictions.
- Misclassifications may be *false positives* or *false negatives.*
- *Cross validation* can help to confirm the accuracy metrics of a model by accounting for possible outliers or overfitting.

- When creating *folds* for performing *v-fold cross validation*, the testing data in each group is *held out*.
- The *ROC AUC* (*Receiver Operating Characteristic Area Under Curve*) calculates how many data points are false positives.
- A *grid search* uses a grid of hyperparameter value combinations to find the best values for your hyperparameters.

Exercise 14-1   Create a classification model

In this exercise, you'll create a classification model to predict whether a health care employee is likely to leave their job. In business terminology, this is known as attrition.

Get and clean the data

1. Create a new R script.
2. Load the tidyverse, tidymodels, rpart.plot, and caret packages.
3. Read the data from the watson_healthcare_modified.csv file that's in the murach_r/data directory.

Prepare the data

4. View the data types for all columns. Note that none of the columns are of the factor type, but many are of the character type.
5. View the number of unique values for all columns of the character type. Note that the Attrition variable only has two values. As a result, this is a binary classification task.
6. Convert the Attrition column from the character type to the factor type. This is necessary because the dependent variable in a classification model must be of the factor type.
7. Drop all columns of the character type that only have a single unique value. This is necessary because having an independent variable that only has a single unique value causes an error if you attempt to tune a classification model.
8. Split the data into testing and training sets.

Create an initial model

9. Create a decision tree model with a depth of 30 and a minimum of 5 data points per node.
10. Fit the model to predict the Attrition variable given all of the remaining variables in the data set.
11. Plot the model. Note that it has many levels.
12. Create predictions for the model.
13. View the confusion matrix for the model as well as the accuracy and kappa metrics.

Tune the model

14. Check the variable importance of the model. Make a note of any variables that have an importance of 0.

15. Create six cross validation folds for the training data.

16. Create an automatic grid of hyperparameters for the model.

17. Use the grid and the folds to tune the model.

18. Display the best hyperparameter values for the model.

Create the final model

19. Create a final model that uses the best hyperparameter values.

20. Fit the final model without using any variables that have an importance of 0.

21. Use the final model to create predictions.

22. View the confusion matrix for the final model.

Section 5

Presentation skills

The chapters in section 3 showed how to create three realistic analyses. However, these chapters didn't show how to present these analyses to a target audience. Now, the lone chapter in section 5 shows how to create HTML documents, PDF files, and slideshows that present the insights gained from an analysis.

Chapter 15

How to use R Markdown to present an analysis

Once you've completed an analysis, you typically want to share it with others. To do that, you can share a document such as an HTML or PDF document with your target audience. Or, if you're giving a presentation, you may want to create a slideshow. To create these types of documents and presentations, you can use a markup language known as R Markdown.

How to get started with R Markdown files

R Markdown files look and work a little differently than regular R scripts. That's why this chapter begins by showing how to create and render an R Markdown file.

How to create an R Markdown file

To work with R Markdown, you can't use a regular R (.R) file. Instead, you use R Markdown (.Rmd) files. The easiest way to create one is to use the RStudio menus as described in figure 15-1 to display the New R Markdown dialog. This dialog lets you enter some basic info about your R Markdown file and select a default output format. This format can be a *document* such as an HTML, PDF, or Word file. Or, it can be for a *presentation*, which is also known as a *slideshow*.

Before you can work with R Markdown files, you need to install the rmarkdown package. By default, RStudio installs and loads this package when you first create an R Markdown file. As a result, you typically don't need to run the install.packages() and library() functions for the markdown package.

However, if you want to output PDF files, you do need to install the TeX component. To do that, you can install and load the tinytex package as shown in this figure. Then, you can run the install_tinytex() function to install the TeX component. If this component installs successfully, you should be able to create PDF files from your R Markdown files.

The New R Markdown dialog

```
New R Markdown

┌─────────────────┐     Title:    [ Chapter 15 examples          ]
│  📄 Document    │
│  🖵 Presentation│     Author:   [ Scott McCoy                  ]
│  Ⓡ Shiny        │
│  🗄 From Template│    Date:     [ 2022-07-19                   ]
│                 │     ☐ Use current date when rendering document
│                 │
│                 │     Default Output Format:
│                 │
│                 │     ◉ HTML
│                 │        Recommended format for authoring (you can switch to PDF
│                 │        or Word output anytime).
│                 │
│                 │     ○ PDF
│                 │        PDF output requires TeX (MiKTeX on Windows, MacTeX
│                 │        2013+ on OS X, TeX Live 2013+ on Linux).
│                 │
│                 │     ○ Word
│                 │        Previewing Word documents requires an installation of MS
│                 │        Word (or Libre/Open Office on Linux).
│                 │
└─────────────────┘

[ Create Empty Document ]                          [ OK ]    [ Cancel ]
```

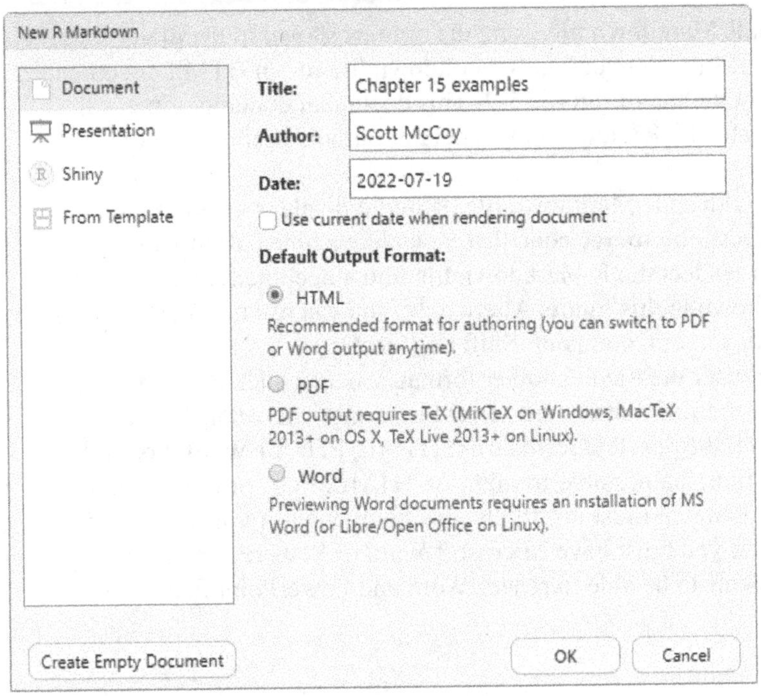

The rmarkdown package

```
install.packages("rmarkdown")
library("rmarkdown")
```

The TeX component for PDF files

```
install.packages("tinytex")
library("tinytex")
install_tinytex()
```

Description

- To use RStudio to create an R Markdown (.Rmd) file, select File→New File→Markdown File from the menu system.

- The New R Markdown dialog lets you create a starting file that renders to a *document* (HTML, PDF, or Word) or a *presentation* (HTML, PDF, or PowerPoint), also known as a *slideshow*. Alternately, you can create a starting file for an interactive Shiny document or presentation.

- By default, RStudio installs and loads the rmarkdown package when you first create an R Markdown file.

- To output PDF files, you must install the TeX component. TeX must be installed when you don't have an R Markdown file open.

- To output Word files, you must have Microsoft Word or Open Office installed.

- To output PowerPoint files, you must have Microsoft PowerPoint or Open Office installed.

Figure 15-1   How to create an R Markdown file

How to render an R Markdown file

If you create an R Markdown file using the options shown in the dialog from the previous figure, RStudio creates an R Markdown file for an HTML document like the one shown at the top of figure 15-2. This document contains some code that was generated by RStudio, and its default output is set to an HTML document.

When working with an R Markdown file, RStudio displays a Knit button in the toolbar just above the source code that looks like a blue ball of yarn. Clicking this button renders the R Markdown file into a document like the HTML document shown in this figure. Alternately, you can use the shortcut keys Ctrl+Shift+K (Windows) or Command+Shift+K (macOS).

If you want to render the file to another format, you can click the down arrow to the right of the Knit button and select the format you want. For an R Markdown document, the possible formats are HTML, PDF, or Word. For an R Markdown presentation, the possible formats are HTML, PDF, or PowerPoint. Remember, though, that you must install the TeX component if you want to render a PDF file, and you must have Microsoft Word or PowerPoint or their Open Office equivalents to be able to render Word and PowerPoint files.

An R Markdown file displayed in RStudio

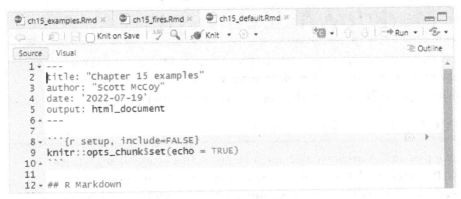

An HTML document after it has been rendered

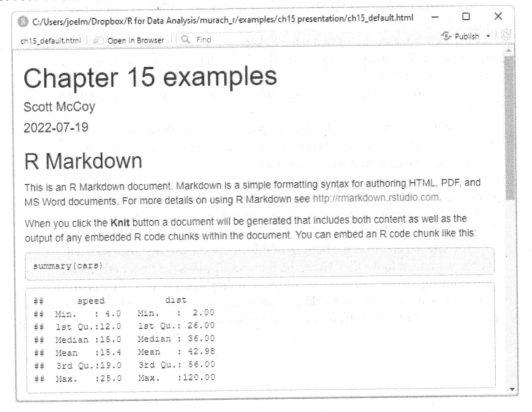

Description

- To render an R Markdown file in the default format, click the Knit button in the toolbar. Or, press Ctrl+Shift+K (Windows) or Command+Shift+K (macOS).
- To render in a different format, click the down arrow to the right of the Knit button and select the format (HTML, PDF, or Word) that you want.

Figure 15-2   How to render an R Markdown file

How to work with R Markdown documents

Now that you know how to create and render an R Markdown file for a document, you're ready to learn more about working with R Markdown documents. Most of these skills also apply to R Markdown presentations.

How to code the YAML header

The top of an R Markdown document begins with a header that's written between two triple-dash (---) sequences. This header sets the options for document output as well as some basic info about the document such as the title, author, and date. Unlike the rest of an R Markdown file, this header is written in a markup language called YAML. As a result, it uses a different syntax than the rest of the file.

Figure 15-3 starts by showing the header that's generated by RStudio when you create an R Markdown file for an HTML document. Then, it shows some header options for an HTML document. For example, you can use the toc and toc_float options to display a floating table of contents as shown by the third example.

In addition, you can use the theme option to change the appearance of the document. There are several themes available by default, and you can even create your own custom theme if you want.

All three options shown in this figure work with an HTML document. However, only the toc option works with PDF and Word documents. HTML documents usually have the most options available for formatting in R Markdown.

The screen capture at the bottom of this figure shows what the rendered HTML document looks like. Here, the browser displays the title, author, and date at the top of the document. In addition, it displays the table of contents in a column on the left side of the document. Since the document includes two headings, the table of contents shows two headings.

A YAML header for an HTML document

```
---
title: "Chapter 15 examples"
author: "Scott McCoy"
date: "2022-07-19"
output: html_document
---
```

Some options for an HTML document

Option	Description
toc	If set to true, the browser displays a table of contents.
toc_float	If set to true, the browser floats the toc in the left column if possible. Otherwise, the browser displays the toc at the top of the document.
theme	Sets the theme for the document. Possible themes include default, bootstrap, cerulean, cosmo, darkly, flatly, journal, lumen, paper, readable, sandstone, simplex, spacelab, united, and yeti.

A YAML header that displays a floating table of contents

```
---
title: "Chapter 15 examples"
author: "Scott McCoy"
date: "2022-07-19"
output:
  html_document:
    toc: true
    toc_float: true
    theme: spacelab
---
```

The rendered HTML document with a floating table of contents

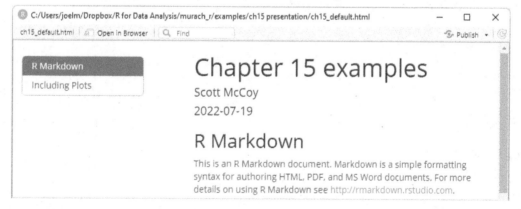

Description

- R Markdown files use YAML for headers.
- The header allows you to specify various information for your document and set some options for how it will be rendered.

Figure 15-3   How to code the YAML header

How to add headings and paragraphs

When you create an R Markdown document, you can use headings to organize your content. This gives your document structure, which makes it easier for your target audience to read and understand your analysis. In addition, RStudio can use the headings to generate a table of contents for the document.

To add headings to a document, you add one or more hash (#) symbols to the beginning of the line as shown in the first example of figure 15-4. The more hash symbols that you add, the lower the level of the heading. For example, one hash symbol creates a level-1 heading, two hash symbols creates a level-2 heading, and so on.

A table of contents organizes headings by level. For example, it puts level-2 headings under level-1 headings. However, it only includes the first three levels of headings in the table of contents. In other words, if you create level-4 headings, RStudio doesn't put them in the table of contents.

To add text to a document, you type in the markdown document like you would in any word processor or text editor. When you want to start a new paragraph, press the Enter key twice before starting a new line.

Although you typically use headings to structure a document, you may sometimes want to add horizontal lines to divide your document into sections. To do that, you can type three dashes as shown at the end of the example in this figure.

The example at the bottom of the figure shows the rendered document. This shows how the first four levels of headings look in the rendered document. In addition, it shows how four paragraphs and a horizontal line look.

Some R Markdown for headings and paragraphs

```
# Heading 1
## Heading 2
### Heading 3
#### Heading 4

To add text, type it anywhere in the document. R handles the line breaks
automatically, so you don't need to worry about them.

To start a new paragraph, press the Enter key twice to leave a blank
line between the two paragraphs.

To add headings, code one or more # symbols before the text for the heading.
The more # symbols that you add, the smaller the heading will be.

To add a horizontal line, code three dashes. This is another way to separate
sections of a document.

---
```

The rendered document

Heading 1

Heading 2

Heading 3

Heading 4

To add text, type it anywhere in the document. R handles the line breaks automatically, so you don't need to worry about them.

To start a new paragraph, press the Enter key twice to leave a blank line between the two paragraphs.

To add headings, code one or more # symbols before the text for the heading. The more # symbols that you add, the smaller the heading will be.

To add a horizontal line, code three dashes. This is another way to separate sections of a document.

Description

- To create a heading, code one or more # symbols before the text for the heading to identify its level. The table of contents includes headings down to the third level.
- You can add a horizontal line with three dashes (---).

Figure 15-4   How to add headings and paragraphs

How to add chunks of code

When you create an R Markdown document, you often use blocks of code known as *chunks* to display data summaries or plots. When you add a chunk of code to a document, it's executed when you render the document. All chunks of code in a document share the same environment. As a result, if you load a package or create a variable in one block, you can use it in another block.

The easiest way to insert a chunk of code into a document is to press Ctrl+Shift+I (Windows) or Command+Shift+I (macOS). When you do that, RStudio inserts three backticks followed by r in curly braces followed by three more backticks. The r identifies the code as R code.

After inserting a code chunk, you can enter the code for the chunk as shown in the first code chunk in figure 15-5. Here, the code uses the print() function to print a message to the console. Since this chunk uses the default options, it displays both the code and the output of the code in the rendered document.

Sometimes, you may want to code a chunk that only shows the output of the code in the rendered document, not the code itself. For example, you typically don't want to display code for a plot or a table, especially for a non-technical audience. If the code doesn't help your target audience understand the analysis, there's no reason to include it in the rendered document. To hide the code but retain the output, you can set the echo parameter to FALSE in the braces for the code block. This is shown in the second code chunk in this figure.

Conversely, you may sometimes want to include the code without showing the output. The most common example of this is code that loads packages. For example, you may want to show your target audience which packages you used in your analysis, but you might not want to display any messages or warnings that are displayed when the library() function loads the packages. To suppress these messages as well as any output from the chode chunk, you can set the result, message, and warning parameters to FALSE as shown in the third code chunk.

Some options for a code chunk

Option	If set to FALSE, don't display...
echo	The code.
results	The results of the code.
message	Any messages displayed by the code.
warning	Any warnings displayed by the code.

The R Markdown with code chunks

```
### Code and its output
```{r}
print("Display the code and the output")
```


Only output
```{r echo=FALSE}
print("Hide the code and display the output")
```


Only code
```{r results=FALSE, message=FALSE, warning=FALSE}
print("Hide the output and display the code")
```
```

The rendered document

Code and its output

```
print("Display the code and the output")
```

```
## [1] "Display the code and the output"
```

Only output

```
## [1] "Hide the code and display the output"
```

Only code

```
print("Hide the output and display the code")
```

Description

- A *chunk* of code begins and ends with three backticks (```). After the first set of backticks, the options for the chunk are coded within braces, after an r.
- To automatically generate the backticks and braces for a code chunk, press Ctrl+Alt+I (Windows) or Command+Alt+I (macOS).
- To specify multiple options, separate each option with a comma.

Figure 15-5   How to add chunks of code

How to run chunks of code

After you insert a chunk of R code into an R Markdown file, you can run it to make sure it works correctly. To do that, you can click on the Run Current Chunk button that's displayed at the top of the chunk. It's a right-facing green arrow in the upper right corner of the chunk.

The example in figure 15-6 shows RStudio after running a code chunk that loads two packages and displays a plot. The plot is displayed in the R Markdown file just below the code chunk.

You can also run the chunk by pressing Ctrl+Shift+Enter (Windows) or Command+Shift+Enter (macOS). This works similarly to running code in an R script by pressing Ctrl+Enter (Windows) or Command+Enter (macOS).

Sometimes, you may want to execute all code in the R Markdown file up to the chunk that you're working on. To do that, you can click the Run All Chunks Above button that's displayed at the top of the chunk that you're working on.

A chunk of code and its output displayed in RStudio

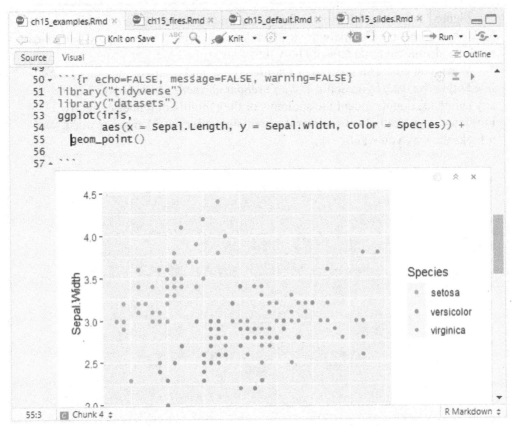

Description

- After you insert a chunk of R code, you can run it to make sure it works correctly.
- To run a chunk of code, click on the Run Current Chunk button that's displayed at the top of the chunk (green right-facing arrow). Or, position the cursor in the chunk and press Ctrl+Shift+Enter (Windows) or Command+Shift+Enter (macOS).
- To run all chunks of code before a chunk, click the Run All Chunks Above button that's displayed at the top of the chunk.

Figure 15-6   How to run chunks of code

How to format text

Figure 15-7 shows how to use R Markdown to format text to enhance the look of your reports. Compare the R Markdown presented in this figure with the rendered document to understand how this works.

When you create an ordered or unordered list, make sure to include a blank line before the list. If you don't, it won't render correctly. Also, if you include any subitems, double indent the subitems or they might not render correctly. Fortunately, you can always render the document after any changes to make sure it looks the way you want.

Some R Markdown for formatting text

```
You can use backticks to format text as `code`.

You can use asterisks to make text  *italic* or **bold**.

You can use use ^ and ~ characters to provide a superscript^2^ or
subscript~2~.

You can use dashes to make lists of unordered items.

- unordered item 1
- unordered item 2
    - unordered subitem 1
    - unordered subitem 2

You can use numbers and letters to make lists of ordered items.

1. ordered item 1
    a. ordered subitem 1
    b. ordered subitem 2
2. ordered item 2
```

The rendered document

You can use backticks to format text as `code`.

You can use asterisks to make text *italic* or **bold**.

You can use use ^ and ~ characters to provide a superscript^2 or subscript$_2$.

You can use dashes to make lists of unordered items.

- unordered item 1
- unordered item 2
    - unordered subitem 1
    - unordered subitem 2

You can use numbers and letters to make lists of ordered items.

1. ordered item 1
    a. ordered subitem 1
    b. ordered subitem 2
2. ordered item 2

Description

- R Markdown provides characters that you can use to apply formatting to text. This works similarly to other markup languages.

Figure 15-7   How to format text

How to create dynamic documents

A *dynamic document* changes depending on a variable in the code. For example, if a company has a sales database that stores sales numbers, it's possible to create a dynamic document that automates sales reports. To do that, you can code a query that pulls the numbers for the current month. Then, you can use R markdown to create a dynamic document that updates itself with the new numbers and month automatically.

To create a dynamic document, you can use backticks to execute inline code. This makes it possible to display the results of code within the YAML header or within text, such as R Markdown headers and paragraphs.

Part 1 of figure 15-8 shows the R Markdown for a dynamic document that displays fire data. In the YAML header, the date option is set to the result that's returned by inline code. Inline code is enclosed in backticks. Within these backticks, an r indicates that the code is R code. Then, this code calls the Sys.Date() function to return the value of the current date. As a result, when you render this document, the date in the header is always set to the current date.

Dynamic documents commonly use the Sys.Date() function to get the current date. That's because most dynamic documents are produced at a regular time interval (daily, weekly, monthly, etc.).

The next section that uses dynamic code begins by using the Sys.Date() function to get the current date and store it in a variable named curDate. Then, it uses the format() function to extract the number for the current month from the current date. Next, for testing, it sets the current year to 2012. That's because the fires data set only contains data up to 2015.

However, if the fires data set contained current data, you could use the statement that's commented out to set the current year to the year that's returned by the Sys.Date() function. In a real-world setting, the fires data set would probably be stored in a database that's updated regularly.

After getting the current month and year, the code filters the data set based on these values. This creates a document that only uses the data for the current month and year. To make that clear, both headings use inline code to display the current month and year. This inline code uses the format() function to extract the name of the current month, not the number.

The R Markdown for a dynamic document

```
---
title: "Automated Report"
author: "Scott McCoy"
date: "`r Sys.Date()`"
output: html_document
---


```{r, echo=FALSE, message=FALSE, warning=FALSE}
library("tidyverse")
library("maps")
fires <- readRDS("../../data/fires_prepared.rds")

curDate <- Sys.Date()
curMonth <- format(curDate, "%m")
# curYear <- format(curDate, "%Y")    # for production
curYear <- 2012                       # for testing

fires_filtered <- fires %>%
  filter(Year == curYear & Month == curMonth & State == "CA")
```


California fire size statistics for `r format(curDate, "%B")` `r
curYear`

```{r}
summary(fires_filtered$Size)
```


All fires in California for `r format(curDate, "%B")` `r curYear`

```{r, message=FALSE, echo=FALSE}
ca_map <- map_data("state", "california")

ggplot(ca_map, aes(x = long, y = lat)) +
  geom_polygon(color = "black", fill = "white") +
  geom_point(data = fires_filtered,
             mapping = aes(x = Longitude, y = Latitude,
                           size = Size, color = Size)) +
  scale_color_gradient2(low = "yellow", mid = "orange",
                        high = "darkred") +
  labs(x = "", y = "") +
  guides(size = "none", color = "none") +
  coord_fixed()
```
```

Description

- A *dynamic document* changes depending on the value of a variable in the code.
- To execute inline R code, code a backtick, an r, a space, the code that you want to execute, and another backtick.

Figure 15-8   How to create dynamic documents (part 1)

Part 2 of figure 15-8 shows the dynamic document after it has been rendered. This shows that the header displays the current date. In addition, it shows that the month and year displayed in the two headings match the statistics and plot that are displayed below these headings. For these headings, the month should be the current month at the time the file is rendered, but the year should be 2012 since the code sets it to this value for testing purposes.

The rendered document

Figure 15-8   How to create dynamic documents (part 2)

How to specify multiple output formats

So far, this chapter has shown how to create an HTML document and set its output options. Now, figure 15-9 summarizes the three output formats for an R Markdown document.

The example in this figure shows how to set the options for three output formats in a single YAML header. This header specifies that the HTML document should have a floating table of contents and use the spacelab theme. It specifies that the PDF document should have a table of contents. And it specifies that the Word document should use the default options, which don't include a table of contents.

Setting the options for the output formats doesn't render the R Markdown document in multiple formats when you click the Knit button. However, it uses the options specified in the YAML header for whichever format you choose to render. For example, if you render the R Markdown document to a Word document, it uses the default options.

Three output options for documents

Option	Description
html_document	Creates an HTML document.
pdf_document	Creates a PDF document.
word_document	Creates a word document.

The YAML header with multiple output formats

```
---
title: 'Chapter 15 examples'
author: "Scott McCoy"
date: "7/7/2022"
output:
  html_document:
    toc: true
    toc_float: true
    theme: spacelab
  pdf_document:
    toc: true
  word_document: default
---
```

Description

- In the YAML header, you can specify multiple output formats and set options for each format.

Figure 15-9   How to specify multiple output formats

The Wildfires document

To give you an idea of how you can use R Markdown to present an analysis, the next few figures show how to create a document for the Wildfires analysis from chapter 10 that you could share with your target audience. Although this document is short, you could easily make it longer to present more of the findings and insights from the Wildfires analysis.

The HTML document displayed in a browser

The document for the Wildfires analysis consists of three main sections: a short introduction, a section about the data, and a section for the analysis. Figure 15-10 shows two screen captures of the HTML document when displayed in a browser. The first screen capture shows the introduction and data sections, and the second screen capture shows the analysis section.

The data section introduces the columns and explains how the data was filtered. It uses R Markdown to create headings, paragraphs, and an unordered list. In addition, it formats the names of the columns as code to make them easier to identify.

The analysis section displays some code that summarizes the values of the Size column, and it presents some of the plots from the Wildfires analysis. Before each plot, the document uses a paragraph to describe the insights that the plot reveals. The analysis is the longest section in the document, which is good. When you create a report like this, your goal is usually to provide insights to your target audience and back them up with visual data.

The introduction and data headings

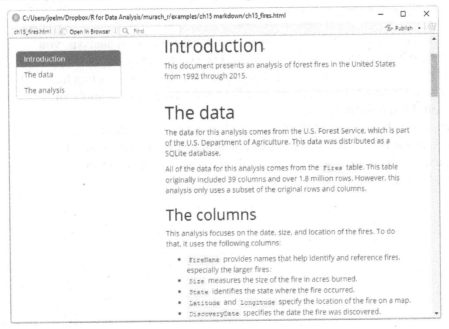

The analysis heading

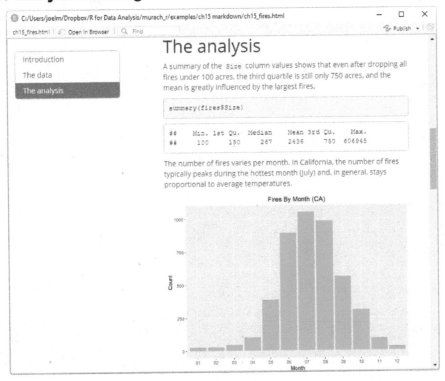

Figure 15-10   The HTML document displayed in a browser

The PDF and Word documents for the same markdown

When you create the R Markdown document for the Wildfires analysis, you can also render it as a PDF or Word document. For example, you might want to create a PDF document to share with colleagues who will only be reading the report. Or, you might want to create a Word file for the document to allow an editor to edit the text for the report.

Figure 15-11 shows that the PDF and Word documents contain the same information as the HTML document. The main differences are the options that are set in the YAML header, which may result in the table of contents or the theme of the document looking different.

The PDF document

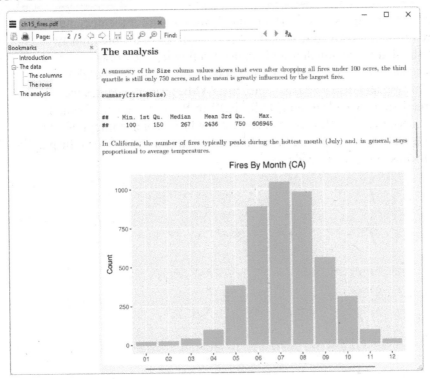

The Word document

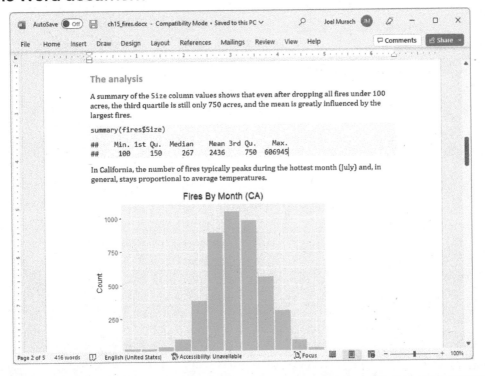

Figure 15-11   The PDF and Word documents for the same markdown

The R Markdown

The three parts of figure 15-12 present the R Markdown for the document. If you compare this markdown code with the screen captures in the previous two figures, you shouldn't have much trouble understanding how it works.

Part 1 presents the YAML header for the document and the R Markdown for the introduction and data sections. Here, the YAML header specifies the options for rendering the code as an HTML, PDF, or Word document. Then, the R Markdown specifies some level-1 and level-2 headings, some paragraphs, an unordered list, and some horizontal lines. In addition, it uses backticks to display the column names in the list as code.

The code for the document

```
---
title: "U.S. Wildfires analysis"
author: "Scott McCoy"
date: "`r Sys.Date()`"
output:
  html_document:
    toc: true
    toc_float: true
    theme: spacelab
  pdf_document:
    toc: true
  word_document:
    toc: true
---


# Introduction

This document presents an analysis of wildfires in the United States
from 1992 through 2015.


---


# The data

The data for this analysis comes from the U.S. Forest Service, which is
part of the U.S. Department of Agriculture. This data was distributed as
a SQLite database.

All of the data for this analysis comes from the `Fires` table. This
table originally included 39 columns and over 1.8 million rows. However,
this analysis only uses a subset of the original rows and columns.

## The columns

This analysis focuses on the date, size, and location of the fires. To
do that, it uses the following columns:

- `FireName` provides names that help identify and reference fires,
especially the larger fires.
- `Size` measures the size of the fire in acres burned.
- `State` identifies the state where the fire occurred.
- `Latitude` and `Longitude` specify the location of the fire on a map.
- `DiscoveryDate` specifies the date the fire was discovered.
- `Year` specifies the year the fire started.
- `Month` specifies the month the fire started.

## The rows

This analysis focuses on fires that burned 100 or more acres. To make
this easy, the analysis dropped all rows that were smaller than 100
acres. This reduced the number of rows to just over 54 thousand. In
addition, this analysis removed a few duplicate rows.

---
```

Figure 15-12   The code for the document (part 1)

Parts 2 and 3 present the R Markdown for the analysis section. To start, the analysis section displays a level-1 header. Then, it uses a code chunk to load the packages needed for the rest of the document. In addition, this code chunk reads the data needed for the rest of the document. This code chunk sets the echo, message, and warning options to FALSE. As a result, it doesn't display this code or any messages or warnings this code might display.

The rest of the document alternates between the paragraphs that describe the results of a code chunk and the code chunk that generates the result. For example, the first paragraph describes the distribution of the values of the Size column. Then, the code chunk displays the summary statistics for this column.

Similarly, the second paragraph describes the plot of the number of fires for each month. Then, the code chunk displays this plot. This code chunk sets the echo option to FALSE so that the document only displays the plot, not its code, which is usually what you want.

The code for the document (continued)

```
# The analysis
```{r echo=FALSE, message=FALSE, warning=FALSE}
library("tidyverse")
library("maps")
fires <- readRDS("../../data/fires_prepared.rds")
```
```

A summary of the `Size` column values shows that even after dropping all fires under 100 acres, the third quartile is still only 750 acres, and the mean is greatly influenced by the largest fires.

```
```{r}
summary(fires$Size)
```
```

In California, the number of fires typically peaks during the hottest month (July) and, in general, stays proportional to average temperatures.

```
```{r, echo=FALSE}
ggplot(fires %>% filter(State == "CA"),
       aes(x = Month)) +
  geom_bar(fill = "darkorange") +
  labs(title = "Fires By Month (CA)", y = "Count") +
  theme(plot.title = element_text(hjust = 0.5))
```
```

Interestingly, the largest fires typically occur in August, September, and October, *after* the summer spike in volume.

```
```{r, echo=FALSE}
ggplot(filter(fires, State == "CA" & Size >= 10000),
       aes(x = Month, y = Size)) +
  geom_boxplot(fill = "darkorange") +
  theme(plot.title = element_text(hjust = 0.5)) +
  labs(title = "Fires >= 10,000 Acres By Month (CA)",
       x = "Month", y = "Fire Size")
```
```

Figure 15-12   The code for the document (part 2)

The rest of the code displays the paragraphs and code chunks for three more plots. Since the code for these plots was taken from the Wildfires analysis presented in chapter 10, you should already be familiar with these plots. To view these plots and better understand the R Markdown for them, we recommend opening the R Markdown file for this document and rendering it. Then, you can compare the markdown with the rendered HTML, PDF, and Word documents.

The code for the document (continued)

In California, the 20 largest fires occurred mostly in the forested and mountainous regions. Further analysis of the relationship between elevation and fire size may be fruitful.

```r
```{r, echo=FALSE}
ca_top_20 <- fires %>%
  filter(State == "CA") %>%
  arrange(desc(Size)) %>%
  head(20)
ca_map <- map_data("state", "california")

ggplot(ca_map, aes(x = long, y = lat)) +
  geom_polygon(color = "black", fill = "white") +
  geom_point(data = ca_top_20,
             mapping = aes(x = Longitude, y = Latitude,
                           size = Size, color = "red")) +
  labs(title = "The 20 Largest Fires in California",
       x = "", y = "") +
  theme(plot.title = element_text(hjust = 0.5)) +
  guides(color = "none") +
  coord_quickmap()
```
```

Analysis of all fires larger than 500 acres confirms this trend as the Central Valley and southeastern deserts have fewer of these fires than the forested areas. This doesn't mean that there aren't fires in the Central Valley. It just means that these fires are smaller than 500 acres, or they aren't recorded in this data set.

```r
```{r, echo=FALSE}
ggplot(ca_map, aes(x = long, y = lat)) +
  geom_polygon(color = "black", fill = "white") +
  geom_point(data = filter(fires, Size > 500 & State == "CA"),
             mapping = aes(x = Longitude, y = Latitude,
                           size = Size, color = Size, alpha = Size)) +
  scale_color_gradient2(low = "yellow", mid = "orange",
                        high = "darkred") +
  coord_quickmap()

```
```

Figure 15-12   The code for the document (part 3)

How to work with R Markdown presentations

Most of the skills for using R Markdown files to create documents also apply to creating presentations, also known as slideshows. As a result, if you understand how to create an R Markdown document, you shouldn't have much trouble creating an R Markdown presentation.

How to start a presentation

One of the main differences between creating a document and a presentation is that you start the R Markdown file by selecting Presentation from the New R Markdown dialog shown in figure 15-1. Then, you can select the default output for the presentation (HTML, PDF, or PowerPoint). Here, there are two types of HTML presentations, ioslides and Slidy. When you click OK, RStudio generates a starting R Markdown file for the slideshow.

The R Markdown file in figure 15-13 has its output option set to an ioslides presentation. However, you can render this document to any of the presentation formats in the same way you could render a document to HTML, PDF, or Word. For an ioslides presentation, RStudio renders the first slide based on the YAML header. After that, it starts a new slide at each level-2 heading. As a result, you can use level-2 headings to start each new slide in the presentation.

The R Markdown in this figure begins with a code chunk that loads the packages and reads the data needed for the rest of the presentation. Then, it uses a level-2 heading to start a slide, and it uses a code chunk to display a plot on that slide. Next, it uses a level-2 heading to start another slide, and it displays some summary data on that slide.

When you create slides, you typically use less text than you do when you create a document. That's because you typically talk about each slide during a presentation. As a result, it isn't necessary to put a long description of the data or plot in the slide. However, the R Markdown works the same as it does for a document. So, you can use it to add headings, bullet points, data, plots, lists, and whatever else you need to make each slide work the way you want for your presentation.

The code for the start of a presentation

```
---
title: "U.S. Wildfires analysis"
author: "Scott McCoy"
date: "7/8/2022"
output: ioslides_presentation
---


```{r echo=FALSE, message=FALSE, warning=FALSE}
library("tidyverse")
library("maps")
fires <- readRDS("../../data/fires_prepared.rds")
```


Fire count by month in California

```{r, echo=FALSE}
ggplot(fires %>% filter(State == "CA"),
       aes(x = Month)) +
  geom_bar(fill = "darkorange") +
  labs(title = "Fires By Month (CA)", y = "Count") +
  theme(plot.title = element_text(hjust = 0.5))
```


Summary of fire sizes >= 100 acres in the U.S.

```{r}
summary(fires$Size)
```
```

Description

- To use RStudio to create a presentation, select File→New File→Markdown
  File from the menu system, select Presentation, and select the type of slideshow
  (HTML, PDF, or PowerPoint).
- RStudio provides for two types of HTML slideshows, ioslides and Slidy.
- If you create an ioslides slideshow, it builds the first slide based on the YAML
  header. After that, you can use the level-2 heading to start each new slide.

Figure 15-13   How to start a presentation

The first two slides of a presentation

Figure 15-14 shows the first two slides of an ioslides presentation after they have been displayed in a browser. To move between slides, you can press the right arrow key to move forward one slide, and you can press the left arrow key to move back one slide.

The first slide displays the info that's in the YAML header. This includes the title of the presentation, the author, and the date. For most presentations, that's a good way to start.

The second slide displays a heading and a plot. This works well because the presenter could easily explain what the plot means to the audience. In general, this plot is clean and easy to interpret. However, there is some duplication between the heading for the slide and the title at the top of the slide, which creates a little extra clutter. As a result, you could make this this slide even more clean by deleting the title from the code that displays the plot.

The first slide

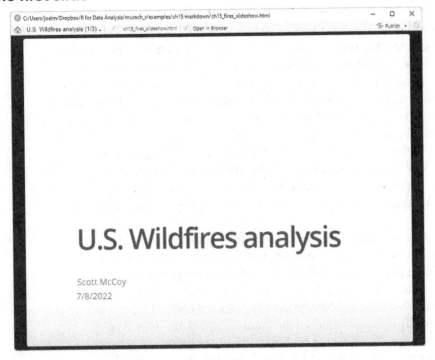

The second slide

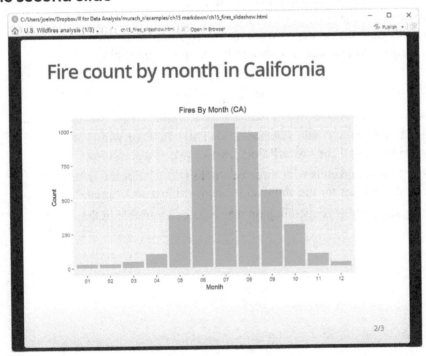

Figure 15-14   The first two slides of a presentation

Perspective

This chapter is designed to give you a solid foundation in using R Markdown to create documents that you can share with your target audience. In addition, it shows how you can use many of the same skills to create the slides for a presentation. However, there's much more to learn about creating documents and presentations. For example, the YAML header provides many more options for controlling the appearance of the rendered output, and R Markdown provides for many other types of formatting.

In addition, there are more packages for working with R markdown documents. For example, you can use the flexdashboard package to create dashboards, which are websites that display your data on a single page. Dashboards are typically used to monitor live data that is continuously updated.

Similar to a dashboard, you can use the Shiny package to create web applications for interactive analysis. Rather than having set plots for dynamic data like a dashboard, Shiny focuses on creating dynamic plots for static data. This typically takes the form of sliders, dials, dropdowns, or other controls that are used to filter the data. The plots are then automatically updated based on the values in the dials. This type of application is rapidly replacing traditional business reporting because it allows a user with no technical knowledge to select the data that they want to see and have it automatically plotted or summarized.

Finally, you can use the bookdown package to develop books with R Markdown. Although this package was designed to create books, it's useful for any long document that you want to break up into multiple files.

Summary

- You can use R Markdown to create a *document* (HTML, PDF, or Word) or a *presentation* (HTML, PDF, or PowerPoint), also known as a *slideshow*.
- A *chunk* of code begins and ends with three backticks (```). After the first set of backticks, the options for the chunk are coded within a set of braces.
- A *dynamic document* changes depending on the value of a variable in the code.

Exercise 15-1   Render and edit the R Markdown file for the Fires analysis

In this exercise, you'll run the R Markdown file presented in this chapter, and edit it to add some plots and information.

Open and render the Fires R Markdown file

1. If you haven't already, install the TeX component so you can render PDF files.

2. Open the R Markdown file named exercise_15-1.R located in this folder:

    `Documents/murach_r/exercises/ch15`

    This file contains the code presented in this chapter and a couple additional plots.

3. Use the Knit button to render to HTML, Word, and PDF files and compare the results.

4. After the code chunk for `summary(fires$Size)`, add another chunk that plots the mean and median fire size for each year in the fires data set. You can copy this code from the Fires analysis. Be sure to include not just the ggplot() statements but also the code preparing the data for the plot, and set echo to FALSE.

5. Add a short explanation for the mean and median plot before the code chunk you just added.

6. Add a new section to the end of the document called "Conclusion." Separate it from the previous Analysis section with a horizontal line and use a level-1 header for the name.

7. Write two short paragraphs describing what you learned from the Fires analysis. Don't worry about exactly what to write as the point here is the formatting, not your text.

8. In the Conclusion section, add a level-2 header called "Recommendations". Then, write some bullet points with your recommendations based on the analysis. These can be as simple as "Be prepared for more fires during the summer" and "Educate Californians about fire safety".

9. Add another level-2 header, this one called "Future Analysis". In this section, use an ordered list to create a short table of contents for a potential future analysis of the fires data.

Exercise 15-2   Create your own R Markdown file for the Polling analysis

In this exercise, you'll create your own R Markdown file for the Polling analysis from chapter 9.

Create a presentation R Markdown file

1.   Open the new R Markdown file dialog, and choose presentation.

2.   Open the ch09_polling.R file found in examples/ch09 for reference. It will be easier to copy in code you already know works than to re-create it.

3.   Add a code chunk with the echo, message, and warning options all set to FALSE that imports and formats the polling data set. The data should be pivoted to its long form.

4.   When you render your slides, if you have some console text in the background, you can also set the results option to FALSE.

Add some slides

5.   Choose three plots from the analysis and create slides from them. Each slide should have a title and a plot.

6.   Create another slide with a heading of "Conclusion". Use bullet points to add some thoughts about the Polling analysis.

Appendix A

How to set up Windows for this book

Before you can learn how to use R for data analysis, you need to install the R language and the RStudio IDE (Integrated Development Environment). This appendix show shows how to do that for Windows. Then, it shows how to install the files for the book. Finally, it shows how to install some additional data analysis packages for R that aren't installed when you install the R language.

How to install R

Figure A-1 shows how to install the R language on your computer. To do that, you can visit the link shown at the top of the figure. Then, you can download and run the installer for the most recent version of the R language. When you run the installer, you can accept the default options from the resulting dialog boxes.

How to install RStudio

Figure A-1 also shows how to install RStudio, the IDE we recommend for this book.

The URL for the latest version of R

https://cran.r-project.org/bin/windows/base/

How to install R

1. Go to the URL shown above. You can also find this page by searching the internet for "R download". This displays the download page for the R language.
2. Click the "Download R for Windows" link to download an exe file for the installer.
3. Double-click the exe file to start the R installer.
4. Respond to the dialog boxes that are displayed when the installer runs. When in doubt, accept the default options.

The URL for RStudio's download page

https://www.rstudio.com/products/rstudio/download/#download

How to install RStudio

1. Go to the URL shown above. You can also find this page by searching the internet for "RStudio Desktop download". This displays the download page for RStudio Desktop.
2. Click the Download button to download an exe file.
3. Double-click the exe file to run the RStudio installer.
4. Respond to the dialog boxes that are displayed when the installer runs. When in doubt, accept the default options.

Description

-   The R programming language is designed for statistical computing and graphics. You can download and install it for free.
-   The Comprehensive R Archive Network (CRAN) is a network of servers around the world that stores the executable files, source code, documentation, and packages for the R language.
-   RStudio is an Integrated Development Environment (IDE) for R.
-   After RStudio has been installed, it's available from the Windows Start menu.

Figure A-1   How to install R and RStudio

How to install the files for this book

Figure A-2 shows how to install the R scripts and data files that you need for this book. This includes the scripts for all of the examples and analyses presented in this book as well as the scripts that you'll need for doing the exercises that are at the end of each chapter.

When you finish the first procedure in this figure, the scripts and data files for this book should be in the folders shown in this figure. Then, you can open these scripts, review the code, run the code, do the exercises, and experiment on your own.

How to install the packages for this book

This book makes use of many R packages for data analysis that aren't installed when you install the R language. To make it easy to install these packages, you can run the script named install_packages.R as shown by the second procedure in figure A-2. After you run it, you can view the messages displayed in the Console pane. These messages are long and hard to read, but if you scroll through them, they should indicate that the packages were installed.

In the messages, you may see some warning messages. Since these messages are warnings, not errors, you can often ignore them. If they bother you or cause problems, you can search the internet to learn more about them. In many cases, you can get rid of warning messages by updating R, RStudio, or the packages that you're using.

Whenever you open a script that attempts to load a package that isn't installed, RStudio prompts you with an Install link that you can click to install the package. As a result, it isn't necessary to run the script shown in this figure. However, running this script is a convenient way to install all of the packages used by this book at once instead of having to install them one by one as you go through the book.

The Murach website

www.murach.com

The folder for the book analyses, examples, and exercises

\Documents\murach_r

The subfolders

Folder	Description
data	The data files that are used by this book.
examples	The examples and analyses presented in each chapter of the book.
exercises	The starting points for the exercises presented at the end of each chapter.
solutions	The solutions to the exercises.

How to install the files for this book

1. Go to www.murach.com.
2. Find the page for *Murach's R for Data Analysis*.
3. If necessary, scroll down to the FREE downloads tab.
4. Click the FREE downloads tab.
5. Click the Download Now button for the zip file. This should download a zip file.
6. Find the zip file on your computer and double-click it. This should extract the files for this book into a folder named murach_r.
7. Move the murach_r folder into your Documents folder.

A script for installing the packages for this book

\Documents\murach_r\install_packages.R

The packages installed by this script

tidyverse, ggforce, ggpubr, RJSONIO, RSQLite, tidymodels, dotwhisker, tinytex, caret, rpart.plot, maps, rmarkdown, markdown, Rcpp

How to install the packages for this book

1. Start RStudio.
2. Open the script file named install_packages.R that's in the murach_r folder.
3. Press Ctrl+Shift+Enter to run the entire script. When you do, RStudio should show the results of running this script in its Console pane. If you get a dialog that asks if you want to restart R prior to install, click Yes.
4. When the script finishes running, read through the messages displayed in the Console pane to make sure that R was able to install these packages. If you get warning messages, you can usually ignore them.

Figure A-2   How to install the files and packages for this book

Appendix B

How to set up macOS for this book

Before you can learn how to use R for data analysis, you need to install the R language and the RStudio IDE (Integrated Development Environment). This appendix show shows how to do that for macOS. Then, it shows how to install the files for the book. Finally, it shows how to install some additional data analysis packages for R that aren't installed when you install the R language.

How to install R

Figure B-1 shows how to install the R language on your computer. To do that, visit the link shown at the top of the figure. Then, you can download and run the installer for the most recent version of the R language. When you run the installer, you can accept the default options from the resulting dialog boxes.

How to install RStudio

Figure B-1 also shows how to install RStudio, the IDE we recommend for this book.

The URL for R's download page

https://cran.r-project.org/bin/macosx/

How to install R

1. Go to the URL shown above. You can also find this page by searching the internet for "R download macos". This displays the download page for the R language.
2. Click the link to download the pkg file for the installer that's right for your computer.
3. Double-click the pkg file to start the R installer.
4. Respond to the dialog boxes that are displayed when the installer runs. When in doubt, accept the default options.

The URL for RStudio's download page

https://www.rstudio.com/products/rstudio/download/#download

How to install RStudio

1. Go to the URL shown above. You can also find this page by searching the internet for "RStudio Desktop download". This displays the download page for RStudio Desktop.
2. Click the Download button to download the dmg file.
3. Double-click the dmg file. In the resulting dialog, drag the RStudio icon into the Applications folder.

Description

- The R programming language is designed for statistical computing and graphics. You can download and install it for free.
- The Comprehensive R Archive Network (CRAN) is a network of severs around the world that stores the executable files, source code, documentation, and packages for the R language.
- RStudio is an Integrated Development Environment (IDE) for R.
- After RStudio has been installed, it's available from the Applications folder.

Figure B-1   How to install R and RStudio

How to install the files for this book

Figure B-2 shows how to install the R scripts and data files that you need for this book. This includes the scripts for all of the examples and analyses presented in this book. In addition, it includes the scripts that you'll need for doing the exercises that are at the end of each chapter.

When you finish the first procedure in this figure, the scripts and data files for this book should be in the folders shown in this figure. Then, you can open these scripts, review the code, run the code, do the exercises, and experiment on your own.

How to install the packages for this book

This book makes use of many R packages for data analysis that aren't installed when you install the R language. To make it easy to install these packages, you can run the script named install_packages.R as shown by the second procedure in figure B-2. After you run it, you can view the messages displayed in the Console pane. These messages are long and hard to read, but if you scroll through them, they should indicate that the packages were installed.

In the messages, you may see some warning messages. Since these messages are warnings, not errors, you can often ignore them. If they bother you or cause problems, you can search the internet to learn more about them. In many cases, you can get rid of warning messages by updating R, RStudio, or the packages that you're using.

Whenever you open a script that attempts to load a package that isn't installed, RStudio prompts you with an Install link that you can click to install the package. As a result, it isn't necessary to run the script shown in this figure. However, running this script is a convenient way to install all of the packages used by this book at once instead of having to install them one by one as you go through the book.

The Murach website

www.murach.com

The folder for the book analyses, examples, and exercises

\Documents\murach_r

The subfolders

Folder	Description
data	The data files that are used by this book.
examples	The examples and analyses presented in each chapter of the book.
exercises	The starting points for the exercises presented at the end of each chapter.
solutions	The solutions to the exercises.

How to install the files for this book

1. Go to www.murach.com.
2. Find the page for *Murach's R for Data Analysis*.
3. If necessary, scroll down to the FREE downloads tab.
4. Click the FREE downloads tab.
5. Click the Download Now button for the zip file. This should download a file named rda1_allfiles.zip.
6. Find the zip file on your computer and double-click it. This should extract the files for this book into a folder named murach_r.
7. Move the murach_r folder into your Documents folder.

A script for installing and loading the packages for this book

\Documents\murach_r\install_packages.R

The packages installed by this script

tidyverse, ggforce, ggpubr, RJSONIO, RSQLite, tidymodels, dotwhisker, tinytex, caret, rpart.plot, maps, rmarkdown, markdown, Rcpp

How to install and load the packages for this book

1. Start RStudio.
2. Open the script file named install_packages.R that's in the murach_r folder.
3. Press Command+Shift+Enter to run the entire script. When you do, RStudio should show the results of running this script in its Console pane. If you get a dialog that asks if you want to restart R prior to install, click Yes.
4. When the script finishes running, read through the messages displayed in the Console pane to make sure that R was able to install these packages. If you get warning messages, you can usually ignore them.

Figure B-2   How to install the files and packages for this book

Index

F

facet_matrix() function, 286-289
facet_wrap() function, 124-125
facet_zoom() function, 268-269
Factor data type, 96-97, 192-193
factor() function, 192-193
False negative, 486-487
False positive, 486-487
Fence, 198-199, 404-405
file.rename() function, 143
Filter rows, 176-177
filter() function, 100-101, 110-111, 176-177
filter_at() function, 418-419
Fit a model, 458-459, 464-465
fit() function, 422-423
Five-number summary, 76-77, 112-113
floor() function, 36-37
for loops, 56-57
format() function, 206-207
Formula, 422-423, 430-431, 440-441, 444-445
FROM clause (SQL), 146-147
fromJSON() function, 150-151
full_join() function, 232-235
Functions, 32-33, 58-59

G

geom_arc() function, 256-257
geom_bar() function, 104-105, 110-111
geom_boxplot() function, 104-105, 112-113
geom_circle() function, 256-257
geom_col() function, 246-247
geom_density() function, 104-105, 116-117
geom_density_2d() function, 104-105, 120-121
geom_ellipse() function, 256-257
geom_errorbar() function, 248-249
geom_function(), 428-429
geom_histogram() function, 104-105, 114-115
geom_label() function, 254-255
geom_line() function, 86-87, 104-107
geom_point() function, 104-105, 108-109
geom_polygon() function, 256-257, 264-269
geom_rect() function, 256-257
geom_segment() function, 254-255
geom_smooth() function, 252-253, 432-433
geom_tile() function, 256-257
Geospatial data, 264-269
Get the data,
    Basketball Shots, 364-365
    phases of data analysis, 6-7
    Polling, 298-305
    Wildfires, 334-337

get_lower_fence() function, 418-419
get_season() function, 368-369
get_upper_fence() function, 418-419
getwd() function, 68-69
ggforce package, 242-243
ggplot() function, 86-87, 102-103
ggplot2 package, 70-71, 86-87, 94-95, 102-103
grep() function, 34-35
Grid of plots, 124-125, 286-289
grid_regular() function, 492-493
Group the data, 216-217
group_by() function, 84-85, 216-217
gsub() function, 34-35
guides() function, 250-251, 276-277

H

head() function, 74-75
heteroscedasticity, 442-443
Histogram, 114-115
Hold out set, 488-489
Home directory symbol (~), 68-69
Hotkeys
    assignment operator, 16-17
    knit .Rmd file, 504-505
    pipe operator, 85
    run code, 12-13
Hyperparameter, 482-483
    of a decision tree, 484-485
    tuning, 482-483

I

I() function, 430-431, 444-445
IDE, 1, 4
identical() function, 236-237
Identify correlations visually, 410-413
if statement, 52-53
ifelse() function, 182-185
Imbalanced classification, 470-471
Independent variable, 400-401
Integrated Development Environment, 1, 4
initial_split() function, 414-415
inner_join() function, 232-235
install.packages() function, 70-71
install_packages script, 542-543, 548-549
install_tinytex() function, 502-503
Installing R
    macOS, 546-547
    Windows, 540-541
Installing RStudio
    macOS, 546-547
    Windows, 540-541

tunable() function, 492-493
tune() function, 492-493
tune_grid() function, 490-493
Tuning hyperparameters, 492-493
Type
 boolean, 16-17
 coercion, 38-39
 conversion, 188-191
 factor, 96-97, 192-193
 numeric, 16-17
 ordered factor, 98-99
 string, 16-17

U

Underfitting, 446-449
ungroup() function, 216-217
Unique number of values, 170-171
Unique values, 168-169
unique() function, 166-167, 176-177
unzip() function, 142-143
update.packages() function. 70-71

V

Value counts, 172-173
Variable, 10-11, 16-17, 78-79
 importance, 482-483
 naming conventions, 18-19
 selection, 482-483
varImp() function, 482-483
vars() function, 124-125, 286-287
Vector, 38-39
vfold_cv() function, 488-489

W

WHERE clause (SQL), 146-147
Wide data, 78-79
Wildfires analysis, 334-359
Working directory, 68-69
Workspace image, 14-15

X

xlim() function, 270-271
xor() function, 176-177

Y

YAML header, 506-507
ylim() function, 270-271